The Theory of Algorithms

Mathematics and Its Applications (*Soviet Series*)

The Theory of Algorithms

by

A. A. Markov
Formerly University of Moscow

and

N. M. Nagorny
Computing Centre of the Academy of Sciences of the U.S.S.R.

Translated by M. Greendlinger

KLUWER ACADEMIC PUBLISHERS
DORDRECHT / BOSTON / LONDON

Library of Congress Cataloging in Publication Data

Markov, A. A. (Andreĭ Andreevich), 1903-
The theory of algorithms.

(Mathematics and the applications ; 23)
Translation of: Teoriía algorifmov.
Bibliography: p.
Includes indexes.
1. Algorithms. I. Nagornyi, N. M. (Nikolai Makarovich) II. Title. III. Series: Mathematics and its applications (Kluwer Academic Publishers). Soviet series ; 23.
QA9.58.M3713 1988 511'.8 88-17582

ISBN 90-277-2773-2

Published by Kluwer Academic Publishers,
P.O. Box 17, 3300 AA Dordrecht, The Netherlands.

Kluwer Academic Publishers incorporates
the publishing programmes of
D. Reidel, Martinus Nijhoff, Dr W. Junk and MTP Press.

Sold and distributed in the U.S.A. and Canada
by Kluwer Academic Publishers,
101 Philip Drive, Norwell, MA 02061, U.S.A.

In all other countries, sold and distributed
by Kluwer Academic Publishers Group,
P.O. Box 322, 3300 AH Dordrecht, The Netherlands.

Printed in The Netherlands

SERIES EDITOR'S PREFACE

Approach your problems from the right end and begin with the answers. Then one day, perhaps you will find the final question.

'The Hermit Clad in Crane Feathers' in R. van Gulik's *The Chinese Maze Murders.*

It isn't that they can't see the solution. It is that they can't see the problem.

G.K. Chesterton. *The Scandal of Father Brown* 'The point of a Pin'.

Growing specialization and diversification have brought a host of monographs and textbooks on increasingly specialized topics. However, the "tree" of knowledge of mathematics and related fields does not grow only by putting forth new branches. It also happens, quite often in fact, that branches which were thought to be completely disparate are suddenly seen to be related.

Further, the kind and level of sophistication of mathematics applied in various sciences has changed drastically in recent years: measure theory is used (non-trivially) in regional and theoretical economics; algebraic geometry interacts with physics; the Minkowsky lemma, coding theory and the structure of water meet one another in packing and covering theory; quantum fields, crystal defects and mathematical programming profit from homotopy theory; Lie algebras are relevant to filtering; and prediction and electrical engineering can use Stein spaces. And in addition to this there are such new emerging subdisciplines as "experimental mathematics", "CFD", "completely integrable systems", "chaos, synergetics and large-scale order", which are almost impossible to fit into the existing classification schemes. They draw upon widely different sections of mathematics. This programme, Mathematics and Its Applications, is devoted to new emerging (sub)disciplines and to such (new) interrelations as exempla gratia:

- a central concept which plays an important role in several different mathematical and/or scientific specialized areas;
- new applications of the results and ideas from one area of scientific endeavour into another;
- influences which the results, problems and concepts of one field of enquiry have and have had on the development of another.

The Mathematics and Its Applications programme tries to make available a careful selection of books which fit the philosophy outlined above. With such books, which are stimulating rather than definitive, intriguing rather than encyclopaedic, we hope to contribute something towards better communication among the practitioners in diversified fields.

Because of the wealth of scholarly research being undertaken in the Soviet Union, Eastern Europe, and Japan, it was decided to devote special attention to the work emanating from these particular regions. Thus it was decided to start three regional series under the umbrella of the main MIA programme.

With the possibility of computing given us by the big machines there came, naturally, a strengthening of interests in concrete methods of calculating all kinds of mathematical objects, i.e. algorithms. And hence interest in what it means to have an algorithm for computing, say, a real number. As long as one is working with concrete algorithms to compute something this is still more or less traditional well-established mathematics except for questions of computational complexity and related matters. This changes drastically as soon as one starts considering questions such as whether an algorithm can exist for a certain class of problems, i.e. questions of algorithmic unsolvability. Thus, for example, there is the recursively unsolvability of the word problem for semigroups: a result obtained independantly and simultaneously by the senior author of this book and E.K. Post (1947). That requires a definition of what an algorithm is precisely, and all of a sudden, and somewhat ironically, we are back into quite foundational questions of both mathematics and computer science. It is perhaps significant that the Russian original of this book (1985) and its venerable precursor of 1954 are both classified under computer science in our library at the CWI in Amsterdam. There are several formalizations of the intuitive concept of algorithm or computability which have been proposed; one of these is Markov's notion of a normal algorithm. All, Turing computability, general recursive functions, λ-definability, normal algorithm, and some others, have been proved equivalent, lending support to Church's thesis that these do indeed capture the intuitive idea in the right way. Which one likes best is a matter of training, taste, and habit. A whole school of algorithmic and computability investigations is based on Markov's notion of a normal algorithm and this is the definitive book on it. In addition, in its final chapter it discusses related topics such as constructive analysis, intuitionistic logic and choice sequences.

The unreasonable effectiveness of mathematics in science ...

Eugene Wigner

Well, if you know of a better 'ole, go to it.

Bruce Bairnsfather

What is now proved was once only imagined.

William Blake

As long as algebra and geometry proceeded along separate paths, their advance was slow and their applications limited.

But when these sciences joined company they drew from each other fresh vitality and thenceforward marched on at a rapid pace towards perfection.

Joseph Louis Lagrange.

Bussum, February 1988

Michiel Hazewinkel

CONTENTS

And the will therein lieth, which dieth not. Who knoweth the mysteries of the will, with its vigor? For God is but a great will pervading all things by nature of its intentness. Man doth not yield himself to the angels, nor unto death utterly, save only through the weakness of his feeble will.

"Ligeia", E. A. Poe

For the English Translation

I am glad that the reader of the English translation will now have the opportunity of becoming acquainted, at least to some extent, with the views of my teacher, A. A. Markov—not through hearsay, but directly. These views provoked a stormy reaction on the part of many of his set-theoretically inclined colleagues. Traces of the prolonged discussions which took place around A. A. Markov's conceptions can be found in this book, but they are merely pale shadows.

The book's somewhat unusual character demanded a somewhat unusual preface to the Russian edition. Let it not stop the reader, even through it might be difficult in certain respects.

Finally, I should like to thank A. A. Markov's last graduate student, M. Dombrovsky, who selflessly helped me perfect the translation, and also the translator, M. Greendlinger, who called my attention to some shortcomings in the Russian edition.

February 28, 1986 N. Nagorny
Moscow

Translator's Preface

When Nikolai Makarovich Nagorny suggested that I translate the Russian original of this book, I had mixed feelings. On the one hand, I felt that my set-theoretical views on the foundations of mathematics might prevent me from doing justice to some of its philosophical excursions. On the other hand, my admiration for Andrei Andreevich Markov as a mathematician, philosopher and person demanded that I try. So did my debt to Andrei Andreevich and all the other participants in the many fascinating free clashes of ideas which helped me better understand not only mathematics, but the USSR as well.

I resolved the above dilemma by enlisting the aid of three constructivists, Elena Greendlinger, Nikolai Nagorny, Mikhail Dombrovsky, and a linguist, Tatyana Greendlinger. Although many of the translation's merits are due to their efforts, for which they deserve my sincerest gratitude, all its deficiencies must be attributed to me alone.

November 20, 1985 — Martin Greendlinger
Shuya

PREFACE

This book is being published at a time when its main author, one of the greatest mathematicians of our time, Andrei Andreevich Markov (1903–1979), is no longer alive. His memory obliges me, his disciple and co-author, to preface the book with some lines in which the reader would be told, at least briefly, about the outstanding contribution made by Markov to the development of the theory of algorithms and the creation of constructive mathematics. In addition, I probably ought to write about his views on mathematics in general, his creative manner and the inimitable originality of his personality.

However, it isn't easy to do this. All the more so, within the limits of a book's preface. It isn't easy if only because it's impossible to simply and uniquely refer Markov's activities to any one specific branch of science. Thus, being a mathematician, he developed in the last period of his activities, devoted to mathematical logic and lasting more than thirty years, his own conception of constructivism—a conception which now occupies a firm place in the foundations of mathematics and which has given rise to a wide range of scientific problems. His views, continuing the ideas of such brilliant thinkers as Brouwer, Weyl, and developing them on a modern basis, raised the discussion on the effectiveness of methods applied in mathematics to a new level. This fully entitles us to characterize him as a profound philosopher of science and a logician with original conceptions.

Neither is it possible to connect Markov's mathematical activities to any single field. One can come across papers of his on axiomatic set theory and the theory of plasticity, published almost simultaneously. Working on the subtlest problems of constructive semantics, he created at the same time a language for describing the functioning of computers. He followed the development of theoretical cybernetics with great attention and made a significant contribution to this branch of science. He also displayed a lively interest in work on mathematical economics and mathematical linguistics, in mathematical problems of biology. The range of questions with which he was preoccupied was so broad, that mention of his results can even be found in textbooks on geophysics.[1]

---

[1]Bursian, V. R.: Бурсиан, В. Р., Теория электромагнитных полей, применяемых в электроразведке V. 1. 2nd edition, Leningrad, (1972). (The Theory of Electromagnetic Fields Applied in Electrical Prospecting. See p. 176, where the most important special case, considered by Markov, of the problem of a point electrode in a layered medium is presented.)

Having already become classical, the work done by Markov during his last period, the period devoted to the foundations of mathematics, the theory of algorithms and mathematical logic, is well known nowadays. But mathematicians of an older generation pay particularly high tribute to his papers of 1940–1946 on free topological groups. Thus, Head of the Soviet Topological School and a Member of the Academy of Sciences, Pavel Sergeevich Alexandrov, repeatedly referred to them as Markov's finest achievement in Mathematics. Even before that, in the early 1930s, Markov published a series of profound papers on the general theory of dynamical systems. It was precisely he who first gave a general definition of a dynamical system that was independent of differential equations.[2] (I regret to say that few people are aware of this nowadays.)

Markov left a permanent imprint on many fields of mathematics. However, those series of papers first published by him were not directly related to mathematics—they were papers on theoretical physics and on celestial mechanics.[3] Moreover, during his early student years, when intending to become a chemist, Markov wrote and published papers on chemistry.

Possessing vast scientific experience of his own, Markov had available an extensive fund of knowledge in the widest variety of scientific fields. By the cast of his mind and the nature of his scientific activity, he was, as N. A. Shanin said in his farewell speech, a naturalist, and his approach to mathematics was the approach of a naturalist, striving to give the development of a theory in a way in which its results would have as real and tangible a meaning as possible. In large measure, this is what explains Markov's sharp rejection of set theory as a basis for constructing mathematics, his striving to build mathematics constructively.

It is not easy to explain in a few words how Markov succeeded in leaving behind such a vast scientific heritage,[4] in spite of his deliberate unhurriedness and thoroughness, sometimes looking almost like sluggishness, his reluctance to be assisted and his gusto for working up details. Neither is it a simple matter to tell about the sources of the aesthetic and stylistic norms to which Markov adhered in his works and required his disciples to observe.

It would be interesting, although difficult, to tell of Markov's extraordinary mode of behaviour, of his observational powers and subtle wit, of his unexpected, at times paradoxical, but invariably apt and precise judgments, which were often presented in lively forms of unforgettable artistry.

Son of a great Russian mathematician who was a member of the Academy of Sciences, Andrei Andreevich Markov (1856–1922), Markov, Jr. belonged by birth to the upper stratum of the Russian intelligentsia, which has given our country many outstanding persons. The best traditions of this environment—a

---

[2]See Linnik and Shanin [1].

[3]Markov was a physicist by education. He graduated from the Physics Department of Leningrad University's Physics and Mathematics Faculty in 1924, after which he worked for a short time at the Institute of Physics and Engineering. He was a graduate student from 1925 to 1928 and then, until 1935, a scientific worker at the Astronomy Institute.

[4]At the present time, the USSR Academy of Sciences has begun preparing a two-volume "Selected Works" of Markov.

belief in the lofty mission of culture, a selfless love for science, adherence to a clear set of moral principles and a plain hierarchy of fundamental values—found their distinct expression in the character and even in the outward appearance of Markov, astonishingly noble, steadfast and democratic, in the highest sense of the word, as he was known and loved by his disciples and everyone else who was close to him in one degree or another.

A fuller account of Markov's life would require a separate treatment, and so it must be postponed. But the best way of honouring my teacher's memory available to me at present consists in trying to complete with maximal care the work begun by him.

* * *

One of the most essential factors determining the significance of mathematics and its place in modern science is the precision and high degree of universality of the means of representation available to it. These means allow us to construct more precise models of the most varied kinds of fundamental scientific concepts, including mathematical concepts. Once such models are constructed, they then become available for analysis with the aid of exact tools, and as a rule, appreciable progress ensues in the study of the concepts described by them.

The development of this type of more precise model was carried out during the second half of the 1930s for one of the most important and widely employed concepts in mathematics—and not only in mathematics—the concept of an ≪Algorithm≫.[5][6]

In mathematical practice it is customary to understand an algorithm as an ≪exact prescription determining a computational procedure which leads from varying initial data to the required result≫. (Markov, [2], p.3). Here the expression ≪computational procedure≫ should, of course, be interpreted in the broad sense, namely, as a discrete process transforming complexes of symbols of definite types.

Euclid's world-famous algorithm for finding the greatest common divisor of two natural numbers is usually mentioned in the literature as a typical example of an algorithm. The rules for performing arithmetical operations on integers written in the decimal system are algorithms which are also known in elementary school. The reader will recall other examples found in his practice.[7]

---

[5] The terms ≪prescription≫, ≪method≫ ≪means≫, etc., are also used in a similar sense in mathematics.

[6] There are two Russian words for "algorithm", "алгоритм" and "алгорифм". The authors use the second, less common but more euphonious, form. (Translator's note).

[7] Sometimes, incidentally, computational recommendations, which turn out on closer analysis not to be algorithms, are referred to in the literature as algorithms (most often simply due to a misunderstanding or a blind following of tradition). The famous ≪method of bisecting a segment≫, ordinarily applied in proving Cauchy's Theorem on a zero of a continuous function with alternating signs, can serve as a typical example. The reasons why this method can not be regarded as an algorithm are given in §64. Nevertheless, it does have a certain algorithmic content, which is just what we clarify in §68.2.

Using the concept of an algorithm, mathematics can be content up to a certain point with a rather vague, inexact formulation of this concept, such as the one given above. In any case, this is how the situation may be as long as the question is the construction of concrete algorithms for solving such and such concrete problems. But as soon as we, having come to conjecture the possible *unsolvability* of any algorithmic problem whatsoever of interest to us,[8] begin trying to prove that it really is unsolvable, we may immediately encounter the necessity of making the concept of an algorithm precise, of standardizing it. In fact, in order to be able to prove the non-existence of an algorithm for solving a problem in case there is no a priori reason *forbidding* its existence,[9] we must understand just what type of object it is whose conjectured existence will be refuted.

The problem of making the general concept of an algorithm precise "was in the air" by the middle of the 1930s because of attempts to get to the bottom of this situation, on the one hand, and extensive research in the field of the foundations of mathematics and mathematical logic, on the other. Papers by Church [1], Kleene [1],[10] Turing [1] and Post [1], in which this problem was solved, were published almost simultaneously. The solutions obtained differed in their technical aspects: Church's λ-conversion, the recursive functions of Herbrand–Gödel–Kleene and Turing machines[11] did not resemble each other outwardly. But the ability of these concepts to simulate each other was soon established, and Church made and justified the claim that the precise definition which was found is adequate, i.e., correctly reflects the essence of the concept of effective computability. This claim was subsequently called "Church's Thesis"—in honour of its author—by Kleene.[12] The age of the theory of algorithms as an exact mathematical discipline may be reckoned from this moment.

The precise definition that was made had an immediate effect: in the very

---

[8]As we now know, such famous algorithmic problems as the decision problem for the predicate calculus, the identity problem for semigroups and for groups, the homeomorphy problem for polyhedra and Hilbert's 10th problem turned out to be unsolvable. The hypothesis of the unsolvability of a given algorithmic problem can arise as a result of systematic failures in persistent attempts to find an algorithm solving it, which is what actually took place in the cases just listed. However, sometimes its unsolvability can be foretold on the basis of general considerations (see, for example, §64.1).

[9]By this we are emphasizing that the question is not about problems of the type "construct an algorithm which, for any three points in a plane, would yield a straight line passing through them". The unsolvability of this kind of problem can, of course, be established even without making the concept of an algorithm precise. In the given example, it follows from the existence of three points which do not lie on a line.

[10]Kleene's work was based on earlier ideas of Herbrand and Gödel.

[11]Finite combinatorial Post processes bear considerable resemblance to Turing machines, in spite of having been defined independently of them.

[12]An interesting account of the history of this question can be found in Kleene's paper [6].

same year of 1936, Church [2] and Turing [1] had already proven the unsolvability of the famous Entscheidungsproblem—the decision problem for classical predicate calculus, which at that time was regarded by Hilbert as the main problem of mathematical logic.

In 1947, the unsolvability of Thue's Problem—the identity problem for semigroups—was established by Markov [5], [6] and Post [2] independently of each other. By the same token, the first example was given of a unsolvable algorithmic problem of a really mathematical (not a mathematical logical) nature.

In 1945, Kleene, using the precise concept of an algorithm, gave an interpretation of intuitionistic arithmetic which radically advanced the development of the foundations of constructive semantics. This paper by Kleene made a significant contribution to the development of constructive logic. Markov noted in conversations with me that it exerted a strong influence on the evolution of his views regarding problems in the foundations of mathematics.[13]

Investigations directed towards clarifying the degree of effectiveness of a series of fundamental concepts and constructions of mathematical analysis were begun with the first steps in the theory of algorithms (Turing [1], [2]). In the course of time, these investigations led to the creation of so-called "constructive mathematical analysis", whose structure is based not on traditional concepts of set theory, burdened with well-known difficulties, but on a much clearer (at least with respect to principles) foundation—on a precise concept of an algorithm and on a constructive understanding of mathematical propositions.[14]

Since then, many remarkable achievements turned out to be related to the concept of an algorithm in mathematics. Side by side with the obtaining of concrete results, certain important general ideas were also developed. In particular, Davis' hypothesis on the coincidence of the recursively enumerable sets with the Diophantine sets, confirmed by Matiyasevich [2], revealed the special place in mathematics occupied by polynomials, i.e., in essence, by elementary arithmetical operations—addition and multiplication. The role of the concept of an algorithm especially increased in connection with the appearance of digital computers and the development of computer science. Realizing, however, the impossibility of anything like a complete coverage[15] of the problems pertaining to this concept, we decided to confine ourselves in this book to presenting only certain facts, those of greatest importance for the general theory of algorithms, accompanying them with a number of applications of this theory to concrete mathematical situations. In addition, considerable attention will be given to a logical analysis of the tools employed.

* * *

Markov's normal algorithms[16] are taken as the basis of the presentation in

---

[13]The reader will find a detailed analysis of these views in the commemorative paper [1] of Nagorny and Shanin.

[14]A number of results and statements of problems characteristic of constructive analysis are given in a special chapter (the ninth) of our book.

[15]The survey article by Uspenskii and Semenov [2], which is a revised version of their survey article [1], may well be recommended to the reader.

[16]Other precise formulations of the concept of an algorithm are not considered h-

this book. They are the precise formulation of the concept of an algorithm found by Markov in the course of this search for as lucid an account as possible of his solution, already mentioned, of Thue's Problem. A mathematician, becoming acquainted with the solution published by Markov and unaware of how it was obtained, is often surprised by the relative simplicity with which this result is achieved. However, this feeling is deceptive. It should not be forgotten that the lucidity and elegance of the conceptually very deep proof of the theorem on the representation of normal algorithms by associative calculi,[17] playing a central role in the range of questions under consideration, emerged as a consequence of the brilliantly chosen definition of a normal algorithm, and that the proof of its original variant, based on the calculus of λ-conversions, was far more unwieldy and complicated.

As was natural to assume, normal algorithms turned out to be equivalent to recursive functions (see Detlovs [1]), and hence also to the other more precise definitions which had been formulated earlier.

In an expanded form, the theory of normal algorithms was first expounded by Markov in his monograph [2],[18] which by now has already become a classic. The concept of a normal algorithm, possessing many merits of a fundamental as well as of a technical nature, turned out to be fruitful and convenient. Having withstood the test of time and proven its viability, it firmly took its place in the modern science of the theory of algorithms, parallel with the concepts of a recursive function and a Turing machine. In particular, it became the main tool of a large scientific school, created and headed by Markov.

Soon after its publication, the monograph [2] was translated into English (twice) and Chinese. However, the Russian edition, being quite small, has long become a bibliographic rarity, and at the present time, all that is actually available to a reader of scientific and educations literature in Russian are summaries of the theory of algorithms, inevitably failing to touch upon many important points. Therefore, the need for a new book on the theory of algorithms began to be felt a long time ago. It increased even more when there arose the necessity of reflecting certain new approaches and results which had appeared in this theory. Moreover, it had become necessary in the course of time to translate into the language of normal algorithms a number of classical facts of the general theory which had not found their way into the monograph [2].

A plan for writing a book which would satisfy this need was often considered by Markov, especially while working on his graded semantic system, which is now frequently called the "Markov Tower". Critically analyzing general problems involved in the possible structure of constructive mathematics, Markov insistently emphasized his dissatisfaction with the fact that constructive logic, on the one hand, and concrete theories of constructive mathematics, on the other,

---

us. The reader can become acquainted with them through the books of Hilbert and Bernays [1], [2], Kleene [5], Maltsev [1], Manin [1], Mendelson [1], Péter [1], Rogers[1], Uspenskii [2], [3].

[17]This is the theorem of §58.1.1 in our book.

[18]The definition of a normal algorithm, examples and a brief sketch of the general theory were published earlier in his paper [1].

are usually presented separately, isolated from each other to a certain extent. He pointed to the necessity of seeking such means of presentation whereby the theory of algorithms, constructive logic and developing branches of constructive mathematics (what he had in mind at first was constructive mathematical analysis) would be treated within the scope of a single theory based on a single language and on a single semantics for this language. It is worth mentioning that the Markov Tower was conceived precisely as the basis for such a theory, uniting the theory of algorithms with constructive logic.

A description of the tower should naturally be preceded by an informal account of the theory of normal algorithms, developed in an appropriate way. Being elaborated, it could then also be used for other purposes (the further choice of material could then depend on personal preferences, the planned size of the book, etc.).

Markov began working on the book three times, but a serious illness prevented him from fully realizing his plan.

One of the manuscript's variants had been developed quite far by Markov. Its profound conception and the consistency with which this conception is realized, makes a striking impression. The complicated, and at the same time transparent, idea of his tower is startling. Markov's unexpected striving to bring his point of view closer—remaining within the bounds of the constructive conception—to the views of Brouwer is no less startling. The brilliantly sustained inductive style of his exposition (we shall comment on this in greater detail later on), replete with genuine artistic mastery, calls for emulation.

This is the variant that I took as the basis of our book. Of course, it allowed for a variety of continuations. Unfortunately, the book acquired its specific outlines when Markov was no longer alive. In making the final decision regarding its plan, it was necessary to take into account the remaining manuscripts, Markov's oral remarks concerning the purpose and style of the book being planned and the views expressed by him in connection with the course on the theory of algorithms that I gave more than twenty times at the University of Moscow and the Moscow Institute of Physics and Engineering on his initiative and using his advice.[19] I was also guided by a number of Markov's published works and, of course, in the first place by his monograph [2], whose structure has a lot in common with that of this book. In order to emphasize these common features, I decided to retain this monograph's title for the newly written book.

Everyone who had studied Markov's works knows how well thought-out they are and how much care he took in writing them, and in this sense the book, had it appeared during his lifetime, would have turned out differently in many respects from the way it now appears before the reader. However, I should like to believe that its present variant would not have given rise to unfavourable criticism of too serious a nature on his part.

* * *

Something should now be said about certain significant differences between this book and the monograph [2].

---

[19]The contents of these lectures were in part used in writing the book.

One of them consists in the fact that considerably more attention is paid here than in [2] to the logical aspect of the presentation. In particular, a number of special sections (§§1–16,[20] 50, 67) devoted to the purely semantic side of the question are introduced into the book. Special attention is paid to the semantics of implication and negation—connectives whose understanding, as the reader will have the possibility of convincing her or himself, gives rise to real difficulties. Since, however, semantics plays a subordinate role in this book, we gradually reduce the number of semantic explanations "to zero", leaving the routine details to the reader, as we move away from the beginning of the book (and, therefore, as the readers' experience grows).

Section 67 is devoted to the consideration of an exceptionally important principle of constructive mathematics—the so-called "constructive selection principle". This principle, first noted and justified by Markov (see [7], for example) and now bearing his name, is characteristic of his constructivism. Among representatives of various trends in the foundations of mathematics, there are marked differences of opinion as to the admissibility of this principle. I therefore fit it into the book where it is first actually needed (thus, the preceding material does not depend on this principle).

Separate remarks and commentaries (§37.5, §47.2, §54.1.3, §56.2.5, §64.1, §§64.5–64.7 etc.), by their nature pertaining to problems in logic and the foundations of mathematics, were inserted by me in the basic text.

Something should be explicitly said about the set-theoretic commentaries in the book (§49 and following), which I introduced on by own initiative, understanding their *helpfulness* for a definite category of readers. Markov's position can be characterized by the consistent *non-use* of set-theoretic ideas, and he probably would not have agreed to the introduction of these commentaries, in order that no false notions about their importance occur to the reader. Acting in this manner, I specifically emphasize the fact that these commentaries are of a purely heuristic nature, that they are not part of the book's framework (and may be omitted in reading), and that no retreat whatsoever from Markov's principled position has, therefore, taken place.

One more special feature of the book is the inductive style of presentation employed in it. The attentive reader will note the book's complete lack of repeated dots followed by such turn of phrase as "etc.", so typical of mathematical literature. On the other hand, one's attention is drawn to the presence of a large number of inductive definitions and proofs by induction. This kind of presentation was planned by Markov with the aim of emphasizing the special role played by the method of induction in elementary semiotics. Consistently putting this plan into practice, I considered it correct and expedient to extend it to the most primitive concepts of semiotics—to the concept of a word, to the relations of graphic equality and difference, and also to the operation of juxtaposing words. In conjunction with the analysis conducted on the relationship between the right and left methods of induction (see §17.3), this led to a convenient and unified system for presenting elementary semiotics, excluding a number of technical complications arising, for example, when the inductions in a construction,

[20]§§1 and 4–16 basically follow the text of Markov's booklet [4].

on the one hand, and in a argument about this construction, on the other, have opposite directions. These difficulties are well known to anyone who has come up against the necessity for a careful presentation of a theory of words. Moreover, the semantic motivation of the induction principle becomes especially natural through such an approach.

The role of induction can also be clearly seen in proofs of the correctness of a number of the concrete normal algorithms given in the book (see §§28–34). These proofs are carried out in great detail. But it would be superficial to consider them excessively detailed: their essence is not only to convince the reader (who, left to her or himself, might be inclined to haste and credulity), but also to carry out, as is now said, a "verification" in accordance with definite rules—actually in the form of a deduction within a certain calculus. With a minimum of skill, such deductions can be obtained on the basis of our proofs without particular effort.

Due to the presence of variable subscripts and repeated dots, the situation is less transparent in the monograph [2], but the idea of verification is undoubtedly already present there, so this monograph may apparently be regarded as the first work in the literature where the problem of verification is developed distinctly and in a sufficiently detailed form, although only in a special case. From the point of view of the problem of verification, §§31 and 32 in our book are particularly instructive. The proofs given in them of the correctness of the doubling and reversing normal algorithms, more sustained in style than those of the monograph [2], demonstrate, in particular, a curious special feature of the application of the induction method (see the remark at the end of §31.2).

As for changes in the technical apparatus of the general theory of normal algorithms, new proofs of the branching (§39) and repeating (§40) theorems were written for this book. The Reduction Theorem (§41.7.1) and its strengthened version (§41.8.1) are proven for the relation of strong equivalence of algorithms, which had not been considered in [2]. The chapter on the universal algorithm was written anew, taking into account the achievements of Ostroukhov and Zharov. Here I was given considerable help by my disciple, Viktor Gavrilovich Zharov, to whom the expression of my gratitude can only be regarded as an agreeable duty for me. The section dealing with the normalization principle was somewhat expanded too.

The material corresponding to Chapter V of [2] was enlarged by me into §§49–50 and 52–56. The additions that were made contain a number of commentaries and necessary supplements to the basic theorems on the impossibility of algorithms, which had at one time been presented in Chapter V of the monograph [2].

Chapter IX is devoted to the examination of a number of questions in constructive analysis, whose intensive development had already been predicted by Markov during the period when he was working on the monograph [2] (see the Introduction to it). This prediction, as we now know, later proved to be correct. Unfortunately, due to lack of space, I had to make the presentation in this chapter somewhat fragmentary and much more schematic than in the others.

For the same reason, the material in Chapter VI of [2] was condensed almost to the minimum. I only took the presentation up through an example of an associative calculus with an unsolvable problem of equivalence to the empty word, and added to this the method of computable invariants, developed by Markov. The problem of recognizing invariant properties of associative calculi is presented in a concise form.

Unfortunately, the book does not contain one of Markov's most remarkable results, for which he was awarded the Chebyshev Prize by the USSR Academy of Sciences—the proof of the unsolvability of the homeomorphy problem for polyhedra. However, making good this omission would have required considerable more space and time than I had at my disposal.

* * *

The task of systematically presenting the theory of algorithms poses a number of difficult problems of both a principled and a technical nature for anyone tackling it. Without going into a detailed analysis of all the problems arising here, we nevertheless note that she or he must somehow decide in what terms the basic concepts of the theory of algorithms will be presented, whether the apparatus of set theory will be involved in these considerations and also what methods of drawing conclusions will be allowed. After the serious analysis of the traditional set-theoretic foundations of mathematics to which they have been subjected since the time of Brouwer and Hilbert, and after the criticism to which traditional Aristotelian logic has been subjected in connection with this, it seems by no means pointless to approach the tackling of the above-mentioned task in such a manner. The difficulties related to the use of set-theoretic ideas are well known. We know that in spite of all the work done by specialists, the so-called "naive" set theory can not be regarded as freed from the threat of the appearance of further paradoxes. It is also hard to ignore the fact that neither can the attempt to make this theory precise through axiomatization be regarded as successful, due to the absence of a consistency proof for the proposed axiom systems. No matter how unpleasant these facts are, the main difficulty of set theory does not even lie in them, but in a simpler and more fundamental circumstance: the substratum of the very concept of a set is unclear. Lacking the possibility of obtaining a precise answer to the question of what a set is, a mathematician is forced to be satisfied with having this concept explained to him by means of examples. But, of course, no one would be satisfied with an arithmetic in which the natural numbers were also introduced only by means of examples!

The questions just raised acquire additional significance in connection with the fact that the concept of an algorithm occupies one of the most central places in investigations on the foundations of mathematics. Hence, in choosing the methods with whose aid the theory of algorithms is to be constructed, it is natural to display maximal restraint. Taking into account what has been said, we shall be even more careful than in [2] that our presentation contains no set-theoretic inclusions and remains within the scope of constructive processes and constructively acceptable methods of reasoning. Along with this, we shall also be careful not to apply specifically constructive methods of drawing conclusions, unacceptable from the set-theoretic point of view, so that in summary we remain working "on neutral territory".

Such an approach to the construction of the theory of algorithms makes it independent of traditional set theory, or to put it even better, of the traditional *doctrine* on sets (Mengenlehre[21]). And perhaps in time this circumstance, and also the experience accumulating in the process of realizing such an approach, will promote the development of the above mentioned doctrine into a real *theory* of sets (Mengentheorie) with an exact definition of its basic concept, and not one introduced by means of examples. In any case, at the present stage of development of the foundations of mathematics, the understanding has already arisen of how a concept of a set, which although standardized, is nevertheless sufficiently flexible, can be precisely defined: we can agree to regard one-parameter conditions, formulated within the scope of any exact language having an exact semantics,[22] as sets. If we do not consider set theory to be an end in itself, then with a bit of fantasy, we can imagine that some day we shall succeed in creating a standardized, but in return, exact set theory, sufficient for the needs of mathematics, just as at one time we succeeded in constructing a standardized, but exact theory of algorithms. It is possible that such a set theory will succeed in preserving the technical merits inherent in this theory in its present state—they are unquestionable, and nobody ever denied them.

* * *

Thematically, the book is subdivided into nine chapters. The chapters consist of sections. The sections are divided into parts, which sometimes have subparts. The sections are numbered consecutively throughout the book.

The system of references adopted in the book is similar to that which was used in [2]. The propositions and formulas in each part (subpart) are numbered separately. Numbers of formulas are written down to their left in parentheses. In numbering an assertion, we include the number of its part (subpart), followed by a decimal point.

A complete reference to an assertion consists of an indication of its section, the number of its part (and also of its subpart, if it exists) and the assertion's number, which are separated from each other by decimal points, whereas the entire reference is enclosed in square brackets. For example, [§18.9.3] denotes a reference to Assertion 3 of Part 9 of §18; in referring to an assertion of the same section, the section's number is omitted. References to section, parts and formulas are made analogously, with the only difference being that no decimal point is placed before the formula's number enclosed in brackets, and that in referring to a formula of the same point (subpoint), its number is omitted. When several references are made simultaneously, they are separated from each other by commas and enclosed in common square brackets.

---

[21]There is actually a great difference between them: as a matter of fact, nobody would ever consider any religious doctrine whatsoever—for example, Christianity—to be a theory. The classical set theorists understood more clearly than we now do that the system of views being developed by them is not a simple collection of definitions, lemmas, theorems, etc., and that it is a definite world outlook, a doctrine. Hausdorff's famous book, translated into Russian with the title "Теория множеств" ("Set Theory"), is entitled "Mengenlehre" in the original. This very same term was employed by Cantor, himself.

[22]About this, see Shanin [1], §7, for example.

Items in the literature are numbered separately for each author (or group of authors).

* * *

N. A. Shanin and B. A. Kushner attentively read the manuscript of the book and made a large number of critical remarks, which on the whole were taken into account by me and helped eliminate many defects in the presentation. I had long discussions with them, as a results of which my final opinions on a number of questions of importance for the book were settled. I heartily thank Nikolai Aleksandrovich and Boris Abramovich for their friendly aid.

I should also like to emphasize that N. A. Shanin's actual influence on the book's final wording is of a far more significant nature than would follow from what was said above. At one time, N. A. Shanin was a direct participant in the formation of many ideas whose reflections can be found in the book. And as for me, as one of its authors, I have been finding out and learning a great deal from Nikolai Aleksandrovich ever since the time when I was a student. In one or another form, this is reflected in that part of the text written by me.

V. I. Khomich and I. D. Zaslavskii read certain parts of the manuscript and made remarks very much to the point. I offer them my kindest thanks for this.

I discussed a number of details of the book's general plan with S. I. Adyan, V. K. Detlovs, A. G. Dragalin and B. A. Kushner. I express my profound gratitude to all of them for their advice, which turned out to be extremely helpful for me.

Sergei Ivanovich Adyan's energetic initiative, which contributed quite a bit to the book's being finished and published, deserves special mention and thanks.

My work on the book was carried out while I was completing my translation of "Foundations of Mathematics"—Hilbert and Bernays' magnum opus. These two pursuits, intertwined in time and essence, interacted with each other and exerted a beneficial influence on each other. On the whole, this constituted a long period of my life. I express my warm and deep gratitude to my family, close friends, colleagues and everyone else who encouraged me during this time.

N. M. Nagorny

CHAPTER I

INTRODUCTION. BASIC PRINCIPLES OF CONSTRUCTIVE SEMANTICS

§1 Constructive Processes and Constructive Objects

1. A simple example of a constructive process is the construction of a row of vertical strokes

$$||||||$$

by writing one such stroke, adding its copy—another stroke—to the right of it, adding one more stroke to those obtained, then one more stroke, then one more and one more. The constructive object depicted three lines above is the result of this constructive process. This constructive object is a material body consisting of paper and dried ink, and the drawing given above is a copy of this constructive object, consisting of paper and dried typographic paint. It, too, is a constructive object, since the preparation of a copy may be considered a constructive act.

We shall call rows of vertical strokes, such as our drawing, including an "empty" row, of which not a single stroke is a part (it can be imagined in the form of a blank sheet of paper), *natural numbers*.

The natural numbers introduced in such a manner are constructive objects.

The natural number represented by our drawing is called ≪six≫ and is sometimes depicted by the Arabic numeral ≪ 6 ≫. The natural number represented by an empty row of strokes is called ≪zero≫, and depicted by the numeral ≪ 0 ≫.

2. Natural numbers are constructed from identical elementary symbols—vertical strokes. Instead of them, one can use elementary symbols of several types in an analogous manner. A constructive process will then consist in writing a certain elementary symbol, in adding another elementary symbol to the right of it, then one more, etc., until the end of the process. The resulting constructive object will in this case be a row of elementary symbols. As in the preceding case, it will be convenient to include the "empty" row of symbols, i.e., a row which does not contain a single symbol, among those we are considering.

Elementary symbols participating in constructions of this kind will be called *letters*.

Such a definition of a letter can, of course, make no claim of exactness. But the peculiarity of our situation will consist precisely in our not defining this concept in any of the traditional ways, for example ≪through the nearest genus and specific difference≫. Whenever we begin one or another concrete investigation, we shall simply agree to regard such and such symbols as elementary, understanding that all other expressions we need will be constructed from them, their component parts. Appropriate agreements are usually arrived at without difficulty. But as for the question ≪What is a letter in general?≫, we do not offer an answer to it which is suitable for all imaginable cases that might appear, since there is no need for such an answer.

The situation is somewhat eased by the circumstance that it is in fact possible to manage with the use of only two different letters, encoding the others with their aid. The reader will later become convinced of this, when getting acquainted with the material presented in parts 1–4 of §41. However, due to purely technical considerations of convenience, we would not want to deal with this at the very beginning.

For the realization of a concrete constructive process, the number of required copies of one and the same letter might not lend itself to any prior estimation. In view of this circumstance, we shall agree to assume that at the beginning of each investigation, we shall have at our disposal certain fixed "standards" of the letters participating in it, and that copies will be made of these standards as the need arises. We shall always regard copying as a constructive act.

Letters from some wide-spread alphabets (Latin, Greek, Russian, etc.,), and also scientific symbols, will ordinarily be used as our letters. A proposal to regard the symbols

$$\mathbf{I},\ a,\ b,\ *,\ \square,\ (,),\ +,\ \#,\ Щ$$

as letters would apparently meet with no objection (here a comma separates one letter from another and is not itself regarded as a letter; parentheses, on the contrary, act in the role of letters).

When choosing letters, we shall always make sure that it will be possible to state with confidence whether any two of them under consideration are *identical* or *different*. Of course, identity of and difference between letters are also conditional and also depend on the agreements made about this matter. Nevertheless, idealizing the situation somewhat, we shall assume these conditions to be strictly fulfilled in what follows.

3. When using letters, we shall, as a rule, be dealing with a finite (and even not a very large) collection of them. This collection can be given in the form of a certain list, constructed in a special way, which we shall call an *alphabet*. We may, for example, agree to write the letters of an alphabet "in a row", separating them from each other by commas and enclosing the entire list in some kind of brackets which have been especially singled out for that purpose—for example, curly.

The following list of letters,

$$\{\mathbf{I},\ a,\ b,\ *,\ \square,\ (,),\ +,\ \#,\ Щ\},$$

having already been considered by us, could serve as an example of an alphabet.

4. Letters occurring in the structure of an alphabet will be called *letters of this alphabet*. Rows of letters of a given alphabet will be called *words in this alphabet*. In particular, an empty row of letters will be called an *empty word* and will be denoted by the capital letter Λ of the Greek alphabet. An empty word is a word in any alphabet.

We especially emphasize the fact that words in alphabets, introduced by us as constructive objects, i.e. as results of certain constructions, will only be treated as such in what follows. We shall not permit ourselves such acts of imagination—regarding them as too strong—when the process of constructing all the words in a given alphabet (in principle incompletable!) is assumed, nevertheless, to have proceeded "to the end", to have been completed, as a result of which a new object of consideration arises—the *set of all words* in the given alphabet, so that the words in the given alphabet are elements of this set. This would be contrary to our basic aim—to present the theory of algorithms in terms of constructive processes.

In everyday life, by words of a given language it is customary to have in mind "meaningful" rows of this language's letters. No requirement whatsoever of meaningfulness is present in our definition. Thus, the row of the English language's letters

papagiglemma,

while not a word of the English language, is a word in the English alphabet in the sense of our definition.

6.The natural numbers, as we have convinced ourselves, may be characterized as words in the alphabet whose only letter is a vertical stroke **I**. Joining a horizontal stroke ≪–≫ and an inclined stroke≪/≫ to this alphabet, we acquire the possibility of constructing rational numbers in a natural way as words in the three-letter alphabet

$$\{\mathbf{I}, -, /\}.$$

In fact, the *integers* can be obtained by adding the words of the form

$$-N$$

where N is a natural number different from Λ, to the natural numbers, and the *rationals*—by adding the words of the form

$$M/N$$

where M is an integer and M is a natural number different from Λ, to the integers.

Rational numbers defined in such a manner are constructive objects.

7. Polynomials with rational coefficients, matrices whose entries are such polynomials, various kinds of equations and systems of equations, a variety of tables and diagrams can also be defined as constructive objects.

The large variety of constructive objects and constructive processes generating them can give rise to a desire to have general definitions of both the former and the latter. Nevertheless, there is no real need for such definitions, since each mathematical theory deals not with constructive objects in general, but with constructive objects of a certain definite kind: for example, with words in a certain alphabet. In all cases with which one has to deal, this kind can be characterized precisely, just as can the type of constructive processes generating such objects.

We shall act in the very same manner: speaking of constructive objects, we shall mean constructive objects of a definite type, which has been characterized precisely. We shall often omit the adjective "constructive" in this context.

§2 Words

1.The linearity of the arrangement of the symbols in the constructive objects with which we deal in mathematics can be violated. Such violations already arise when we begin using subscripts and exponents. Even greater deviations from linearity arise when matrices are written as arrays. Nevertheless, one usually succeeds in overcoming all such violations without any change in the essence of the matter. Thus, for example, we may agree to write out the system of words

$$(P, Q, R, S, T, U)$$

composed of the words P, Q, R, S, T, U in the form of a word,

$$\alpha P\alpha Q\alpha R\alpha S\alpha T\alpha U\alpha,$$

and the array

$$\begin{pmatrix} P, Q \\ R, S \\ T, U \end{pmatrix}$$

in the form of a word, $\beta P\alpha Q\beta R\alpha S\beta T\alpha U\beta$, where α and β denote any two distinct letters which have not been used so far. Such a joining of "new" letters to an alphabet opens up the possibility of writing out more complicated configurations, composed of objects already written out as words in this alphabet, in a row.

Devices of a similar kind make it possible to represent the objects under consideration as words. Hence, there will be no essential loss in generality if we agree to assume in the future that words in alphabets are *constructive objects of the general kind*.

2. Taking this special role of words into account, we shall go into their definitions in somewhat greater detail.

First of all, it seems advisable to organize the process of generating words in such a way that the case of the empty word turns out to be included in the general scheme. To this end, we shall assume that the first step in the construction of a word always consists in writing the empty word (nothing is written; a blank sheet of paper is taken), then adding to its right the first letter of the word (strictly speaking, this letter is simply written down), then adding the next letter to (the right of) what has been obtained, etc., until the end of the process.

Furthermore, the definition of a word should, of course, be given a "working" character, which would insure the convenience of its use in various kinds of formulations and constructions in what is to follow.

Taking what was said into account, we shall formulate this definition in the following form, in which we shall use it from now on.

Let A be any alphabet whatsoever. We shall call constructive objects *words in the alphabet* A if they are obtained as a result of the unfolding of constructive processes carried out on the basis of the following rules:

a) we assume the empty word Λ to be a word in the alphabet A;
b) if a constructive object P has already turned out to be a word in the alphabet A, then we also assume the constructive object $P\xi$, where ξ is any letter of A, to be a word in the alphabet A.

(Here $P\xi$ denotes the object obtained as a result of adding the letter ξ to the right of P; in particular, $\Lambda\xi$ denotes ξ.)

Words obtained as a result of processes ending with an application of Rule b) will be called *non-empty*. In what follows, we shall convince ourselves that every non-empty word in A can be represented uniquely in the form $P\xi$, where P is a word and ξ is a letter.

3. Definitions similar to the one just formulated are customarily called *inductive*. As a rule, an inductive definition is used for the presentation of some *type of constructive objects*. It usually has the following structure: one part of the definition declares objects of a given type to be such and such concrete constructive objects, (this is Part a) in our case), while the other parts speak of how new objects of the given type are constructed from those which are already available. (This is Part b) in our case.)

Another method, slightly different from the one just given, is often employed for describing constructive processes. We have in mind the method based on the concept of a *calculus*. This concept is one of the most important in modern mathematics, and similarly to the way a precise concept of an algorithm is used for standardizing the idea of a *prescription*, the concept of a calculus, having been made precise in an appropriate manner, can be used for standardizing our inexact, vague idea of *permission*. Without going into detail about a precise definition of a calculus, we shall only note that a concrete calculus is usually constructed in the following way. Certain constructive objects—the *axioms* of this calculus—are fixed. *Inference rules*, allowing us to construct certain new objects, starting with those already obtained, are specified. It is permitted to carry out *derivations*, starting with axioms and using inference rules. An ordered list of objects, each of whose terms is either one of the calculus' axioms or else obtained from any number of the preceding terms according to one of the inference rules, is called a derivation in the given calculus. A derivation having a given object as its last term is called a *derivation of this object*. An object is considered *derivable* in a given calculus if it is possible to construct its derivation in this calculus.

A derivation of a given object may be regarded as its *construction*. Since any initial segment of a derivation is also a derivation, every derivation contains, along with a construction of its "main" object (the derivation's final term), constructions of all the "auxiliary" objects used in it during the course of its unfolding.

There sometimes arises a need to consider so-called *derivations with analysis*, i.e. derivations containing additional information about the "origin" of its terms. If some term of the derivation is taken as an axiom of the calculus, then it is indicated in the analysis just which axiom this is. Otherwise, the inference rule and the preceding terms to which it is applied are indicated. Agreeing to write derivations "in a column", we can write down derivations with analysis "in two columns", placing the required explanation alongside each of the derivation's terms.

In a number of cases, it is convenient to consider derivations with certain restrictions. Thus, formalizing our given definition of words in the form of a calculus, we find it natural to consider derivations in which the axiom (empty word) occurs only once (and, hence, comes first), while each remaining term is obtained by applying an inference rule (transition from P to $P\xi$) to its immediate predecessor. Derivations possessing this property are just the ones that account for constructions unfolding according to our original inductive definition.

4. Comparing words in some fixed alphabet A, we can come across two words composed of the same letters arranged in the same way. But it can also happen that one of the words being compared turns out longer than the other, or that being of the same length, these words will differ in their letters at certain positions. In the former case, we shall call the words *graphically equal*, while in the latter—*graphically distinct*. With the aim of making these definitions more precise, we introduce two binary relations between words in A—the relation $\risingdotseq_{\mathrm{A}}$ of graphical equality of words in A and the relation $\neq_{\mathrm{A}}$ of graphical distinctness of words in A, assuming that:

P.1. The statement

$$\Lambda \risingdotseq_{\mathrm{A}} \Lambda$$

is true.

P.2. The statements

$$P\xi \risingdotseq_{\mathrm{A}} \Lambda, \quad \Lambda \risingdotseq_{\mathrm{A}} P\xi,$$

where P is a word in A and ξ is a letter of this alphabet, are false.

P.3. The statement

$$P\xi \risingdotseq_{\mathrm{A}} Q\eta$$

where P and Q are words in A and ξ and η are letters of this alphabet, is true if ξ and η are identical and the statement $P \risingdotseq_{\mathrm{A}} Q$ is true, and false if ξ and η are different or the statement $P \risingdotseq_{\mathrm{A}} Q$ is false.

P.4. The statement

$$P \neq_{\mathrm{A}} Q$$

is true if the statement $P \risingdotseq_{\mathrm{A}} Q$ is false, and conversely, is false if the latter is true.

It is not difficult to understand that these relations could have also been defined independently of each other, but it is technically more convenient to first define the relation $\fallingdotseq_{\mathrm{A}}$, and then express the relation $\not\fallingdotseq_{\mathrm{A}}$ in terms of it.

In case the alphabet A in question is either clear or of no concern, the subscript A in the symbols $\fallingdotseq_{\mathrm{A}}$ and $\not\fallingdotseq_{\mathrm{A}}$ will be omitted.

Note that by virtue of P.3, the question as to the graphic equality of two non-empty words is reduced to an analogous question for words of smaller length, so that we have available a method allowing us to determine for any two words P and Q in the alphabet A, which of the statements

$$P \fallingdotseq_{\mathrm{A}} Q, \quad P \not\fallingdotseq_{\mathrm{A}} Q$$

is true.

5. The *operation of juxtaposing words* will play an important role in the sequel. Its application to words P and Q in an alphabet A will consist in adding a word which graphically equals Q to the right of a word which graphically equals P, obtaining as a result a word called the *juxtaposition of the words* P and Q. We shall temporarily denote the juxtaposition of the words P and Q by $[P, Q]_{\mathrm{A}}$. In the future, when the associative law will be established for this operation, we shall replace the notation $[P, Q]_{\mathrm{A}}$ by a simpler one. We shall now give a definition of this operation which is more convenient for utilization, somewhat more precise and takes into account the inductive nature of our definition of words. Namely, having fixed an alphabet A, we set

$$[P, \Lambda]_{\mathrm{A}} \rightleftharpoons P$$
$$[P, Q\xi]_{\mathrm{A}} \rightleftharpoons [P, Q]_{\mathrm{A}}\,\xi$$

for any two words P and Q in A and any letter ξ of this alphabet (here the symbol $\rightleftharpoons$ means equality by definition). This is the definition of the operation of juxtaposition from which we shall start out in what follows.

The first equality allows one to find the juxtaposition of the word P and the empty word for any P. The second uniquely reduces finding the juxtaposition of P with a word $Q\xi$ to finding the juxtaposition of P with the word Q "preceding" $Q\xi$. Therefore, the juxtaposition of P and Q can be constructed for any P and Q.

As above, let us agree to omit the subscript in the operation symbol in cases not subject to doubt.

6. The attentive reader has already noted that in parts 4 and 5 we make essential use of a certain primitive elementary operation—the *operation of right juxtaposition of a letter to a word*. The role of this operation will later (§17.3) be analyzed in greater detail.

7. Alphabets were defined above (§1.3) as collections of symbols, which we agree to regard as letters. It is convenient to depict such a collection as a word composed of the alphabet's letters, taken in a definite order, for example, in the order of their succession. In what follows, we shall often act in precisely this fashion. Nevertheless, the order of the letters in an alphabet will not, as a rule, be of interest to us. Bearing this in mind, we shall call two alphabets consisting

of the same letters *equicomposed*, and all our investigations involving alphabets will be carried out up to an equicomposition of the alphabets figuring in them.

8. We introduced a number of basic concepts in the *semiotics* of objects having linear structure, in the theory of words. This theory considers words as complexes of symbols, taken by themselves, regardless of what these complexes mean (as a rule, symbols do mean something in mathematics). We shall now interrupt our account of the theory of words until §17, and shall consider a number of questions of principle which are going to play a very important role in what follows.

§3 Languages. Statements

1. We convey information to each other, using a definite "language". A *language* is a means of conveying information.

Speaking of language, philologists for the most part have in mind oral language—speech. But in this book we shall understand language as a graphical means of conveying information—written language. With such a means, information is conveyed in the form of a collection of words in some definite alphabet—in the form of a text.

Therefore, a language is a means of conveying information with the aid of constructive objects.

2. Information in a given language, conveyed by a sender to an addressee, should be understood by the latter, preferably uniquely. This is usually accomplished by dividing a text into pieces, called ≪statements≫.

The concept of a ≪statement≫ is defined in each language in its own way.

3. The situation regarding this concept is simplest of all in certain artificial languages. Rules for constructing statements admit precise formulations in such languages. Some of these rules directly qualify definite words as statements. Other rules codify the building up of new statements from those which have already been constructed. Those and only those objects which can be constructed in accordance with the given rules are considered to be *statements*. The rules for constructing statements form the *syntax* of the artificial language under consideration.

Our understanding of statements is further determined on the basis of the syntax. But we want statements to express something, to state something. (We are now excluding imperative and interrogative statements from consideration.) This is achieved by means of definite agreements on the understanding of statements—by means of so-called *semantic agreements*.

Having clarified with the aid of semantic agreements what a given statement expresses, we can pose the question of its truth, the question as to whether what the statement expresses is really the case. This question can be answered in various ways, but the current state of our knowledge may prevent it from being answered at all. A statement is *right(true)* if what it expresses really is the case;[1]

[1] What we are saying should be made precise. This can be done differently in different situations.

it is *wrong(false)* if what it expresses is not the case.

The semantic rules form the *semantics* of a given language. The language itself may be regarded as consisting of two parts—syntax and semantics.

The sketch of the construction of an artificial language just considered does not exhaust all cases. There are artificial ≪statement≫ languages, constructed along more complicated lines. For example, the formation of the grammatical category of constructive objects called statements can be carried out simultaneously with constructions of some other categories of objects. Statements being precisely defined independently of the meaning attributed to them is a characteristic feature of all these languages. Statements turn out to be constructive objects in all languages of this kind.

4. The situation is much less satisfactory in the so-called "natural" languages. Such a language can have a "grammar", attempting to bring order to speech and writing. This grammar also has its syntax and its semantics, but the former and the latter are usually very involved and inexact. The semantics often interferes with the syntax, which is far from being conducive to conceptual clarity.

We are forced, nevertheless, to make use of the natural languages which we have been taught since childhood. In particular, in this book we use the English language,[2] enriched only slightly with special designations and terms. We apply and study certain statements of this language. Here the term ≪statement≫ is being used in a somewhat imprecise sense., The word ≪statement≫ is a synonym of the word ≪assertion≫. A statement tells about something. It can be true, but it can be false. Incidentally, the concept of a statement in the English language[2] can probably be defined by means of a more or less exact syntax, after which our understanding of statements can be determined by means of more or less exact semantic agreements. Hence, "natural" languages can apparently be regarded as bad "artificial" languages.[3]

5. Fortunately, we shall not need a natural language in its entirety for the construction of the theory of algorithms. It will be sufficient to make use of a relatively narrow, restricted fragment of this language, having, due to its restrictedness, a precise logical structure, which in turn allows us to develop relatively precise principles for understanding statements formulated by means of this fragment. In developing these principles, account will be taken of the special, constructive nature of the objects under consideration.

If so desired, this fragment could be transformed into an artificial language with an exact syntax and an exact semantics. In fact, this was even done,[4] and the results were presented by Markov in [18]–[27]. However, teaching the reader this language would require considerable effort, and so we leave it for the future.

---

[2]Here the authors refer to the Russian language. (Translator's note).

[3]The quotation marks here are justified by the fact that "natural" languages are artificial: in fact, their grammars were also invented, i.e. created artificially, by somebody. On the other hand, many artificial special languages arose quite naturally for the sake of practical needs.

[4]For the Russian language (Translator's note).

§4 Abstraction of Potential Feasibility

1. Carrying out constructive processes, we often come up against obstacles connected with a lack of time, space and material. One usually succeeds in somehow by-passing these obstacles. However, our constructive possibilities really are limited, and there are no grounds for supposing that the obstacles caused by their restrictedness can always be obviated. Rather to the contrary, it seems that modern physics and cosmology testify to the impossibility in principle of surmounting such obstacles.

Nevertheless, in what follows we shall not take these obstacles into account in our arguments about constructive processes and constructive objects. We shall argue as through these obstacles did not exist, i.e., as if at any moment we had at our disposal the space and the time and the material needed for carrying out the next step of the constrictive process under consideration.

Acting in this manner, we shall disregard the restrictedness of our possibilities concerning space, time and material. This disregarding is customarily called the *abstraction of potential feasibility.*

2. The abstraction of potential feasibility allows us to consider arbitrarily long constructive processes and arbitrarily large constructive objects. Their feasibility is potential: they would be feasible in practice, had we available sufficient space, time and material. Invoking the abstraction of potential feasibility, we assume the following statement to be true:

2.1. *To each word in a given alphabet, any letter of this alphabet can be added to the right (left), and this yields another word in the same alphabet.*

In particular,

2.2. *One can be added to each natural number, and this yields another natural number.*

3. Applications of the abstraction of potential feasibility in what follows will to a great extent be implicit. Having in mind the potential feasibility of a process or object, we shall simply say that this process or object is feasible, omitting the word ≪potentially≫.

4. Every application of the abstraction of potential feasibility is an imaginative act. This is equally true for any of the abstractions inherent in mathematics or some other abstract science. Classical[5] mathematics invokes abstractions going much farther than those of constructive mathematics. In particular, it makes use of the abstraction of actual infinity, i.e., allows itself to reason about "infinite sets" as about given non-constructive "objects". The difference between "classicists" and "constructivists" consists in their being willing to accept abstractions of different sorts.

[5]For *brevity*, this is what we call mathematics based on set-theoretic ideas.

§5 Abstraction of Identification

1. In considering two letters of some alphabet, we state that they are identical or that they are different. For example, we see that the second and fourth letters of the word ≪state≫ are identical, but the first and last letters of this word are different. Moreover, the identity of and difference between letters are determined "by sight".

Recognition of the identity of or difference between letters is easily effected for an acceptable alphabet. For this it is only required that the differences between non-identical letters considerably exceed the small differences between identical letters.

2. In reasoning about the letters of an alphabet, we usually invoke the *abstraction of identification*, i.e., we allow ourselves to speak of two identical letters as of one and the same letter. For example, we may say that the letter ≪t≫ occurs twice in the word ≪state≫, instead of saying that two letters, identical with the letter ≪t≫, occur in this word.

3. The abstraction of identification is applicable, of course, not only to the letters of an alphabet, but also to constructive objects of other kinds, as long as it makes sense to speak about some sort of identity of these objects and it is natural to consider these objects from the point of view of identity, i.e., ignore differences between identical objects. In particular, this abstraction is applicable when considering words in a given alphabet, since the identity of words is defined in a natural way as their graphic equality.

We shall usually apply the abstraction of identification to letters and to words without any mention of this. On the contrary, we shall note the cases when this abstraction is not being applied to letters or to words.

4. Applying the abstraction of identification in what follows, we shall regard it only as a convenient means of expression which shortens formulations. Starting with such shortened formulations, we can always easily recover the longer formulations which avoid the abstraction of identification.

Invoking the abstraction of identification, one may assert that the number ≪six≫ is unique.

5. Every constructive process whose result is identical to a constructive object will be called a *construction* of this object.

The constructive processes we consider will always be regarded as consisting of a sequence of elementary constructive acts—the steps of the process. Not only the writing of a new elementary symbol, but the elimination of an old symbol (for example, by its being erased from the blackboard) may appear as a step of a constructive process. A step may also consist in the comparison of any two elementary symbols which were previously obtained, in order to establish their identity or difference. The result of such a step may be the obtaining of a third elementary symbol, depending on whether those compared are identical or different.

§6 Existence of a Constructive Object

In mathematics, results related to our ability to carry out constructive processes are often formulated as *existence theorems* asserting that there are objects satisfying certain requirements. In constructive mathematics, we understand by this that the construction of such objects is potentially feasible, i.e., that we possess a method for their construction.

This constructive understanding of existence theorems differs from how they are understood in classical mathematics. True, classicists do not like to explain their understanding of such theorems. However, it follows from the way they employ the word ≪exists≫ that their understanding of this word differs from the constructive understanding. They consider it possible, for example, to assert the existence of a constructive object satisfying a given requirement as soon as they succeed in refuting the proposition that no object satisfies this requirement by means of a ≪reductio ad absurdum≫. Moreover, a method for constructing the desired object may even be unknown. Therefore, the constructive understanding of statements about existence differs from the classical.

§7 Disjunctions

1. Constructive mathematics also differs from classical mathematics in its understanding of *disjunctions*, i.e., statements formed from two statements by joining them with the connective ≪or≫, as, for example:

> ≪there exists a prime number less than two or there exists a prime number greater than two≫

The statements joined by the connective ≪or≫ in the construction of a disjunction are called respectively, the *first* and the *second terms* of this disjunction.

In constructive mathematics, a disjunction is understood as the feasibility of singling out its true term i.e., as the potential feasibility of a constructive process yielding one of the disjunction's terms, which will be true. Thus, in the example just given, the disjunction's second term can be singled out as true, and so the disjunction under consideration is true.

2. The constructive understanding of disjunctions just described differs from that in classical mathematics. True, classicists do not explain their understanding of disjunctions. However, they regard a disjunction as true as soon as they succeed in refuting the proposition that neither of its terms is true. The ability to find a true disjunctive term is not required for this.

3. Three-term disjunctions, four-term disjunctions, etc., can be constructed analogously to the way in which this was done for two-term disjunctions. For example, a four-term disjunction is constructed from four statements, joined by means of three connectives ≪or≫. These statements are called the *terms of the disjunction.* Such a disjunction is also understood as the feasibility of singling out its true term.

In the case when we have at our disposal a list of all the constructive objects of the kind we are interested in and, moreover, this list contains more than one name, a statement on the existence of a constructive object satisfying a given requirement turns out to be equivalent, as is easily seen, to the many-term disjunction, each of whose terms asserts that one of the objects in the list satisfies the requirement introduced, where all objects in the list appear in the disjunction in the sense indicated.

§8 The Problem of Creating Constructive Mathematical Logic

1. It follows from what was said in §§6 and 7 that constructive mathematics is in need of a special logic, different from the logic of classical mathematics. As a matter of fact, at least two of the usual six logical connectives, ≪and≫, ≪or≫, ≪if..., then≫, ≪it is not the case that ≫,[6] ≪for every≫, ≪there exists... such that≫, are understood differently in constructive than in classical mathematics. A different understanding of logical connectives naturally requires a different use of them, different rules of operation—in one word, a different logic.

2. One should not, of course, be surprised by the idea of the non-uniqueness of logic. As a matter of fact, why should all our reasoning, regardless of what we are considering, be governed by the very same laws? There are no grounds for such an assumption.

The first to persistently emphasize this idea was one of the greatest mathematicians of our century, Brouwer, who proposed his own special approach to the construction of mathematics, opposed to Cantor's set-theoretical conception and called intuitionistic by its author. Brouwer convincingly showed that the construction of intuitionistic mathematics requires a special, intuitionistic logic, different from classical logic. Heyting discussed this matter in detail in Chapter I of the book [1].

3. We shall attempt to accomplish the task of developing *constructive mathematical logic* (i.e., the logic of constructive mathematics) by means of a suitable interpretation of traditional logical phrases, more precisely, of the interpretation of statements formed with the aid of these phrases. This interpretation should correspond to the aim of constructive mathematics—the study of constructive processes and constructive objects.

We have already dealt with this task for the connectives ≪or≫ and ≪there exists...such that≫ in §§6 and 7. The connective ≪and≫ offers no difficulties in this respect. The most common way of understanding it is suitable for constructive mathematics.

Statements formed from two statements, joined by the connective ≪and≫, are called *conjunctions*. The two statements which are joined by the connective ≪and≫ in the construction of a conjunction are called the *first* and *second terms* of this conjunction. We shall understand a conjunction as an assertion of the truth of both its terms.

---

[6]It would be more precise to speak of the connective ≪not≫. However, the systematic use of phrases with this word leads to awkward constructions.

Many-term conjunctions are treated analogously.

It remains to indicate acceptable constructive interpretations for the other three connectives: ≪for every≫, ≪if..., then≫ and ≪it is not the case that≫.

§9 Universal Statements

1. An important role in mathematics is played by *universal statements*, i.e., statements beginning with the words ≪every≫, ≪for every≫, ≪whatever be≫, etc., How can they be understood constructively?

2. The problem is simply solved when there is a list of all objects of the kind to which the universal statement refers. In this case, the statement that every object of this kind satisfies a given requirement can be understood as a many-term conjunction, each of whose terms asserts this about one object in the list, where all objects in the list appear in the conjunction in the sense indicated.

We are implicitly assuming here that the list of objects of the kind under consideration contains more than one name. But if this list contains only one name, we shall then, of course, understand or universal statement as the statement:

≪the unique object named in the list satisfies the given requirement≫

Finally, it is desirable to agree on a definite way of understanding universal statements about objects of a given kind in the case when their list is empty, i.e., when there are no objects of this kind. We shall then understand our statement as the trivial truth $\Lambda = \Lambda$ (independently of what the requirement which an object should satisfy consists of).

With such an understanding of universal statements in those cases when there is a list of objects of the kind under consideration, the following holds. When adding a single new object to this kind, the statement that every object of the extended kind satisfies the formulated requirement turns out equivalent to the conjunction of the previous universal statement (pertaining to the initial kind) and the statement that the added object satisfies this requirement.

3. Everything said in Part 2 pertained to the case when there is a list of objects of the kind under consideration. There may, however, be no such list, or it may even to unfeasible because there are "too many" of these objects.

Fortunately, there is another possible way of understanding universal statements constructively. For certain objects of the kind under consideration, it may be established by means of suitable arguments that they satisfy the formulated requirements. These arguments may be of such a nature that the possibility of developing analogous arguments for any other object of the kind under consideration will be clear. A conviction will then grow up in us that whatever the object of this kind given to us, we shall be able to prove (make obvious through arguments) that this object satisfies the requirement presented. We are claiming that precisely this situation prevails when we say that every object of the given kind satisfies the given requirement.

Thus, we have arrived at the following understanding of statements to the effect that every constructive object of a given kind satisfies a given requirement. Such a statement claims our ability to prove for any given object of the kind under consideration, that it satisfies the requirement presented.

The source of our confidence in being able to carry out the required proof, regardless of which object we are given, may be our limited experience with such proofs, provided that this experience has made it clear how we should act in an arbitrary case. This ability of ours to become convinced of the truth of universal statements on the basis of limited experience is what we shall call our *intuition of the universal.*

4. In the case when it is asserted that any word in a certain alphabet A satisfies a given requirement, it is possible to make extensive use of a method based on the special, inductive nature of the definition of words. It consists in the following. First we show that

1) the empty word Λ satisfies the given requirement.

Then we show that

2) if some word P in A satisfies the given requirement, then whatever be the letter ξ of the alphabet A, the word $P\xi$ also satisfies this requirement.

We may then assert that every word in A satisfies the formulated requirement.

In fact, tracing the process of constructing a word, we see that at each step of this process, some word satisfying the formulated requirement is built. In particular, this also holds for the last step, when the word of interest to us is obtained.

We shall call this method the *method of right induction on the construction of a word*, or simply *right induction.*

5. In case the alphabet A consists of the single letter **I**, right induction is the ordinary arithmetical ≪method of complete induction≫. In this case, the words in A are the natural numbers, the number zero coincides with the empty word and the method itself consists in the following.

Suppose we want to prove that every natural number satisfies a certain condition F with the natural variable N. For this we prove that:

1) zero satisfies F,

2) for every N, the following implication holds: if N satisfies F, then $N\mathbf{I}$ also satisfies F.

Then every natural number satisfies F.

As an example of how this method is applied, we shall prove with its aid the statement

(1) ≪whatever be the natural number N, we have the graphic equality

$$[\mathbf{I}, N] \asymp [N, \mathbf{I}] \gg .$$

In fact, for $N \asymp \Lambda$, we have[7]

$$\begin{aligned}[\mathbf{I}, \Lambda] &\asymp \mathbf{I} \\ &\asymp \Lambda\mathbf{I} \\ &\asymp [\Lambda, \Lambda]\mathbf{I} \\ &\asymp [\Lambda, \Lambda\mathbf{I}], \\ &\asymp [\Lambda, \mathbf{I}],\end{aligned}$$

i.e.,

(2) $$[\mathbf{I}, \Lambda] \asymp [\Lambda, \mathbf{I}].$$

Now let N be an arbitrary natural number, and assume that we have the equality

(3) $$[\mathbf{I}, N] \asymp [N, \mathbf{I}].$$

Then

$$\begin{aligned}[\mathbf{I}, N\mathbf{I}] &\asymp [\mathbf{I}, N]\mathbf{I} \\ &\asymp [N, \mathbf{I}]\mathbf{I} \qquad (3) \\ &\asymp [N, \Lambda]\mathbf{I}\mathbf{I} \\ &\asymp N\mathbf{I}\mathbf{I} \\ &\asymp [N\mathbf{I}, \mathbf{I}]\end{aligned}$$

i.e.,

(4) $$[\mathbf{I}, N\mathbf{I}] \asymp [N\mathbf{I}, \mathbf{I}].$$

Consequently, if (3) holds, then (4) also holds. In combination with (2), this allows us to conclude that the statement (1) is true.

As another example, we shall establish the truth of the statement

(5) $\ll$whatever be the natural number N, we have the graphic equality

$$[\mathbf{I}, [N, \mathbf{I}]] \asymp [[\mathbf{I}, N], \mathbf{I}] \gg .$$

For this we shall make use of the truth of (1).

In fact, when $N \asymp \Lambda$ we have

$$\begin{aligned}[\mathbf{I}, [\Lambda, \mathbf{I}]] &\asymp [\mathbf{I}, [\Lambda, \Lambda]\mathbf{I}] \\ &\asymp [\mathbf{I}, \Lambda\mathbf{I}] \\ &\asymp [\mathbf{I}, \mathbf{I}] \\ &\asymp [[\mathbf{I}, \Lambda], \mathbf{I}],\end{aligned}$$

[7]See the first paragraph in part 5 of §2.

i.e.,

$$(6) \qquad [\mathbf{I}, [\Lambda, \mathbf{I}]] \asymp [[\mathbf{I}, \Lambda], \mathbf{I}].$$

Now let N be an arbitrary natural number, and assume we have the equality

$$(7) \qquad [\mathbf{I}, [N, \mathbf{I}]] \asymp [[\mathbf{I}, N], \mathbf{I}].$$

Then

$$\begin{aligned} [\mathbf{I}, [N\mathbf{I}, \mathbf{I}]] &\asymp [\mathbf{I}, [N\mathbf{I}, \Lambda]\mathbf{I}] \\ &\asymp [\mathbf{I}, N\mathbf{I}\mathbf{I}] \\ &\asymp [N\mathbf{I}\mathbf{I}, \mathbf{I}] && [(1)] \\ &\asymp [[N\mathbf{I}, \Lambda,]\mathbf{I}, \mathbf{I}] \\ &\asymp [[N\mathbf{I}, \mathbf{I}], \mathbf{I}] \\ &\asymp [[\mathbf{I}, N\mathbf{I}], \mathbf{I}] && [(1)] \end{aligned}$$

i.e.,

$$(8) \qquad [\mathbf{I}, [N\mathbf{I}, \mathbf{I}]] \asymp [[\mathbf{I}, N\mathbf{I}], \mathbf{I}].$$

Further, taking into consideration (6) and the fact that (8) follows from (7), we conclude, using right induction, that (5) is true.

6. The method of right induction just presented requires certain clarifications.

First of all, we must explain what meaning we put in the term ≪requirement imposed on a word≫. Leaving the detailed discussion of this question for the future, we shall present in the next section enough clarification for an initial understanding.

Besides, as we shall later be convinced in §13, Part 2) of our method, dealing with the inductive step, requires special analysis. In what sense should the phrase ≪if... ,then≫ (implication) appearing in this part be understood? Should it not be understood differently in different situations? Postponing the discussion of the given problem until §§13–15, we note that the following property of implications was used in justification of our method:

6.1. *Starting from the truth of the two statements*

$$A$$

and

$$\ll \textit{if } A, \textit{then } B \gg,$$

it is possible to infer the truth of the statement

$$B$$

(the so-called ≪rule of modus ponens≫).

7. We shall also apply the method of complete induction in other forms, often employed in mathematics. For example, in order to show that every natural number N satisfies a condition F, we shall sometimes prove that for an arbitrary natural number N, the following implication is true:

≪if all numbers less than N satisfy F, then N also satisfies F≫ .

Having proved this implication, we shall also have the right to regard as proven the assertion of interest to us that every natural number N satisfies F.

Our justification for this form of the method of complete induction is the usual one. Here we need the property of implication consisting in the fact that

7.1. *Every implication with a false premise is true.*

Carrying out concrete proofs by induction, we at first shall, as a rule, indicate in precisely what sense implication is being used in the given argument. However, to the degree that the reader will accumulate experience, we shall shift this task more and more onto her or him, with no special stipulation to this effect.

8. We do not have sufficient evidence for considering the method of induction to be exhaustive. Let us note, however, that with its aid it is possible to establish practically all universal statements of interest to us which are found in the elementary theory of words.

§10 Variables. Predicates

1. It is advisable to preface our examination of the two remaining logical connectives, ≪it is not the case that≫ and ≪if..., then≫, with a few words about ≪variables≫ and ≪predicates≫.

As is well known, it is convenient to make use of literal designations in mathematics. We shall also apply them.

Certain letters singled out in the text will denote arbitrary constructive objects of a definite kind; of exactly what kind should be clear from the context. Moreover, such a letter will usually preserve its individuality throughout an entire passage in the sense that in all its occurrences it will denote one and the same object (while distinct letters may denote one and the same object, as well as distinct objects). Letters used in such a fashion will be called *free variables.* The employment of free variables is a certain extension of ordinary language.

When a free variable will be used for denoting an arbitrary object of a given kind, the objects, themselves of this kind, will be called the *admissible values* of this free variable.

In particular, the admissible values of a free variable may be the statements of some fixed language. We shall call such free variables *propositional.*

The admissible values of a free variable may be the words in a given alphabet A. We shall call such free variables *verbal variables in the alphabet* A.

We shall call verbal variables in the alphabet **I** *natural.*

The admissible values of a free variable may be the letters of a given alphabet. Such free variables will be called *literal.*

In what follows, the Latin capital letters A, B, C will generally appear as free propositional variables; the Latin capital letters M, N—as free natural variables; the Latin capital letters P, Q, R—as free verbal variables; the Greek small letters ξ, η, ς—as free literal variables.

2. The employment of free variables gives rise to such passages as[8]

(1) ≪ N is odd ≫,

(2) ≪ M is perfect ≫,

(3) ≪ M is less than N ≫ .

Replacing the free natural variables in them by their admissible values—natural numbers—we obtain, in particular, the statements

(4) ≪ **III** is odd ≫,

(5) ≪ **IIIIII** is perfect ≫,

(6) ≪ **IIIIII** is less than **III** ≫ .

Passages yielding statements as a result of replacing free variables by their admissible values are called *predicates.*

The passages (1), (2) and (3) can serve as examples of predicates.

One distinguishes *one-place, two-place, three-place,* etc., predicates depending on how many different free variables occur in them.

The predicates (1) and (2) are one-place, while (3) is a two- place predicate.

Predicates whose variables are all verbal and literal variables in an alphabet A will be called *verbal predicates in the alphabet* A.

Given a one-place predicate, the result of replacing all occurrences of its free variable by the same admissible value is a statement.

For example, taking the one-place predicate (2) and replacing the natural variable in it by the natural numbers **I**, **II**, **III**, etc., we get the statements

≪ **I** is perfect ≫,

≪ **II** is perfect ≫,

≪ **III** is perfect ≫ ...

Some of them turn out to be true, such as, for example, the statement

≪ **IIIIII** is perfect ≫ .

---

[8] The reader will probably recall that a natural number is called *perfect* if it equals the sum of its divisors which differ from it. For example, the number 6 is perfect, since its divisors are the numbers 1, 2, 3 and 6, while $6 = 1 + 2 + 3$. 24 perfect numbers have been found so far. They all turned out to be even. The decimal notation for the largest of them contains more than twelve thousand digits. At the present time, the existence of perfect odd numbers is unknown.

Therefore, the given one-place predicate singles out certain natural numbers from among the natural numbers, and may be regarded as a requirement imposed on natural numbers. In the given case, the numbers satisfying this requirement are the perfect numbers.

It is clear that one-place predicates, in general, may be regarded as requirements imposed on a constructive object of a given kind.

In an analogous manner, it is possible to regard a two-place predicate as a requirement on a pair of objects of given kinds, a three-place predicate, as a requirement imposed on a triple of objects of given kinds, etc.

Free variables whose admissible values are verbal predicates in a given alphabet will be called *free predicate variables* in the given alphabet. In what follows, the Latin capital letters F and G will, as a rule, be used as free predicate variables.

3. For the formulation of universal and existential statements, variables of a different type are used- - - so-called *bound variables.* For example, in order to express the fact that adding one to a natural number always yields a number distinct from the former, we write

(1) $$\ll \text{given any } N,\ N \neq N+1 \gg;$$

in order to express the fact that there exists a perfect number, we write

(2) $$\ll \text{there exists an } N \text{ such that } N \text{ is perfect} \gg .$$

Both these passages are statements, but they are not predicates with the free variable N. Replacing the letter N in these passages by some natural number, we would not get a statement, but a meaningless passage such as

$$\ll \text{given any } 5,\ 5 \neq 5+1 \gg,$$
$$\ll \text{there exists a } 6 \text{ such that } 6 \text{ is perfect} \gg .$$

The letter N in the passages (1) and (2) is called a bound variable.

Bound variables appear in universal statements and in existential statements. The former may have the form

(3) $$\ll \text{given any } X,\ F \text{ holds} \gg,$$

where F is a verbal predicate with the free variable X; the latter may have the form

(4) $$\ll \text{there exists an } X \text{ such that } F \gg$$

with the same F.

The statement (3) asserts that every admissible value of the variable X satisfies the predicate F; the statement (4) asserts that an admissible value of the variable X satisfying F can be constructed.

The variable X is free in the predicate F, but all its occurrences in the statements (3) and (4) are bound.

With the employment of bound variables, the statement

$\ll$ there exists an odd perfect number $\gg$

can be formulated as follows:

(5) $\ll$ there exists an N such that N is odd and N is perfect $\gg$;

the statement

$\ll$every natural number is even or imperfect$\gg$

can be formulated as follows:

(6) $\ll$ given any N, N is even or N is imperfect $\gg$.

§11 Direct Negation. Decidable Statements

1. Studying our ability to carry out constructive processes, we formulate the results of this study in the form of certain statements asserting that at the present time we are able to construct certain objects, we have mastered certain general methods, etc. It is natural to ask what the negations of statements of this kind can look like, negations also saying something about our constructive abilities.

A naive answer says that negation of the statement

(1) $\ll$at the present time we are able to construct an object satisfying the requirement... $\gg$

is the statement

(2) $\ll$at the present time we are unable to construct an object satisfying the requirement... $\gg$;

that negation of the statement

(3) $\ll$at the present time we possess a general method for ... $\gg$

is the statement

(4) $\ll$at the present time we do not possess a general method for... $\gg$

This answer must, however, be rejected because statements of the form (2) and (4) may well turn out to be true today and false tomorrow. Today we may in fact be unable to solve a given constructive problem, but may learn how to do so tomorrow. Incidentally, it is fitting to impose a conservation requirement on mathematical truths: a truth established at a given moment should remain such tomorrow, the day after tomorrow, etc. True statements of the form (1) and (3) satisfy this requirement,[9] but this can not, generally speaking, be said about true statements of the form (2) and (4). Hence, true statements of the form (2) and (4) may not always be regarded as mathematical.[10]

We see, therefore, that a naive treatment of negation can take us beyond the bounds of mathematics. Desiring to stay within these bounds, we should be concerned about acquiring a suitable positive understanding of negation, i.e., about an understanding for which the negation of a statement would also testify to some sort of ability on our part. In this connection, we shall naturally require that the negation of a statement always be incompatible with it, i.e. that the conjunction of a statement with its negation be false.

2. The problem of understanding negation positively is solved simply in certain cases. For example, this is how matters stand with statements about the identity of and difference between letters of a given alphabet. Denying that ξ and η are identical means asserting that ξ and η are different; denying that ξ and η are different means asserting that ξ and η are identical. Both these statements tell of our ability to perform certain constructive acts. The requirement of a statement's incompatibility with its negation is obviously fulfilled here. Furthermore, the *law of the excluded middle* is observed, i.e., the disjunction whose second term is the given statement, and whose first, its negation, is true. This disjunction should, of course, be understood constructively. It means that we are capable of indicating which of the statements—the given one or its negation—is true. Therefore, we are capable of determining whether the given statement is true.

In general, in the case when statements A and B are such that their conjunction is false, but their disjunction is true, we shall say that B is a *direct negation* of A. We shall say that a statement A is decidable when we have succeeded in selecting its direct negation.

In case a statement A is *decidable*, we have a method for determining whether A is true. We determine this by establishing the truth of the disjunction

$$\ll A \text{ or } B \gg,$$

where B is a direct negation of A.

3. It is easy to prove the following theorems on decidable statements and their direct negations:

3.1. *The conjunction of two decidable statements is a decidable statement. The disjuction of direct negations of two statements is a direct negation of their conjunction.*

---

[9] A fault-finding reader will detect excessive optimism here. For it has happened that people have forgotten how to do something. Let the reader imagine that effective measures have been taken to prevent such forgetting.

[10] The reader will find some concretization of these considerations in §12.1.

3.2. *The disjunction of two decidable statements is a decidable statement. The conjunction of direct negations of two decidable statements is a direct negation of their disjunction.*

3.3. *Any direct negation of a decidable statement is a decidable statement. Every decidable statement is a direct negation of its direct negation.*

4. A good example of a decidable statement is the statement obtained from the predicate §10.2 (2) by replacing the natural variable by some natural number. A direct negation of this statement is the statement obtained from the predicate

(1) $\ll M$ is imperfect $\gg$

by replacing the natural variable by the same natural number. The incompatibility of these two statements is obvious, while their disjunction is based on our having a method for determining whether a given natural number is perfect: for this one must make up a list of all its proper divisors (which is obviously possible), find their sum and compare it with the given natural number.

We shall say that a one-place predicate G with free variable X is a *direct negation of the one-place predicate* F with the same variable, if the replacement of X by any admissible value of this variable turns F and G into statements which are direct negations of each other.

The predicate (1) is a direct negation of the predicate §10.2 (2). The predicate

(2) $\ll M$ is even or M is imperfect $\gg$

is a direct negation of the predicate

(3) $\ll M$ is odd and M is perfect $\gg$ [3.1]

We shall say that a one-place predicate F is *decidable* if its direct negation is available.

The predicates (1)–(3) and §10.2 (2) are decidable.

Let us assume that a one-place predicate F with a free variable X is decidable and that G is its direct negation. Then the statement

$\ll$ for every X, F or G holds $\gg$

is true. This means that we are capable of proving any disjunction

$\ll A$ or $B \gg$,

where A and B are obtained from F and G, respectively, by replacing the variable X by any admissible value of this variable. But the proof of this disjunction yields a method for recognizing one of the statements A and B as true. Moreover, by virtue of the incompatibility of A and B, the statement A will be true only in case it is precisely the one which will be recognized as such. Therefore, in case a predicate F with a free variable X is decidable, we possess a general method of recognizing, for any allowable value of the free variable X, whether the result of replacing X in F by this admissible value is true.

5. It is clear that a decidable statement is false if we have succeeded in constructing its true direct negation.

The following theorems are easily proven:

5.1. *Every true statement is decidable. Every false statement is its direct negation.*

5.2. *Every false statement is decidable. Every true statement is its direct negation.*

5.3. *Every decidable statement is true or false.*

6. Theorems 5.1–5.3 show that in every sufficiently expressive language, a decidable statement has a great many direct negations. It is sometimes desirable to have a simple formal method of constructing, starting with any given decidable statement of a given language, one of its direct negations. Such a method actually is available in many cases.

7. The two-place predicates of identity among the letters of a given alphabet and of graphic equality among the words in a given alphabet are examples of decidable predicates. Employing the symbol $\leftrightharpoons$ for expressing the graphical equality among words (and, in particular, letters), we can write down these predicates in the form

$$\xi \leftrightharpoons \eta$$

and

$$P \leftrightharpoons Q$$

where the admissible values of the literal variables ξ and η are the letters of the given alphabet, while those of the verbal variables P and Q are the words in it.

The two-place predicates of graphic difference between letters and words are their direct negations. Employing the difference symbol $\not\leftrightharpoons$ we can write them down in the form

$$\xi \not\leftrightharpoons \eta$$

and

$$P \not\leftrightharpoons Q.$$

The law of the excluded middle holds for the predicates of graphic equality and difference:

7.1. *Whatever be the letters ξ and η of the alphabet* A, *the disjunction*

$$\ll \xi \leftrightharpoons \eta \text{ or } \xi \not\leftrightharpoons \eta \gg$$

is true.

7.2. *Whatever be the words P and Q in the alphabet* A, *the disjunction*

$$\ll P \leftrightharpoons Q \text{ or } P \not\leftrightharpoons Q \gg$$

is true.

The following statements are also true.

7.3. *Whatever be the letters* ξ *and* η *of the alphabet A, the conjunction*

$$\ll \xi = \eta \text{ and } \xi \neq \eta \gg$$

is false.

7.4. *Whatever be the words* P *and* Q *in the alphabet* A, *the conjunction*

$$\ll P = Q \text{ and } P \neq Q \gg$$

is false.

The statements 7.1–7.4 express the fact that the predicates $\xi = \eta$ and $\xi \neq \eta$ are direct negations of each other, as are the predicates $P = Q$ and $P \neq Q$.

§12 Semidecidable Statements. Strengthened Negation

1. Let us now consider a statement about the existence of a word in a given alphabet satisfying a given requirement expressed by a decidable predicate, i.e., a statement of the form

(1) $\ll$ there exists an X such that $F \gg$,

where X is a free verbal variable and F, a decidable one-place predicate with this variable.

We shall call statements of this kind *semidecidable*.

The statement §10.3(5):[11]

(2) $\ll$ there exists an N such that N is odd and perfect $\gg$

is an example of a semidecidable statement. If its direct negation were at our disposal, we would have a method allowing one to determine whether this statement is true. However, we do not have such a method. At the present time, not only do we not know whether an odd perfect number exists, but we do not even know how we might find this out. This means that at the present time we are unable to construct a direct negation of this statement and, therefore, can not assert that it exists.

But what meaning can be given to a negation of the statement (2) under consideration?

Let us recall what meaning this statement, itself, has. It means that at the present time, we possess a method of constructing an odd and perfect natural number. It would seem that this statement's negation should be understood as follows:

(3) $\ll$at the present time, we do not possess a method of constructing an odd and perfect natural number.$\gg$

[11]The question of this statement's truth is one of the most ancient mathematical problems.

The negation of our statement (2), thus understood, is true at the moment when these lines are being written: at the present time (1987), we really are unable to construct an odd and perfect natural number. However, we have no guarantee that after a certain period of time, say, by 2087, a method of constructing such a number will not have been found, that the statement (3) will not have ceased being true. Therefore, the proposed understanding of (2) would force us to deal with statements which are true at a given moment, but not guaranteed from a future refutation. This sort of statement should hardly be included in mathematics. Remaining within the realm of mathematics, we are compelled to reject the statement (3) as the way to understand the statement (2).

Let us now note that the statement (3) would be guaranteed from refutation if the statement §10.3(6):

(4) $\ll$ given any N, N is even or imperfect $\gg$

were to be proven. In fact, then the substitution of any natural number for the free natural variable in the predicate §11.4(2) would yield a true statement, and since this predicate is a direct negation of the predicate §11.4(3), the substitution of any natural number for the free variable in the predicate §11.4(3) would yield a false statement. In this case, we would never have a method of constructing an odd and perfect natural number.

This suggests that as a negation of (2), we take (4) which would guarantee the truth of (3) "for ever" and would turn out to be true for ever, itself, as soon as its truth would be established. This is precisely how we propose that the negation of the statement (2) be understood.

Investigations analogous to those just carried out can obviously be performed for any semidecidable statement. They suggest that we regard the statement

(5) $\ll$ given any X, G holds $\gg$,

where G is a direct negation of F, to be a direct negation of the semidecidable statement (1).

The statement (5) will be called a ***strengthened negation*** of the statement (1). In particular, the statement (4) is a strengthened negation of the statement (2).

2. For this definition of a strengthened negation of a semidecidable statement, the question of the law of the excluded middle naturally arises, i.e., the question as to whether the disjunction

(1) $\ll A$ or $B \gg$,

where A is a semidecidable statement and B, its strengthened negation, is always true. We have no basis for a positive answer to this question. It is not difficult to see that in case the disjunction (1) is true, B is a direct negation of A, since A and B are obviously incompatible. A is decidable in this case. But we have seen that the decidability of the statement 1(2) has not been established. Proving the truth of the disjunction (1) in the case when the role of A is played by the statement 1(2) asserting the existence of an odd perfect natural number is still an unsolved problem.[12]

3. The term ≪semidecidable statement≫ is justified by the following considerations. There is a general method establishing the truth of every true semidecidable statement 1(1). It consists in systematically testing in turn the words in the given alphabet and deciding for each of them, with the aid of the general method indicated in §11.4, whether the statement obtained from F by replacing X by this word is true. In case the statement (1) is true, i.e., when a word for which the result of the replacement is true can be constructed, we come across such a word in the course of our testing, and then break off the process.

§13 Material Implication

1. By *implications* we mean statements formed from two statements joined by the connective ≪if..., then≫, where the first statement, called the *implication's premise*, is placed between ≪if≫ and ≪then≫, while the second, called the *implication's conclusion*, is placed after ≪then≫.

The statement

(1) ≪if there exists an odd and perfect natural number, then there exists an odd and perfect natural number greater than ten≫.

can serve as an example of an implication.

2. A naive explanation of the meaning of an implication consists in the following.

Every implication asserts that having acknowledged the premise to be true, we shall have to acknowledge the conclusion to be true.

This explanation gives rise to the following questions. How can we assume the premise to be true (but what if it is false, which, for example has not been ruled out yet for the implication 1(1))? What force makes us acknowledge the conclusion to be true (but what if we find it so unpalatable that we would rather take back our acknowledgement of the premise than agree with the conclusion)? Can we not, nevertheless, somehow resist this force? Since it is not easy to answer these questions, we must recognize the given explanation of the meaning of an implication to be insufficient.

3. The following explanation of the meaning of an implication is accepted in *classical* mathematical logic: an implication asserts the same thing as the disjunction whose first term is the negation of the implication's premise and whose second is its conclusion. For example, the implication 1(1) is understood in this interpretation as the disjunction

≪it is not the case that there exists an odd and perfect natural number

---

[12]For comparison see Hilbert and Bernays [1] **1**, p.61

or

there exists an odd and perfect natural number greater than ten≫.

The understanding of an implication is therefore reduced to the understanding of a disjunction and a negation. The constructive understanding of a disjunction was spoken of above [§7]. But as for negation, we considered two interpretations of it, which are successfully applied in the appropriate cases [§§11 and 12]. Now which understanding of a negation should be chosen if we accept the interpretation of an implication just indicated?

In the case when an implication's premise is a semidecidable statement, it seems natural to understand its negation as strengthened. But then we would have to understand the implication

(1) ≪if there exists an odd and perfect natural number, then there exists an odd and perfect natural number≫

as the disjunction

(2) ≪given any N, N is even or N is imperfect

or

there exists an N such that N is odd and N is perfect≫

[§12.1, §11.4], whose truth has not yet been established. Therefore, we would have to regard the truth of the implication (1), whose conclusion coincides with its premise, as unestablished. Since this diverges too sharply from the ordinary usage of the connective ≪if..., then≫, we can not recognize the given interpretation of an implication as universally suitable.

There are, however, cases when the reduction of an implication to a negation and a disjunction, which we have considered, does not lead to a sharp divergence from ordinary word usage, and when this reduction seems natural. These are the cases of implications with decidable premises.

In these cases, we shall understand an implication as a disjunction whose first term is a direct negation of the implication's premise, and whose second is its conclusion. An implication so understood will be called a *material implication*.

Thus we call the implication

$$\ll \text{if } A, \text{ then } B \gg,$$

understood as the disjunction

$$\ll C \text{ or } B \gg,$$

where C is a direct negation of A, the material implication with premise A and conclusion B.

According to this definition, a material implication has a decidable premise.

4. As is not difficult to see, the rule of modus ponens and the rule of the chain inference are valid for material implication:

4.1. *Having a true material implication with a true premise, we may infer that the conclusion of this implication is true.*

4.2. *Having true material implications:*

$$\ll \text{if } A, \text{then } B \gg$$

and

$$\ll \text{if } B, \text{then } C \gg,$$

we may infer that the material implication

$$\ll \text{if } A, \text{then } C \gg$$

is true.

The following theorems on material implication are also easily proven:

4.3. *Having a true statement A and a decidable statement B, we may infer the truth of the material implications*

$$\ll \text{if } B, \text{ then } A \text{ and } B \gg$$

and

$$\ll \text{if } B, \text{ then } B \text{ and } A \gg .$$

4.4. *Having true material implications*

$$\ll \text{if } A, \text{ then } B \gg \tag{1}$$

and

$$\ll \text{if } C, \text{ then } D \gg, \tag{2}$$

we may infer the truth of the material implications

$$\ll \text{ if } A \text{ and } C, \text{ then } B \text{ and } D \gg$$

and

$$\ll \text{if } A \text{ or } C, \text{ then } B \text{ or D} \gg .$$

4.5. *Having a true material implication* (1), *where B is decidable, we may infer the truth of the material implication* (2), *where C is a direct negation of B and D is a direct negation of A.*

4.6. *The material implication*

$$\ll \text{if } A, \text{ then } A \gg,$$

where A is any decidable statement, is true.

The definition of a material implication immediately implies the truth of the following assertions:

4.7. *Every material implication with a true conclusion is true.*

4.8. *Every material implication with a false premise is true.*

5. In view of the non-uniqueness of a direct negation [§11.6], a material implication with a given premise and a given conclusion will also fail to be unique. In languages in which a standard direct negation has been singled out syntactically, material implication can also be standardized syntactically.

6. The following obvious sufficient condition for a material implication's decidability will be used in the sequel:

6.1. *For the decidability of a material implication, it is sufficient that its conclusion be decidable.*

Let us prove the following proposition, also used in the sequel:

6.2. *To establish the truth of a material implication, it is sufficient to prove the truth of its conclusion under the assumption that its premise is true.*

In fact, suppose we are able to prove the truth of the conclusion B of the material implication 4(1) under the assumption that its premise A is true. A is a decidable statement, i.e., we possess a method for recognizing the truth or falsity of A. Let us apply this method. If, as a result, it turns out that A is true, then we shall be able to prove the truth of B. But if it turns out that A is false, then a negation C of A will be true. Therefore,we shall always find the true term of the disjunction

$$\ll C \text{ or } B \gg,$$

i.e., we shall establish the truth of the material implication 4(1).

§14 Strengthened Implication

1. Let us now consider implications with a semidecidable premise, i.e., statements of the form

(1) $\ll$ if there exists an X such that F, then $A \gg$,

where F is a decidable predicate with the free variable X and A is a statement.

How are such implications to be understood?

The premise of the implication (1) means that at the present time, we possess a method for constructing a word in the alphabet under consideration, satisfying the predicate F. Let us try to interpret the implication (1) in the spirit of material implications as the disjunction

(2) $\ll$ at the present time we do not possess a method for constructing a word in the given alphabet, satisfying the predicate F, or $A \gg$.

One finds that this disjunction can be true at a given moment, but is not insured against refutation in the future. This will be the case when A is false and we are unable to construct a desired word now, but shall be able to do so later.

As we have already noted, statements of this kind should hardly be included in mathematics. Therefore, one must reject the disjunction (2) as the interpretation of the implication (1).

Note, however, that the statement (2) would be insured against refutation if the statement

(3) $\ll$ given any X, if F, then $A \gg$

were proven, where the implication obtained from the phrase

(4) $\ll$ if F, then $A \gg$

after the replacement of the variable X by any word in our alphabet should be understood as a material implication.

In fact, assume that the statement (3) has been proven. This means that we are able to prove every material implication obtained from (4) by replacing the variable X by an arbitrary word in our alphabet. Determine whether a method for constructing a word satisfying the predicate F is known at the present time. If not, then the first term of the disjunction (2) will be true. But if the desired method is known, then apply it for the construction of a required word. Substitute this word in place of X in the predicate F, which yields a true statement B. The material implication

$\ll$ if B, then $A \gg$,

obtained from (4) when X is replaced by the word just found, is also true. Consequently, A is true in the given case [§13.4.1]. Therefore, we have a method for singling out a true term of the disjunction (2), i.e., this disjunction is true. Thus we proved that establishing the truth of the statement (3) insures the truth of the statement (2) for ever. Moreover, the statement (3), itself, will not be "spoiled" in the course of time.

All this suggests that precisely the statement (3) be considered as the interpretation of the implication (1). We shall call the implication (1), understood in this way, the *strengthened implication* with the premise

$\ll$ there exists an X such that $F \gg$

and the conclusion A.

According to the definition, a strengthened implication always has a semidecidable premise.

2. The rule of modus ponens is valid for a strengthened implication:

2.1. *Having a true strengthened implication with a true premise, we may infer the truth of this implication's conclusion.*

In fact, the truth of the strengthened implication 1(1) means the truth of the statement 1(3). Assuming this implication's premise also to be true and arguing as above, we convince ourselves of the truth of the conclusion of the strengthened implication 1(1).

It is not difficult to see that the rule of the chain inference is valid for a strengthened implication, i.e., that we have the theorem:

2.2. *Having true strengthened implications* $\ll$if A, then $B \gg$ *and* $\ll$if B, then $C \gg$, *we may infer the truth of the strengthened implication* $\ll$if A, then $C \gg$.

3. A statement can simultaneously be a material implication and a strengthened implication. This is the case if and only if it has the form:

$$\ll \text{if there exists an } X \text{ such that } F, \text{ then } A \gg,$$

where the predicate F is decidable and such that the statement $\ll$there exists an X such that $F \gg$ is also decidable. In this case, two ways of understanding the given statement arise: the first is when the statement is understood as a material implication; the second is when it is understood as a strengthened implication. Fortunately, these two statements are compatible with each other in the following sense: as soon as the statement under consideration is true in one of its understandings, it is also true in the other. Let us prove this.

Let B denote the statement

$$\ll \text{if there exists an } X \text{ such that } F, \text{ then } A \gg .$$

Suppose B turned out to be true when regarded as a material implication. Let U be some admissible value of the variable occurring here. Determine whether it satisfies the predicate F, which is possible in view of this predicate's decidability.

If it turns out that U satisfies F, then there exists an X such that F, and by virtue of the truth of the material implication B, A is true [§13.4.1], and hence the material implication

$$\ll \text{if } U \text{ satisfies } F, \text{ then } A \gg$$

is true. This material implication is also true in the case when U does not satisfy F, since its premise is false in this case. Therefore, whatever be X, the material implication

$$\ll \text{if } F, \text{ then } A \gg$$

is true. This means that B is true when regarded as a strengthened implication.

Suppose, on the other hand, that the statement B is true when regarded as a strengthened implication, i.e., that the statement

$$\ll \text{given any } X, \text{ if } F, \text{ then } A \gg$$

is true. Determine whether the decidable statement

$$\ll \text{there exists an } X \text{ such that } F \gg \tag{1}$$

is true. If it turns out to be true, then taking a U for which F is satisfied, we see that both of the following statements are true: F and

$$\ll \text{if } F, \text{ then } A \gg .$$

Hence A is true in the given case [§13.4.1] and the statement B, regarded as a material implication, is true. It is also true in this sense when the statement (1), being the premise of the statement B, is false. Therefore, the statement B is true when regarded as a material implication. Q.E.D.

§15 Deductive Implication

1. The cases which we have considered of implications with decidable or semidecidable premises obviously do not exhaust all possibilities. It is therefore appropriate to continue our search for possible interpretations of an implication.

The usual explication of the meaning of an implication, mentioned in §13.2, seemed insufficiently clear to us because it was related to a proposal "to assume" something true and then "be forced" to acknowledge that something else is true. It is not possible, however, to make all this precise? Can we not regard the implication

$$\ll \text{if } A, \text{ then } B \gg$$

as expressing the possibility of deriving B from A in accordance with definite, fixed, formal rules, which would be reasonable? Then, in establishing the truth of an implication, we need not concern ourselves with either the meaning or the truth of its premise or its conclusion. Of course, for this it would be necessary that the premises and conclusions of the implications under consideration be stated in some formal language having an exact syntax. It is also natural to require that this language have a precise semantics, i.e., that the meaningful expressions of this language—they can be called statements or formulas—have a clearly defined meaning.

Of course, we should be allowed to draw any true statements of the given language into the process of deriving B from A, in correspondence with the usual practice of mathematical reasoning. This, however, is equivalent to regarding every true statement of the given language as derivable from A. Therefore, among the rules of inference it is natural to have the rule

> *every true statement (of the given language) is regarded as derivable from A.*

It is also natural to have the rule

> *A is regarded as derivable from A.*

Other inference rules may have the form

(1) *in case C is derivable from A, while E is related to C in such and such a way, E is regarded as derivable from A,*

or

(2) *in case C and D are derivable from A, while E is related to C and D in such and such a way, E is regarded as derivable from A,*

or some even more complicated analogous form. By relations between statements, what is meant here are purely formal syntactic relations, which can be detected by merely analyzing these statements as certain phrases, without appealing to their meaning. Rules of inference may, however, have the following more complex form:

(3) *in case every statement of some precise form $\mathfrak{S}$ is derivable from A, a statement E, related to statements of the form $\mathfrak{S}$ in such and such a way, is also regarded as derivable from A.*

For the application of rules of this form, it is necessary to know that every statement of a given form is derivable from A. We can acquire confidence in this, as in a universal statement's truth, on the basis, for example, of our limited experience in deriving certain statements of this form [cf. §9] with the aid of other inference rules.

Rules of inference characterize a derivation from A as a process. They permit us to keep getting new statements derivable from A. The order in which we apply inference rules depends, of course, on us. Hence, the process of deriving statements from A is not uniquely determined. Our confidence in the feasibility of any of certain other processes of deriving statements from A may be needed at certain stages of this process. This confidence can be based on our intuition of the universal.

2. We required above that rules of inference be "reasonable". It is appropriate to give this requirement the following meaning. Each of the inference rules should be such that being applied to true formulas, it yields only true formulas.

For example, in each of the rules of the form 1(1), the relation between C and E should guarantee the truth of E in case C is true; in every rule of the form 1(2), the relation among D, C and E should guarantee the truth of E in case D and C are true; in each rule of the form 1(3), the relation between $\mathfrak{S}$ and E should guarantee the truth of E in case every statement of the form $\mathfrak{S}$ is true.

When this requirement of reasonableness (or semantic suitability) is observed, every statement derived from a true statement will be true.

3. Having a formal language in which the statements under consideration are formulated and a sufficiently powerful system of semantically suitable rules of inference, we can define the meaning of an implication in the following manner: an implication expresses the derivability of its conclusion from its premise with a given collection of inference rules. An implication so understood will be called a *deductive implication.*[13]

A drawback in the definition of a deductive implication can be seen in its being tied to a definite system of inference rules and dependent on this system. However, certain systems of inference rules considered by the first author ([15, [21]) are natural in this respect. In any case, the concepts of deductive implication based on them are worthy of attention.

---

[13]The difficulties arising here are discussed by Hilbert and Bernays [2]. See, in particular, **II**, pp. 371, 372.

4. Another understanding of negation can be defined through deductive implication. This understanding corresponds to what is customarily called"reductio ad absurdum" in mathematics. Having an appropriate formal language and a semantically suitable system of formal inference rules for statements formulated in this language, we can define the reduction of a statement A to an absurdity as deriving a definite statement—"the standard absurdity"—from A. Just what statement is regarded as the standard absurdity in this connection, is quite inconsequential for our purposes. It is only important that its falsity be obvious.

The negation of a statement A (formulated in the given language) will then be understood as the deductive implication

$$\ll \text{if } A, \text{ then } f \gg,$$

where the standard absurdity is denoted by f. A negation understood in this way will be called a *reductional negation.*

Since a reductional negation is defined in terms of deductive implication, much of what was said above about deductive implication can also be said about a reductional negation.

Establishing the truth of a reductional negation of some statement does not require going into the meaning of this statement. A statement whose reductional negation has been established is false. Of course, one's understanding of a reductional negation depends on the chosen system of inference rules. But this does not prevent us from being interested in a reductional negation corresponding to a definite natural system of inference rules.

§16 The Idea of a Graded Semantic System

1. The interpretation of implication as deductive forces us to construct special formal languages for constructive mathematical logic. Moreover, one such language is insufficient.

In fact, having a formal language suitable for constructing statements of a definite form, we can be in a position to introduce, as described above, deductive implications with premises and conclusions of this form. However, these deductive implications, themselves, will not be expressed by formulas of this language. But it is natural for us to want to regard these deductive implication, too, as statements which can be combined with the aid of logical connectives. In particular, we may want to consider implications whose premises and conclusions will be deductive implications which have already been constructed. If we also want to interpret such implications as deductive, we shall obviously need a new formal language, capable of expressing deductive implications which connect statements of our previous language.

This leads us to the following idea of a graded construction of constructive mathematical logic.

The formal languages we need are built one after the other by means of successive extensions. An application of a logical connective to formulas of any of these languages will not always yield a formula of the same language, but always does yield a formula of the language of the next stage.

The carrying out of this plan is briefly described in Markov's notes [18]–[24]. A very significant role is played in it by the concept of a normal algorithm and other concepts related to the consideration of words in a given alphabet. In this monograph, which can serve as an introduction to the given plan, we present, in particular, this introductory material. To a large degree, it is covered by Markov's monograph [2]. However, the presentation is now more rigorous and systematic, and also better adapted to the construction of a graded semantic system of constructive logic.

At the beginning of the book, we shall try to carefully analyze the meaning of the implications and negations applied. To the degree that the reader's experience accumulates, we shall decrease the number of semantic explanations, so as not to make the presentation unwieldy. In each separate case, however, the reader can, if she or he so desires, extract details of interest from the proof. Sometimes the reader will probably want to dwell on considerations of "common sense". We shall not hinder her or him in doing this, since in the final analysis semantics should make common sense precise, but not abolish it; and as for the places where common sense can let one down, we more or less completely "fenced them in" with appropriate explanations.

CHAPTER II

SEMIOTICS OF LINEARLY ARRANGED CONSTRUCTIVE OBJECTS

§17 Words (Continuation of §2)

1. During the course of §§17–21, we shall assume that a certain alphabet A[1] is fixed. Words in it will be called simply words, and its letters, simply letters. We shall denote words by the verbal variables P, Q, R, S, letters, by the literal variables ξ, η, ς.

2. We introduced the operation of juxtaposing words in §2.5. We agreed to denote the result of applying this operation to the words P and Q by means of $[P, Q]$. It is natural to expect that the operation so introduced is subject to the associative law, i.e., for any words P, Q and R, the graphic equality

$$(1) \qquad [P, [Q, R]] \asymp [[P, Q], R]$$

is fulfilled. We shall show that this is really the case.

To this end, let us fix the words P and Q, consider the property of words R ≪to satisfy the graphic equality(1)≫ and establish by means of right induction on R [§9.4] that any word R possesses this property.

In fact, the empty word Λ possesses it, since

$$\begin{aligned} [P, [Q, \Lambda]] &\asymp [P, Q] \\ &\asymp [[P, Q], \Lambda]. \end{aligned}$$

Now let R be an arbitrary word satisfying (1), and let ξ be an arbitrary letter. Then

$$\begin{aligned} [P, [Q, R\xi]] &\asymp [P, [Q, R]\xi] \\ &\asymp [P, [Q, R]]\xi \\ &\asymp [[P, Q], R]\xi \quad [(1)] \\ &\asymp [[P, Q], R\xi], \end{aligned}$$

[1] As variables denoting alphabets, we shall make use of straight capital letters of the Russian alphabet, А, Б, В, etc., resorting to subscripts when convenient.

i.e.,

(2) $$[P, [Q, R\xi]] \rightleftharpoons [[P, Q], R\xi].$$

Therefore, the empty word possesses the indicated property, and if any word R possesses it, then, whatever be the letter ξ, the word $R\xi$ possesses it. Hence, the indicated property is possesses by any word R, and so the equality (1) holds for any P, Q and R. Q.E.D.

Note that the predicate of graphic equality is decidable [§11.7], and so the implication ≪if (1), then (2)≫ may be understood as a material implication for any P, Q, R and ξ.

Taking into account the associative law just established, we shall now—as usual—write PQ instead of $[P, Q]$, PQR instead of $[P, [Q, R]]$, etc.

It is obvious that PQR is the word obtained by writing down in order a word (graphically) equal to P, a word equal to Q and a word equal to R. We shall call the word PQR the juxtaposition of the words P, Q, and R, the word $PQRS$, the juxtaposition of P, Q, R and S, etc.

We see, therefore, that the associative law of addition for natural numbers is a special case of the general associative law for the juxtaposition of words.

3. In formulating our definition of words in a given alphabet [§2.2], we made use in an essential way of having at our disposal a certain initial elementary operation—the operation of ≪right juxtaposition≫ of this alphabet's letters to words. Its presence gave us the possibility of unfolding constructive processes in constructing words in the direction of ≪left to right≫. We justified the method of right induction with a reference to just such a means of generating words. It is natural to ask: what will happen if we replace the operation of right juxtaposition by the analogous operation of ≪left juxtaposition≫, or make use of both these operations jointly, alternating their application in an arbitrary fashion? What will take place if we complement (or replace) the method of right induction by the analogous ≪method of left induction≫?

3.1. In order to answer the first of these questions, let us consider a calculus S_{A} which depends on the alphabet A. The objects derivable in this calculus will be called lr-words in the alphabet A. As the only axiom of S_{A}, we shall take the "empty" lr-word, which we shall also denote by Λ. We shall have two rules of inference:

R) it is permitted to pass from the lr-word U to the lr-word $U\xi$;

L) it is permitted to pass from the lr-word U to the lr-word ξU (here ξ is an arbitrary letter of the alphabet A, while $U\xi$ and ξU denote the results of adding the letter ξ to the right and left, respectively, of the lr-word U).

A derivation with analysis in S_{A} will be called *normal* if the axiom occurs in it only once (in the first step) and if every lr-word which is a term in it is derived from its immediate predecessor. We shall call a derivation with analysis in S_{A} *right-sided* if there are no references to the rule L) in its analysis.

It is easy to show by induction on the length of a derivation that the terms of any right-sided normal derivation in S_{A} are words in A.

Let ξ be an arbitrary letter. Consider the operation φ_ξ on right-sided normal derivations, defined in the following manner. Taking an arbitrary derivation

W of the indicated type, we add ξ to the left of each of its terms and write down the empty word Λ above the list of words obtained in this way. Denote the list so obtained by W'. It is easy to prove by induction on its length that it is a derivation in S_A and that passing from any of its terms to the immediate successor can be carried out in accordance with the rule R). Furthermore, the axiom occurs only once in W'. Consequently, giving W' the corresponding analysis, we can convert it into a right-sided normal derivation. We shall take precisely this derivation as $\varphi_\xi(W)$.

It is easy to see that if the final term of W is P, then the final term of $\varphi_\xi(W)$ will be ξP.

The following assertion holds:

3.1.1. *For every derivation in S_A, a right-sided normal derivation S_A with the same final term can be constructed.*

In fact, taking an arbitrary derivation W in S_A, constructing any analysis for it and moving back from the final to the initial word in accordance with the analysis, we obtain a normal derivation W_1 with the same final term. If, moreover, it turns out that there is no reference to the rule L) in the analysis of W_1, then this derivation will be the desired one. But if there will be such references, then we take the initial segment of the derivation W_1 up to the first of them (inclusive). This segment will have the form

W_2
ξP (by the rule L) from the preceding formula),

where W_2 is a right-sided normal derivation and ξ is the letter added by the rule L) to the final term of the derivation W_2, which, therefore, is P. Apply the operation φ_ξ to W_2. According to the remark made above, the final term of the right-sided normal derivation $\varphi_\xi(W_2)$ will be ξP, and, replacing the initial segment under consideration of W_1 by $\varphi_\xi(W_2)$, we decrease the number of references to the rule L) in the derivation W_1 by one. Repeating the indicated procedure the necessary number of times, we arrive at a right-sided normal derivation W_3 in S_A, whose final term coincides with that of the derivation W_1, and hence also of W. Q.E.D.

Taking into account the fact that the objects derived by means of right-sided normal derivations are words, we may assert that

3.1.2. *Every lr-word is a word.*

3.2. As was already noted, along with the method of *right* induction, one can formulate the analogous method of *left* induction on the construction of words, consisting in the following. Suppose we want to prove that every word in an alphabet A satisfies a predicate F. For this we prove that:

1) Λ satisfies F

and

2) whatever be the word P, if P satisfies F, then every word ξP, where ξ is an arbitrary letter, satisfies F.

In a number of cases, it is convenient to make use of precisely this method of left induction.

The question of the relationship between the methods of right and left induction is of some definite interest. Is it possible, for example, to replace every proof carried out with the use of the method of left induction by a corresponding proof which uses only right induction? The answer to this question turns out to be affirmative. However, in order to obtain it, we must consider an operation on words as a preliminary—the operation of their *reversal.*

The result of applying this operation to a word P will be denoted by $[P^{\smile}$. We give the values of this operation by means of the following defining equalities:

$$[\Lambda^{\smile} \rightleftharpoons \Lambda,$$

$$[P\xi^{\smile} \rightleftharpoons \xi[P^{\smile}$$

(here $\rightleftharpoons$ denotes the symbol for equality by definition, while P and ξ are, respectively, an arbitrary word in and an arbitrary letter of the alphabet A).

REMARK: In mathematics, such definitions of operations are called *recursive.* The first of the defining qualities allows us to find the reversal of the empty word. The second uniquely reduces finding the reversal of a non-empty word $P\xi$ to finding the reversal of the word P "preceding" it, so that both equalities taken together give us in a natural way a method, a determined process, allowing us to start with any word P in A and obtain a word $[P^{\smile}$ in A, such that the equalities

$$[\Lambda^{\smile} \asymp \Lambda$$

and

$$[P\xi^{\smile} \asymp \xi[P^{\smile}$$

will be satisfied (here P denotes an arbitrary word in A, and ξ, and arbitrary letter of this alphabet).

Later, when we will be acquainted with the concept of a *verbal algorithm* in §26, we shall understand that this method is a certain verbal algorithm in the alphabet A. A reader acquainted with recursive functions will see in our definition a simple example of "verbal" *primitive recursion.*

A definition analogous to the one just given was already met with above (see §2.5). In what follows we shall meet with other recursively defined operations. In this connection we shall always have in mind the clarification just made.

The following assertions hold:

3.2.1. $[\xi^{\smile} \asymp \xi.$

3.2.2. $[PQ^{\smile} \asymp [Q^{\smile}[P^{\smile}.$

3.2.3. $[[P^{\smile\smile} \asymp P.$

In fact,

$$\begin{aligned}[\xi^{\smile} &\asymp [\Lambda\xi^{\smile}\\ &\asymp \xi[\Lambda^{\smile}\\ &\asymp \xi\Lambda\\ &\asymp \xi.\end{aligned}$$

Thus we have proven 3.2.1.

We shall prove 3.2.2. in the following way: let us fix P and show by means of right induction on Q that for any Q, the graphic equality $[PQ^{\smile} \mathrel{\overline{\underline{\times}}} [Q^{\smile}[P^{\smile}$ holds. In fact, it is satisfied for $Q \mathrel{\overline{\underline{\times}}} \Lambda$:

$$\begin{aligned} [P\Lambda^{\smile} &\mathrel{\overline{\underline{\times}}} [P^{\smile} \\ &\mathrel{\overline{\underline{\times}}} \Lambda[P^{\smile} \\ &\mathrel{\overline{\underline{\times}}} [\Lambda^{\smile}[P^{\smile}. \end{aligned}$$

Now let Q be arbitrary, and let the indicated equality hold. Then for any letter ξ,

$$\begin{aligned} [PQ\xi^{\smile} &\mathrel{\overline{\underline{\times}}} \xi[PQ^{\smile} \\ &\mathrel{\overline{\underline{\times}}} \xi[Q^{\smile}[P^{\smile} \\ &\mathrel{\overline{\underline{\times}}} [Q\xi^{\smile}[P^{\smile}. \end{aligned}$$

Q.E.D.

We shall also prove 3.2.3 using the method of right induction. In fact, when $P \mathrel{\overline{\underline{\times}}} \Lambda$ the equality $[[P^{\smile\smile} \mathrel{\overline{\underline{\times}}} P$ holds:

$$\begin{aligned} [[\Lambda^{\smile\smile} &\mathrel{\overline{\underline{\times}}} [\Lambda^{\smile} \\ &\mathrel{\overline{\underline{\times}}} \Lambda. \end{aligned}$$

Let it hold for some P. Then for any letter ξ,

$$\begin{aligned} [[P\xi^{\smile\smile} &\mathrel{\overline{\underline{\times}}} [\xi[P^{\smile\smile} \\ &\mathrel{\overline{\underline{\times}}} [[P^{\smile\smile}[\xi^{\smile} && [3.2.2] \\ &\mathrel{\overline{\underline{\times}}} P\xi && [3.2.3,\ 3.2.1]. \end{aligned}$$

Q.E.D.

Now let F be any property of words, and suppose we have proven by the method of *left* induction that any word possesses this property. We shall show that this assertion can be proven by the method of *right* induction applied to some other property of words.

In fact, consider the property G defined in the following way:

$$G(P) \rightleftharpoons F([P^{\smile}).$$

Then, since $[[P^{\smile\smile} \mathrel{\overline{\underline{\times}}} P$, P's having F is equivalent to $[P^{\smile}$'s having G. Besides, since $[\Lambda^{\smile} \mathrel{\overline{\underline{\times}}} \Lambda$, Λ's having F is equivalent to Λ's having G.

By assumption, we proved that for any word P, the implication

> ≪if P possesses the property F, then every word ξP also possesses it, where ξ is an arbitrary letter≫

holds. Consequently, the following implication holds for any P:

> ≪if $[P^{\smile}$ possesses the property F, then every word $\xi[P^{\smile}$ also possesses it, where ξ is an arbitrary letter (i.e., every word $[P\xi^{\smile}$)≫.

But then the implication

≪if P possesses the property G, then every $P\xi$ also possesses it, where ξ is an arbitrary letter≫

is true for any P. Since Λ possesses the property G, using the method of *right* induction we conclude that any word P possesses the property G, and so the word $[P^{\smile}$ also possesses it, which is equivalent to P's having F. Q.E.D.

Therefore, the method of left induction reduces to that of right induction.

4. The next proposition follows immediately from the definition of the graphic equality of words [§2.4]:

4.1. *If $P\xi \risingdotseq Q\eta$, then $P \risingdotseq Q$ and $\xi \risingdotseq \eta$.*

The analogous proposition is also true:

4.2. *If $\xi P \risingdotseq \eta Q$, then $P \risingdotseq Q$ and $\xi \risingdotseq \eta$.*

In fact, let $\xi P \risingdotseq \eta Q$. Then $[\xi P^{\smile} \risingdotseq [\eta Q^{\smile}$. But then $[P^{\smile}\xi \risingdotseq [Q^{\smile}\eta$ [3.2.2, 3.2.1] and so $[P^{\smile} \risingdotseq [Q^{\smile}$ and $\xi \risingdotseq \eta$. But it follows from $[P^{\smile} \risingdotseq [Q^{\smile}$ that $[[P^{\smile\smile} \risingdotseq [[Q^{\smile\smile}$ and, therefore, $P \risingdotseq Q$ [3.2.3]. Q.E.D.

Thus, every non-empty word admits a unique representation in the form $P\xi$ and a unique representation on the form ξP. The letter ξ in the unique representation a non-empty word Q in the form $P\xi$ will be called the *last letter of the word Q*; the letter ξ in the unique representation of a non-empty word Q in the form ξP will be called the *first letter of the word Q*.

4.3. *Every non-empty word has a unique first letter and a unique last letter.*

5. The statement $PR \risingdotseq QR$ is decidable [§11.7]. We shall prove the material implication

5.1. *If $PR \risingdotseq QR$, then $P \risingdotseq Q$.*

Fix the words P and Q. As for the word R, let us agree to say that it is *right cancellable* if the material implication

(1) $$\ll \text{if } PR \risingdotseq QR, \text{ then } P \risingdotseq Q \gg$$

is true. We shall show that every word is right cancellable.

The word Λ possesses this property [§13.4.6, §2.5]. Let us prove the material implication

(2) ≪ if R is right cancellable, then $R\xi$ is also right cancellable ≫ .

In fact, assume that R is right cancellable, i.e., that the material implication (1) is true. We have the true material implication

(3) $$\ll \text{if } PR\xi \risingdotseq QR\xi, \text{ then } PR \risingdotseq QR \gg \qquad [4.1].$$

Hence, we have the material implication

$$\ll \text{if } PR\xi \risingdotseq QR\xi, \text{ then } P \risingdotseq Q \gg \qquad [(3), (1), \S13.4.2],$$

meaning that $R\xi$ is cancellable. By the same token, we have proven the truth of the material implication (2). Applying the method of right induction, we convince ourselves of the truth of Theorem 5.1.

Analogously, we have the material implication

5.2. *If* $RP \leftrightharpoons RQ$, *then* $P \leftrightharpoons Q$.

In fact, let $RP \leftrightharpoons RQ$. Then $[RP^{\smile} \leftrightharpoons [RQ^{\smile}$, i.e., $[P^{\smile}[R^{\smile} \leftrightharpoons [Q^{\smile}[R^{\smile}$ [3.2.2]. But then $[P^{\smile} \leftrightharpoons [Q^{\smile}$ [5.1] and, therefore, $[[P^{\smile\smile} \leftrightharpoons [[Q^{\smile\smile}$, i.e., $P \leftrightharpoons Q$ [3.2.3]. Q.E.D.

We call Theorems 5.1 and 5.2 cancellation laws.

6. The following theorems are obvious:

6.1. $P\xi Q \not\leftrightharpoons \Lambda$.

6.2. *If* $P \not\leftrightharpoons \Lambda$, *then* $PR \not\leftrightharpoons \Lambda$ *and* $RP \not\leftrightharpoons \Lambda$ [6.1].

6.3. *If* $PR \leftrightharpoons \Lambda$, *then* $P \leftrightharpoons \Lambda$ *and* $R \leftrightharpoons \Lambda$ [6.2].

7. Finally, we should also introduce several concepts concerning alphabets, which will be helpful in the future.

First of all, we agree to say that the alphabet Б is an *extension* of the alphabet A if every letter of A is also a letter of Б. In this case, we shall sometimes write A⊂Б.

Let us introduce, moreover, the *difference* between the alphabets A and Б, which we will denote by means of (A\Б). We shall formulate the definition of (A\Б) inductively. Let A and Б be any alphabets, and let ξ be a letter which is not a letter of A. Then we set

$$(\Lambda\backslash Б) \rightleftharpoons \Lambda,$$

$$(A\xi\backslash Б) \rightleftharpoons \begin{cases} (A\backslash Б) & \text{if } \xi \text{ is a letter of Б,} \\ (A\backslash Б)\xi & \text{if } \xi \text{ is not a letter of Б.} \end{cases}$$

It is not difficult to show that (A\Б) consists of those and only those letters which are letters of A and are not letters of Б.

After this, we can introduce the *union* (A∪Б) of alphabets A and Б, setting

$$(A \cup Б) \rightleftharpoons A(Б\backslash A).$$

It is easy to verify that A∪Б consists of those and only those letters which are letters of at least one of the alphabets A and Б.

The symbols ⊂, \ and ∪ contain allusions to set-theoretic analogies. However, we should understand them, of course, in accordance with the definitions just formulated, without any appeal to set theory.

§18 Initial and Terminal Segments of Words

1. We say that a word P is an *initial segment of a word* Q *if* there exists a word X such that $Q \leftrightharpoons PX$; we say that a word P is a *terminal segment of a word* Q if there exists a word X such that $Q \leftrightharpoons XP$.

According to these definitions, we have:

1.1. *The two-place predicates*

$$\ll P \text{ is an initial segment of } Q \gg$$

and

$$\ll P \text{ is a terminal segment of } Q \gg$$

are semidecidable.

Let us prove the strengthened implication

1.2. *If Q is an initial segment of $P\xi$, then Q is an initial segment of P or $Q \rightleftharpoons P\xi$.*

In view of 1.1, the statement 1.2 may in fact be regarded as a strengthened implication, i.e., as the following universal statement:

> ≪given any X, if $P\xi \rightleftharpoons QX$, then Q is an initial segment of P or $Q \rightleftharpoons P\xi$ ≫.

For its proof, it is sufficient to prove the material implication

(1) ≪if $P\xi \rightleftharpoons QR$, then Q is an initial segment of P or $Q \rightleftharpoons P\xi$ ≫

for any words P, Q and R and any letter ξ [§14.1, §9.3].

Assume that

$$P\xi \rightleftharpoons QR. \tag{2}$$

We have $R \not\rightleftharpoons \Lambda$ or $R \rightleftharpoons \Lambda$ [§11.7.2].

If $R \not\rightleftharpoons \Lambda$, then there are an S and an η such that

$$R \rightleftharpoons S\eta. \tag{3}$$

We then have

$$P\xi \rightleftharpoons QS\eta \qquad [(2), (3)]$$

and so $P \rightleftharpoons QS$ [§17.4.1]. Therefore, in the given case Q is an initial segment of P.

But if $R \rightleftharpoons \Lambda$, then $Q \rightleftharpoons P\xi$ [(2)].

Therefore in both cases we have the disjunction

(4) ≪ Q is an initial segment of P or $Q \rightleftharpoons P\xi$ ≫ .

We proved it assuming that the equality (2) is true. Consequently, the material implication (1) holds [§13.6.2]. Q.E.D.

2. It is easy to perceive the truth of the following assertions:

2.1. *Every word is an initial and terminal segment of itself.*

2.2. *The empty word is an initial and terminal segment of every word.*

2.3. *The empty word is the unique initial (the unique terminal) segment of the empty word* [2.1, 2.2, §17.6.3].

2.4. *Every initial segment of the word P is an initial segment of the word PQ; every terminal segment of the word P is a terminal segment of the word QP.*

The latter follows from the associative law for juxtaposing words.

2.5. *The word Q is an initial segment of the word $P\xi$ if and only if the disjunction* 1(4) *holds* [1.2, 2.4, 2.1].

By virtue of 2.5, if we have available a list of all the initial segments of a word P, we can then obtain a list of all the initial segments of the word $P\xi$ by adding this word to the former list. Making use of and taking into account the fact that the list of words whose unique element is Λ is the list of all the initial segments of the word Λ [2.3], we convince ourselves of the possibility of constructing a list of any word's initial segments, i.e., of the truth of the assertion

2.6. *Given any word, a list of all its initial segments can be constructed.*

Analogously, one can prove

2.7. *Given any word, a list of all its terminal segments can be constructed.*

By virtue of 2.6 we have

2.8. *The two-place predicate*

$$\ll P \text{ is an initial segment of } Q \gg$$

is decidable.

In fact, in order to find out whether a word P is an initial segment of a word Q, it is sufficient, having constructed a list of all the initial segments of Q, to compare P with each of this list's elements, determining whether P coincides with this element.

Analogously to the way in which 2.8 was proven, we can prove the assertion

2.9. *The two-place predicate*

$$\ll P \text{ is a terminal segment of } Q \gg$$

is decidable.

3. It is now possible to construct and prove material implications whose premises are statements of the forms: $\ll X$ is an initial segment of $Y \gg$, $\ll X$ is a terminal segment of $Y \gg$ and conjunctions of statements of these forms.

3.1. *If X is an initial segment of Y and Y is an initial segment of Z, then X is an initial segment of Z.*

3.2. *If X is a terminal segment of Y and Y is a terminal segment of Z, then X is a terminal segment of Z.*

3.3. *If X is in initial (terminal) segment of Y and Y is an initial (terminal) segment of X, then $X \leftrightharpoons Y$.*

Theorems 3.1 and 3.2 are proven on the basis of the definitions of the predicates $\ll$ X is an initial segment of Y $\gg$ and $\ll$ X is a terminal segment of Y $\gg$ with the aid of the associative law for juxtaposing words. Let us prove the first of the material implications 3.3 (the one which pertains to initial segments). Suppose that X is an initial segment of Y and Y is an initial segment of X. Then words P and Q can be constructed such that

(1) $$Y \leftrightharpoons XP.$$

and

(2) $$X \leftrightharpoons YQ$$

We have

(3) $Y \leftrightharpoons YQP$ [(1), (2)],

(4) $\Lambda \leftrightharpoons QP$ [(3), §17.5.2],

(5) $Q \leftrightharpoons \Lambda$ [(4), §17.6.3],

$X \leftrightharpoons Y$ [(2), (5)].

This equality has been proven under the assumption that X is an initial segment of Y and Y is an initial segment of X. Consequently, we have the material implication

> $\ll$if X is an initial segment of Y and Y is an initial segment of X, then $X \leftrightharpoons Y$ $\gg$

The second part of Theorem 3.3 (pertaining to terminal segments) is proven analogously.

We shall prove the theorem

3.4. *If X and Y are initial segments of Z, then X is an initial segment of Y or Y is an initial segment of X.*

Let us agree to say during the course of this proof that Z is *orderly* if whatever be its initial segments X and Y, one of them is an initial segment of the other. We must prove that every word is orderly [2.3, 2.1]. Let Z be an orderly word and let ξ be a letter. We shall prove that the word $Z\xi$ is orderly. Let X and Y be initial segments of $Z\xi$. Then we have the disjunctions:

(6) $\ll$ X is an initial segment of Z or $X \leftrightharpoons Z\xi$ $\gg$,

(7) $\ll$ Y is an initial segment of Z or $Y \leftrightharpoons Z\xi$ $\gg$.

If the first terms of both of the disjunctions (6) and (7) hold, then X is an initial segment of Y or Y is an initial segment of X, since the word Z is orderly. If the second term of the disjunction (6) holds, then Y is an initial segment of X, since Y is an initial segment of $Z\xi$. Finally, if the second term of the disjunction (7) holds, then X is an initial segment of Y, since X is an initial segment of $Z\xi$. Therefore, in all cases X is an initial segment of Y or Y is an initial segment of X. Q.E.D.

We call Theorem 3.4 the *theorem on two initial segments.*

Analogously we have the *theorem on two terminal segments:*

3.5. *If X and Y are terminal segments of Z, then X is a terminal segment of Y or Y is a terminal segment of X.*

4. We shall say that X is a *proper initial segment* of Y if X is an initial segment of Y and $X \neq Y$; we shall say that X is a *proper terminal segment* of Y if X is a terminal segment of Y and $X \neq Y$.

4.1. *X is an initial segment of Y if and only if X is a proper initial segment of Y or $X = Y$* [§11.7.2].

4.2. *X is a terminal segment of Y if and only if X is a proper terminal segment of Y or $X = Y$* [§11.7.2].

Let us prove the following lemma, understanding the negation in it as a strengthened negation:

4.3. *$P\xi$ is not an initial segment of P.*

In fact the lemma's assertion means that there does not exist a word X such that

$$P = P\xi X. \tag{1}$$

The predicate of graphic equality is decidable, and its direct negation is the predicate of graphic difference [§11.7]. Hence, the statement: given any X,

$$P \neq P\xi X \tag{2}$$

is a strengthened negation of our assertion. Let us prove this.

In fact, let P, ξ, X be arbitrary, and let (1) be true. Then

$$\Lambda = \xi X \qquad [§17.5.1],$$

which is false by virtue of the definition of graphic equality [§2.4]. Consequently, (1) is false, and so (2) is true by virtue of the definition of graphic difference [§2.4]. Q.E.D.

Now let us prove the theorem

4.4. *X is a proper initial segment of $P\xi$ if and only if X is an initial segment of P.*

Let X be a proper initial segment of the word $P\xi$. Then X is an initial segment of $P\xi$ and $X \neq P\xi$. Hence, X is an initial segment of P or $X = P\xi$. The second term of the last disjunction is easily eliminated, and so X is an initial segment of P. Now suppose, on the other hand, that X is an initial segment of P. Then $X \neq P\xi$ [4.3]. Nevertheless, X is an initial segment of $P\xi$ [2.4]. Therefore, X is a proper initial segment of $P\xi$.

The following modifications of the theorem on two initial segments hold:

4.5. *If X and Y are initial segments of Z, then X is an initial segment of Y or Y is a proper initial segment of X* [3.4, 4.1, 2.1].

4.6. *If X and $Y\xi$ are initial segments of Z, then X is an initial segment of Y or $Y\xi$ is an initial segment of X* [4.5, 4.4].

Analogously to the way in which we proved Theorems 4.3–4.6, we can prove the following four theorems:

4.7. *The word ξP is not a terminal segment of the word P.*

4.8. *X is a proper terminal segment of ξP if and only if X is a terminal segment of P.*

4.9. *If X and Y are terminal segments of Z, then X is a terminal segment of Y or Y is a proper terminal segment of X.*

4.10. *If X and ξY are terminal segments of Z, then X is a terminal segment of Y or ξY is a terminal segment of X.*

4.11. *PQ is an initial segment of PR if and only if Q is an initial segment of R.*

The next two theorems deal with initial segments of the juxtaposition of two words.

4.12. *Every initial segment of the word PQ either is an initial segment of P or else has the form PR, where R is an initial segment of Q* [4.5, 4.11].

4.13. *Every initial segment of the word PQ either is a proper initial segment of the word P or else has the form PR where R is an initial segment of Q* [4.5, 4.11].

Analogous statements can be proven for terminal segments of the juxtaposition of two words.

4.14. *The empty word is the unique proper initial segment of a one-letter word* [4.4, 2.3, 2.2].

4.15. *Every non-empty initial segment of the word ξP has the form ξQ, where Q is an initial segment of P* [4.13, 4.14].

We obviously have the following two true statements:

4.16. *The word P is an initial segment of the word PQ.*

4.17. *No word is a proper initial segment of itself.*

It is easily proven that

4.18. *If P is a proper initial segment of Q and Q is an initial segment of R, then P is a proper initial segment of R* [§17.6.2, §17.5.2].

4.19. *If Q is an initial segment of P, then P is not a proper initial segment of Q* [4.18, 4.17].

5. Looking over lists of two words' initial segments, we can determine whether these two words have a non-empty common initial segment. We shall say that the words P and Q are *relatively prime on the left* if there does not exist a non-empty word which is an initial segment of P as well as of Q.

5.1. *The predicate*

$$\ll X \text{and } Y \text{ are relatively prime on the left} \gg$$

is decidable.

Instead of $\ll X$ and Y are relatively prime on the left$\gg$, we shall sometimes say $\ll X$ is relatively prime on the left to $Y \gg$.

5.2. *The empty word is relatively prime on the left to every word* [2.3].

5.3. *If $\xi \neq \eta$, then ξP is relatively prime on the left to ηQ* [4.15, §17.4.2].

5.4. *If $\xi \eqcirc \eta$, then ξP is not relatively prime on the left to ηQ.*

This follows directly from the definition of relative primality on the left.

5.5. *If X is relatively prime on the left to Y and $Y \neq \Lambda$, then X is relatively prime on the left to $Y\xi$* [5.2, 5.4, 5.3].

We shall prove the theorem

5.6. *Whatever be the words P and Q, we can construct R, S and T such that*

$$P \eqcirc RS, \tag{1}$$
$$Q \eqcirc RT \tag{2}$$

and S is relatively prime on the left to T.

Let us fix the word P. We shall say (during the course of this proof) that the word Q is *proper* if one can construct words R, S and T satisfying the conditions (1) and (2) and such that S is relatively prime on the left to T. It is required to prove that every word (in the alphabet under consideration) is proper.

The empty word is proper, since, setting by definition

$$R \rightleftharpoons \Lambda,$$
$$S \rightleftharpoons P,$$
$$T \rightleftharpoons \Lambda,$$

we see that the conditions imposed on R, S and T are satisfied for $Q \eqcirc \Lambda$ [5.2].

Assume that the word Q is proper; we shall prove that every word $Q\xi$, where ξ is any letter, will then also be proper.

According to the definition of proper, one can construct words R, S and T satisfying the conditions (1) and (2) and such that S is relatively prime on the left to T. If it turns out that $T \neq \Lambda$, then S is relatively prime on the left to $T\xi$ [5.5] and, setting

(3) $$R_1 \rightleftharpoons R,$$
(4) $$S_1 \rightleftharpoons S,$$
(5) $$T_1 \rightleftharpoons T\xi$$

by definition, we see that

(6) $$P = R_1S_1 \quad [(1), (3), (4)],$$
(7) $$Q\xi = R_1T_1 \quad [(2), (3), (5)]$$

and that S_1 is relatively prime on the left to T_1. Therefore, in this case the word $Q\xi$ is proper [(6), (7)]. Now let

(8) $$T = \Lambda.$$

We then have

(9) $$Q = R \quad [(2), (8)].$$

If $S = \Lambda$, then S is relatively prime on the left to ξ [5.2] and, setting

(10) $$R_1 \rightleftharpoons R,$$
(11) $$S_1 \rightleftharpoons \Lambda,$$
(12) $$T_1 \rightleftharpoons \xi$$

by definition,we see that the equalities (6) and (7) hold [(1), (10), (11), (2), (8), (12)] and that S_1 is relatively prime on the left to T_1. Therefore, the word $Q\xi$ is also proper in this case.

Finally, if $S \neq \Lambda$, then there exist an η and a V such that

(13) $$S \neq \eta V.$$

If $\eta = \xi$ here, then we have

(14) $$Q = R \quad [(2), (8)]$$

and, setting

(15) $$R_1 \rightleftharpoons Q\xi,$$
(16) $$S_1 \rightleftharpoons V,$$
(17) $$T_1 \rightleftharpoons \Lambda$$

by definition, we see that the conditions (6) and (7) are satisfied [(1), (14), (13), (15), (14), (17)] and that S_1 is relatively prime on the left to T_1 [(17), 5.2]. Consequently, $Q\xi$ is proper.

But if $\eta \neq \xi$, then we again have (14) and, setting

$$R_1 \rightleftharpoons Q, \tag{18}$$

$$S_1 \rightleftharpoons \eta V, \tag{19}$$

$$T_1 \rightleftharpoons \xi \tag{20}$$

by definition, we see that the conditions (6) and (7) are satisfied [(1), (14), (13), (18)–(20)] and that S_1 is relatively prime on the left to T_1 [(19), (20), 5.3]. Consequently, $Q\xi$ is proper and Theorem 5.6 is proven.

5.7. *If the conditions* (1) *and* (2) *are satisfied and* S *is relatively prime on the left to* T, *then every common initial segment of the words* P *and* Q *is an initial segment of the word* R [4.12, §17.5.2].

5.8. *Whatever be the words* P *and* Q, *there exists a unique triple of words* R, S *and* T *satisfying the conditions* (1) *and* (2) *and such that* S *is relatively prime on the left to* T [5.7, 3.4, §17.5.2].

Given any words P and Q, the word R in the unique triple of words R, S and T satisfying the conditions (1) and (2) and such that S is relatively prime on the left to T, will be called the greatest common initial segment of the words P and Q.

5.9. *Whatever be the words* P *and* Q, *there exists a unique greatest common initial segment of these words* [5.8].

5.10. *Every common initial segment of the words* P *and* Q *is an initial segment of their greatest common initial segment* [5.7].

5.11. *Every initial segment of the greatest common initial segment of two words is their common initial segment* [2.4].

5.12. *A word* R *is a common initial segment of the words* P *and* Q *if and only if it is an initial segment of their greatest common initial segment* [5.10, 5.11].

6. Now consider the case when our alphabet consists of the letter **I**. The words in this alphabet are the natural numbers [§1.1]. We shall say that the natural number M is *majorized* by the natural number N if M is an initial segment of N; we shall say that the natural number M is *less* than the natural number N if M is a proper initial segment of N. In what follows, the notation

$$M \leqslant N$$

will mean that M is majorized by N, while the notation

$$M < N$$

will mean that M is less than N.

The following statements are special cases of statements found in previous parts of this section (the letters M, N, K, L, H are used as natural variables).

6.1. $M \leqslant M$.

6.2. $\Lambda \leqslant M$.

6.3. *If* $M \leqslant \Lambda$, *then* $M \asymp \Lambda$.

6.4. *If* $M \leqslant N$, *then* $M \leqslant NL$.[2]

6.5. $M \leqslant N|$ *if and only if* $M \leqslant N$ *or* $M \asymp N|$.

6.6. *If* $M \leqslant N$ *and* $N \leqslant L$, *then* $M \leqslant L$.

6.7. *If* $M \leqslant N$ *and* $N \leqslant M$, *then* $M \asymp N$.

6.8. $M \leqslant N$ *if and only if* $M < N$ *or* $M \asymp N$.

6.9. $M < N|$ *if and only if* $M \leqslant N$.

6.10. $MN \leqslant ML$ *if and only if* $N \leqslant L$.

6.11. *If* $M < |$, *then* $M \asymp \Lambda$.

6.12. *It is not the case that* $M < M$ [4.17].

6.13. *If* $N \leqslant M$, *then it is not the case that* $M < N$ [4.19].

6.14. *If* $M < N$ *and* $N \leqslant L$, *then* $M < L$ [4.18].

The next two statements pertain to relative primality of natural numbers on the left.

6.15. *If* $M \neq \Lambda$ *and* $N \neq \Lambda$, *then* M *and* N *are not relatively prime on the left* [5.4].

6.16. *If* M *and* N *are relatively prime on the left, then* $M \asymp \Lambda$ *or* $N \asymp \Lambda$.

We have the obvious lemma

6.17. *Every initial segment of a natural number is a natural number; every terminal segment of a natural number is a natural number.*

It is now easy to prove

6.18. $M \leqslant N$ *or* $N \leqslant M$ [5.6, 6.17, 6.16].

Further, we prove that

6.19. $M < N$ *or* $N < M$ *or* $M \asymp N$ [6.18, 6.8].

In the sequel, we shall also make use of the fact that

---

[2] We call the reader's attention to the fact that here NL means the *juxtaposition* of the words N and L, i.e., in essence, the *sum* of the natural numbers N and L. Concerning *multiplication* of natural numbers, see §20.

6.20. $M \leqslant MN$ [4.16].

7. We shall sometimes apply the following *method of induction on the initial segments of a word.*

In order to prove that every initial segment of the word P satisfies the verbal one-place predicate F, we prove that

1) Λ satisfies F,
2) for every word X and every letter ξ, we have the implication

≪if $X\xi$ is an initial segment of P and X satisfies F, then $X\xi$ satisfies F≫.

Then every initial segment of the word P satisfies F.

In fact, let us construct the one-place verbal predicate G

(1) ≪if X is an initial segment of P, then X satisfies F≫.

Λ satisfies G, since Λ satisfies F.

Let X satisfy G and let ξ be a letter. Assume that $X\xi$ in an initial segment of P. Then X is an initial segment of P [3.1]. Since X satisfies G, X satisfies F. But then, by the second property of the predicate F, the word $X\xi$ satisfies F. We arrived at this conclusion assuming that $X\xi$ is an initial segment of P. Consequently, the implication

≪if X is an initial segment of P, then $X\xi$ satisfies F≫

holds. It means that $X\xi$ satisfies G. We arrived at this conclusion assuming that X satisfies G. Consequently, the implication

≪if X satisfies G, then $X\xi$ satisfies G≫

is true [§13.6.2]. Here X is any word in the alphabet under consideration and ξ is any of its letters. Applying the method of right induction on the construction of a word, we convince ourselves that every word (in the alphabet under consideration) satisfies the predicate G. But this just means that every initial segment of the word P satisfies the predicate F [(1)].

In the special case when the alphabet under consideration consists of the single letter $\mathbf{I}$, the method of induction on the initial segments of a word turns into the following *method of restricted arithmetical induction.*

In order to prove that every natural number majorized by the natural number N satisfies the one-place arithmetical predicate F, we prove that:

1) Λ satisfies F,
2) for every natural number M, we have the implication

≪if $M\mathbf{I} \leqslant N$ and M satisfies F, then $M\mathbf{I}$ satisfies F≫.

Then every natural number majorized by the number N satisfies the predicate F.

8. If the word X is an initial segment of the word Z, then the word Y, such that $XY \rightleftharpoons Z$, is unique [§17.5.2]. We shall call this unique word the *terminal complement of the word X in the word Z*, and shall denote it by

(1) $$(X \leftharpoonup Z).$$

According to this definition, we have

8.1. *If X is an initial segment of Z, then the terminal complement of X in Z is a terminal segment of Z.*

8.2. *The expression* (1) *makes sense if and only if X is an initial segment of Z.*

8.3. $(X \leftharpoondown XY) \asymp Y$.

8.4. *If X is an initial segment of Z, then*

$$X(X \leftharpoondown Z) \asymp Z. \tag{2}$$

We shall prove

8.5. *If $X\xi$ is an initial segment of Z, then*

$$(X \leftharpoondown Z) \asymp \xi(X\xi \leftharpoondown Z). \tag{3}$$

In fact, then X is an initial segment of Z [3.1], and we have the equality (2) and the analogous equality

$$X\xi(X\xi \leftharpoondown Z) \asymp Z. \tag{4}$$

By virtue of (2) and (4), the equality (3) holds [§17.5.2].

The following statements are easily proven:

8.6. $(\Lambda \leftharpoondown Z) \asymp Z$.

8.7. $(Z \leftharpoondown Z) \asymp \Lambda$.

8.8. *If X is an initial segment of Z, then*

$$(X \leftharpoondown ZY) \asymp (X \leftharpoondown Z)Y.$$

9. If the word X is a terminal segment of the word Z, then the word Y, such that $YX \asymp Z$, is unique [§17.5.1]. We shall call this unique word the *initial complement of the word X in the word Z,* and shall denote it by

$$(X \rightharpoondown Z). \tag{5}$$

We have

9.1. *If X is a terminal segment of Z, then the initial complement of X in Z is an initial segment of Z.*

9.2. *The expression* (5) *makes sense if and only if X is a terminal segment of Z.*

9.3. $(X \rightharpoondown YX) \asymp Y$.

9.4. *If X is a terminal segment of Z, then*

$$(X \rightharpoondown Z)X \asymp Z.$$

9.5. *If ξX is a terminal segment of Z, then*

$$(X \rightharpoondown Z) \asymp (X\xi \rightharpoondown Z)\xi.$$

9.6. $(\Lambda \rightharpoondown Z) \asymp Z.$

9.7. $(Z \rightharpoondown Z) \asymp \Lambda.$

9.8. *If X is a terminal segment of Z, then*

$$(X \rightharpoondown YZ) \asymp Y(X \rightharpoondown Z).$$

10. We shall introduce the *operation of subtraction* for natural numbers as a special case of the formation of a terminal complement. Namely, for $M \leqslant N$ we set

$$(N - M) \rightleftharpoons (M \leftharpoondown N)$$

We have

10.1. *If $M \leqslant N$, then $(N - M)$ is a natural number* [8.1].

10.2. *The expression $(N - M)$ makes sense if and only if $M \leqslant N$* [8.2].

10.3. $(MN - M) \asymp N$ [8.3].

10.4. *If $M \leqslant N$, then*

$$M(N - M) \asymp N \quad [8.4].$$

10.5. *If $M\mathbf{I} \leqslant N$, then*

$$(N - M) \asymp \mathbf{I}(N - M\mathbf{I}) \quad [8.5].$$

10.6. $(N - \Lambda) \asymp N$ [8.6].

10.7. $(N - N) \asymp \Lambda$ [8.7].

10.8. *If $M \leqslant N$, then*

$$(NL - M) \asymp (N - M)L \quad [8.8].$$

10.9. $N\mathbf{I} \asymp \mathbf{I}N$ [§9.5(1)].

10.10. *If $M \leqslant N$, then $M(N - M) \asymp (N - M)M$.*

PROOF: During the course of this proof we shall say that N is *proper* if for every M such that $M \leqslant N$, the equality

(1) $$M(N - M) \asymp (N - M)M$$

holds. It is required to prove that every natural number is proper.

Λ is proper [6.3, 10.7].

Let us assume that N is proper and prove the propriety of $N\mathbf{I}$.

Let $M \leqslant N\mathbf{I}$. Then $M \leqslant N$ or $M \asymp N\mathbf{I}$ [6.5]. If $M \leqslant N$, then the equality (1) holds, by virtue of which we have

$$
\begin{aligned}
M(N\mathbf{I} - M) &\asymp M(N - M)\mathbf{I} && [10.8]\\
&\asymp (N - M)M\mathbf{I} && [(1)]\\
&\asymp (N - M)\mathbf{I}M && [10.9]\\
&\asymp (N\mathbf{I} - M)M && [10.8].
\end{aligned}
$$

Therefore, when $M \leqslant N$ we have the equality

$$M(N\mathbf{I} - M) \asymp (N\mathbf{I} - M)M.$$

It also holds when $M \asymp N\mathbf{I}$ [10.7]. Consequently, the natural number $N\mathbf{I}$ is proper. By the same token, we have proven the (material) implication

$\ll$if N is proper, then $N\mathbf{I}$ is proper$\gg$ [§13.6.2].

Applying now the method of arithmetical induction, we convince ourselves of the truth of statement 10.10.

We shall now prove the commutativity of addition for natural numbers.

10.11. $$MN \asymp NM$$

$$
\begin{aligned}
MN &\asymp M(MN - M) && [10.3]\\
&\asymp (MN - M)M && [6.20,\ 10.10]\\
&\asymp NM && [10.3].
\end{aligned}
$$

11. In many cases, especially when natural and rational numbers will be appearing not as the main objects under consideration, but as their numerical characteristics, we shall employ the generally accepted arithmetical symbolism, using the sign $\ll = \gg$ to denote the equality of numbers, the signs $\ll + \gg$ and $\ll - \gg$ for denoting the operations of addition and subtraction, etc. In those cases when no collisions will arise, we shall not make any special stipulations.

§19 Length of a Word. Projection of a Word on an Alphabet

1. The *length* $[X^{\delta}$ of the word X can be inductively defined[3] by the equalities

(1) $$[\Lambda^{\delta} \rightleftharpoons \Lambda,$$

(2) $$[X\xi^{\delta} \rightleftharpoons [X^{\delta}\mathbf{I}.$$

The following statements are proven by the method of induction:

[3]See the remark in §17.3.2 (pp.39–40).

1.1. *The length of every word is a natural number.*

1.2. $[XY^\delta \rightleftharpoons [X^\delta[Y^\delta$.

1.3. $[XYZ^\delta \rightleftharpoons [X^\delta[Y^\delta[Z^\delta$ [1.2].

1.4. *If X is an initial segment of Y, then* $[X^\delta \leqslant [Y^\delta$ [1.2].

1.5. *If $X \not\rightleftharpoons \Lambda$, then* $[X^\delta \not\rightleftharpoons \Lambda$ [(2)].

1.6. $[X^\delta \rightleftharpoons \Lambda$ *if and only if* $X \rightleftharpoons \Lambda$ [(1), 1.5].

1.7. *If X is a proper initial segment of Y, then* $[X^\delta < [Y^\delta$ [1.2, 1.6, §17.5.2].

2. Using the method of restricted arithmetical induction, it is not difficult to prove the statement

2.1. *Whatever be the word X and the natural number M majorized by the length of X, there exists an initial segment Y of X, such that*

$$[Y^\delta \rightleftharpoons M. \tag{1}$$

Let us also prove

2.2. *Whatever be the word X and the natural number M less than $[X^\delta$, there exists a proper initial segment Y of X, such that*

$$[Y^\delta \rightleftharpoons M.$$

In fact, under these conditions $M \leqslant [X^\delta$ [§18.6.8], and so there exists an initial segment Y of the word X, satisfying the condition (1) [2.1]. Since $M < [X^\delta$, we have

$$[Y^\delta < [X^\delta$$

and so, according to the definition of a proper initial segment,

$$[Y^\delta \not\rightleftharpoons [X^\delta$$

and, consequently, $Y \not\rightleftharpoons X$. Therefore, Y is a proper initial segment of X. Q.E.D.

We shall prove

2.3. *If $X \rightleftharpoons YXZ$, then $Y \rightleftharpoons \Lambda$ and $Z \rightleftharpoons \Lambda$.*

Let $X \rightleftharpoons YXZ$. Then

$$\begin{aligned} [X^\delta &\rightleftharpoons [Y^\delta[X^\delta[Z^\delta \quad [1.3] \\ &\rightleftharpoons [X^\delta[Y^\delta[Z^\delta \quad [\S 18.10.11], \end{aligned}$$

from which we obtain in turn

$$\begin{aligned} [Y^\delta[Z^\delta &\rightleftharpoons \Lambda \quad [\S 17.5.2], \\ [Y^\delta &\rightleftharpoons \Lambda \quad [\S 17.6.3], \\ [Z^\delta &\rightleftharpoons \Lambda \quad [\S 17.6.3], \\ Y &\rightleftharpoons \Lambda \quad [1.6], \\ Z &\rightleftharpoons \Lambda \quad [1.6]. \end{aligned}$$

Further, we have

2.4. *If* $X \neq \Lambda$, *then* XY *is not an initial segment of* Y [2.3].

3. We shall call the word obtained from a word P in an alphabet A by deleting all letters which are not letters of an alphabet Б, the *projection of the word* P *onto the alphabet* Б. We shall denote the projection of P onto Б by the symbol

$$[P^{Б}.$$

More precisely:

$$[\Lambda^{Б} \rightleftharpoons \Lambda,$$

$$[P\xi^{Б} \rightleftharpoons \begin{cases} [P^{Б}\xi, & \text{if } \xi \text{ is a letter of the alphabet Б,} \\ [P^{Б}, & \text{if } \xi \text{ is not a letter of the alphabet Б.} \end{cases}$$

These equalities are just what will serve as the inductive definition[4] of the projection of a word onto an alphabet. With their aid, it is easy to prove the following two statements:

3.1. *The projection of a word in an alphabet* A *onto an alphabet* Б *is a word in the projection of the alphabet* A[5] *onto the alphabet* Б.

3.2. $[PQ^{Б} \asymp [P^{Б}[Q^{Б}$.

4. The length of a word P's projection onto an alphabet Б is expressed in the following way:

$$[[P^{Б\delta}.$$

The value of this expression can be found for any P.

We have the statements:

4.1. $[[PQ^{Б\delta} \asymp [[P^{Б\delta}[[Q^{Б\delta}$ [1.2, 3.2].

4.2. *If* $Q \asymp RPS$, *then* $[[P^{Б\delta} \leqslant [[Q^{Б\delta}$ [4.1].

4.3. *If* ξ *is a letter of the alphabet* Б, *then* $[[\xi^{Б\delta} \asymp \mathsf{I}$.

4.4. $[[\Lambda^{Б\delta} \asymp \Lambda$.

4.5. *If the alphabets* A *and* Б *have no letters in common, then*

$$[P^{Б} \asymp \Lambda$$

for every word P *in the alphabet* A [3.1].

[4]See the remark in §17.3.2.

[5]In this context, the alphabet A is regarded as a word (see §2.7). Its projection onto the alphabet Б is again an alphabet.

§20 Multiplication of a Word by a Natural Number

1. Let us define inductively[6] the *product* $(P \times N)$ *of the word* P *in the alphabet* A *by the natural number* N with the aid of the equalities

$$(1) \qquad (P \times \Lambda) \rightleftharpoons \Lambda,$$

$$(2) \qquad (P \times N\mathsf{I}) \rightleftharpoons (P \times N)P.$$

The following statements are easily proven.

1.1. $(P \times N)$ *is a word in the alphabet* A *under consideration* [induction on N for a fixed P].

1.2. $(P \times MN) \asymp (P \times M)(P \times N)$ [(1), (2), induction on N for fixed P and M].

1.3. $(MN \times L) \asymp (M \times L)(N \times L)$ [(1), (2), §18.10.11, induction on L for fixed M and N].

1.4. $((P \times M) \times N) \asymp (P \times (M \times N))$ [(1), (2), 1.2, induction on N for fixed P and M].

1.5. $(\Lambda \times N) \asymp \Lambda$ [induction on N, (1), (2)].

1.6. $(\mathsf{I} \times N) \asymp N$ [induction on N, (1), (2)].

1.7. $[(\xi \times N)^{\delta} \asymp N$ [(2), §19.1(2), induction on N].

1.8. $(M \times N) \asymp (N \times M)$ [induction on N for a fixed M, 1.5, (2), 1.6, 1.3].

2. In what follows, we shall often use the notation ξ^N instead of $(\xi \times N)$, where ξ is a letter and N is a natural number.

§21 The Least Number Theorem

1. Let us prove the following lemma:

1.1. *Every word in the alphabet* O**I** *has the form* RS, *where* R *is a word in the alphabet* O *and* S *is either empty or else has a stroke,* **I**, *as an initial segment.*

Let W be a word in the alphabet O**I**. We construct a word V, defining it in the following manner:

$$(1) \qquad V \rightleftharpoons (\mathrm{O} \times [W^{\delta}).$$

V is a word in the one-letter alphabet O [§20.1.1]. Moreover,

$$(2) \qquad [V^{\delta} \asymp [W^{\delta} \quad [(1), §20.1.7].$$

---

[6] See the remark in §17.3.2.

According to Theorem §18.5.6, we can construct words R, S and T in the alphabet OI, such that

$$W \leftrightharpoons RS, \tag{3}$$

$$V \leftrightharpoons RT \tag{4}$$

and S and T are relatively prime on the left.

Since V is a word in the alphabet O, R and T are also words in this alphabet [(4)]. We have $[R^\delta[S^\delta \leftrightharpoons [R^\delta[T^\delta$ [(2)–(4), §19.1.2], from which

$$[S^\delta \leftrightharpoons [T^\delta \quad [\S 17.5.2]. \tag{5}$$

If $T \leftrightharpoons \Lambda$, then $S \leftrightharpoons \Lambda$ [§19.1.6, (5)] and the first of the possibilities specified in Lemma 1.1 is realized. But if $T \not\leftrightharpoons \Lambda$, then $S \not\leftrightharpoons \Lambda$ [§19.1.6, (5)]. In this case, the non-empty word T in the alphabet O has the letter O as an initial segment, while the non-empty word S in the alphabet OI has a letter distinct from O as an initial segment, in view of the relative primality of the words S and T on the left [§18.5.4]. Therefore, in the given case S has the letter I as an initial segment, and the second of the possibilities indicated in Lemma 1.1 is realized. In view of (3) and the fact that R is a word in O, Lemma 1.1 is proven.

2. Let F be a one-place arithmetical predicate, and let M be a natural number. We shall say that M is the *least number satisfying* F if M satisfies F, but every natural number less than M does not satisfy F.

We shall prove the following Least Number Theorem:

2.1. *Given any decidable one-place arithmetical predicate F and any natural number satisfying it, the least number satisfying F can be found.*

Let F be a decidable one-place arithmetical predicate, and let the natural number N satisfy F. Construct the *characteristic operator of the predicate* F, defining it by the following equalities:

$$[M^\chi \rightleftharpoons \begin{cases} \mathsf{I}, & \text{if } M \text{ satisfies } F, \\ \mathsf{O}, & \text{if } M \text{ does not satisfy } F. \end{cases} \tag{1}$$

Here $[M^\chi$ denotes the result of applying[7] the characteristic operator of the predicate F to the number M. Further, construct the *evolute of the predicate* F, defining it by the following recursive equalities:[8]

$$[\Lambda^\epsilon \rightleftharpoons \Lambda, \tag{2}$$

$$[M\mathsf{I}^\epsilon \rightleftharpoons [M^\epsilon[M^\chi. \tag{3}$$

Here $[M^\epsilon$ denotes the result of applying the evolute of the predicate F to the natural number M. The characteristic operator of the predicate F is obviously

[7] The decidability of F allows one to find the value of this result for a given M.

[8] See again the remark in §17.3.2. Taking the experience acquired by the reader into account, we shall limit the number of references to this remark in the future.

applicable to every natural number and changes it into a letter of the alphabet OI. The evolute of the predicate F is also applicable to every natural number and changes it into a word in the alphabet OI.

The following equality is easily proven by the method of arithmetical induction:

$$[[M^{\varepsilon\delta} \leftrightharpoons M. \tag{4}$$

We shall prove that for every natural number M and every initial segment T of the word $[M^{\varepsilon}$, we have the equality

$$[[T^{\delta\varepsilon} \leftrightharpoons T. \tag{5}$$

During the course of the proof, we shall say that the natural number M is *proper* if (5) holds for every initial segment T of $[M^{\varepsilon}$. The predicate ≪ M is proper≫ is obviously decidable. We shall prove that every natural number is proper.

The empty word Λ is proper [§18.2.3, §19.1(1), (2)]. Assume that the word M is proper, and let us then prove that the word $M\mathsf{I}$ is also proper.

Let T be any initial segment of $[M\mathsf{I}^{\varepsilon}$, i.e., of the word $[M^{\varepsilon}[M^{\chi}$. Since $[M^{\chi}$ is a one-letter word, T either is an initial segment of $[M^{\varepsilon}$ or else coincides with $[M^{\varepsilon}[M^{\chi}$ [§18.1.2]. In the first case, (5) holds, since M is proper. In the second case, this equality also holds, since in this case

$$\begin{aligned} [T^{\delta} &\leftrightharpoons [[M^{\varepsilon\delta}[[M^{\chi\delta} && [\S 19.1.2] \\ &\leftrightharpoons M\mathsf{I} && [(4), (1)], \end{aligned}$$

and hence

$$\begin{aligned} [[T^{\delta\varepsilon} &\leftrightharpoons [M\mathsf{I}^{\varepsilon} \\ &\leftrightharpoons [M^{\varepsilon}[M^{\chi} && [(3)] \\ &\leftrightharpoons T. \end{aligned}$$

Applying the method of arithmetical induction, we conclude that every natural number is proper.

Let us now conclude the proof of Theorem 2.1.

Since $[N\mathsf{I}^{\varepsilon}$ is a word in the alphabet OI, we can apply to it Lemma 1.1, by virtue of which there exist words R and S, such that

$$[N\mathsf{I}^{\varepsilon} \leftrightharpoons RS, \tag{6}$$

where R is a word in the alphabet O, while S either is empty or else has the letter I as an initial segment.

We shall now show that the length of the word R is the required least number satisfying the predicate F. Since N satisfies the predicate F, we have

$$[N^{\chi} \leftrightharpoons \mathsf{I}. \tag{7}$$

Hence,

$$(8) \qquad RS \Bumpeq [N^{\varepsilon}[N^{\chi} \quad [(6), (3)]$$
$$\Bumpeq [N^{\varepsilon}\mathbf{I} \quad [(7)].$$

Consequently, RS, unlike R, is not a word in the alphabet O. Therefore,

$$S \neq \Lambda$$

and, according to the alternative for S ($S \Bumpeq \Lambda$ or S has $\mathbf{I}$ as an initial segment), S has $\mathbf{I}$ as an initial segment, i.e., there exists a word U such that

$$(9) \qquad S \Bumpeq \mathbf{I}U.$$

We therefore have

$$[N\mathbf{I}^{\varepsilon} \Bumpeq R\mathbf{I}U \quad [(6), (9)].$$

Consequently, $R\mathbf{I}$ is an initial segment of $[N\mathbf{I}^{\varepsilon}$. By what was proven above, it follows from this that

$$(10) \qquad [[R\mathbf{I}^{\delta\varepsilon} \Bumpeq R\mathbf{I}.$$

On the other hand,

$$[[R\mathbf{I}^{\delta\varepsilon} \Bumpeq [[R^{\delta}\mathbf{I}^{\varepsilon} \quad [\S 19.1.2]$$
$$(11) \qquad \Bumpeq [[R^{\delta\varepsilon}[[R^{\delta\chi} \quad [(3)].$$

Therefore,

$$(12) \qquad R\mathbf{I} \Bumpeq [[R^{\delta\varepsilon}[[R^{\delta\chi} \quad [(10), (11)].$$

Here $[[R^{\delta\chi}$ is a one-letter word. Consequently,

$$[[R^{\delta\chi} \Bumpeq \mathbf{I} \quad [\S 17.4.1],$$

but this means that the natural number $[R^{\delta}$ satisfies the predicate F [(1)]. It remains to prove that no smaller number satisfies F.

Let L be a natural number such that

$$L < [R^{\delta}.$$

Then there exists a proper initial segment Y of the word R such that

$$(13) \qquad [Y^{\delta} \Bumpeq L$$

[§19.2.2]. There exists a T such that

$$(14) \qquad R \Bumpeq YT.$$

$T \neq \Lambda$ since $Y \neq R$. Y and T are words in the alphabet O, since R is a word in this alphabet. Since $T \neq \Lambda$, there exists a word U such that

$$T \asymp \mathrm{O}U. \tag{15}$$

We have

$$\begin{aligned} [N\mathsf{I}^{\epsilon} &\asymp RS && [(6)] \\ &\asymp YTS && [(14)] \\ &\asymp Y\mathrm{O}US && [(15)]. \end{aligned}$$

Therefore, $Y\mathrm{O}$ is an initial segment of the word $[N\mathsf{I}^{\epsilon}$. Hence

$$\begin{aligned} Y\mathrm{O} &\asymp [[Y\mathrm{O}^{\delta\epsilon} \\ &\asymp [[Y^{\delta}\mathsf{I}^{\epsilon} && [\S 19.1\ (2)] \\ &\asymp [[Y^{\delta\epsilon}[[Y^{\delta\chi} && [(3)]. \end{aligned}$$

Here $[[Y^{\delta\chi}$ is a one-letter word. Consequently,

$$[[Y^{\delta\chi} \asymp \mathrm{O} \quad [\S 17.4.1],$$

i.e.,

$$[L^{\chi} \asymp \mathrm{O} \quad [(13)].$$

This means that L does not satisfy the predicate F. Q.E.D.

§22 Pairs of Words

1. In this section, we shall as before consider an alphabet A to be fixed and, unless stipulated to the contrary, shall understand ≪words≫ to mean words in the alphabet A. Besides, we shall also hold fixed a letter α—an arbitrary letter which is not a letter of A. We shall employ the letters P, Q, R and S as verbal variables in A, and the letters X and Y, as verbal variables in the alphabet Aα. Let us prove the statement

1.1. *If $P\alpha X \asymp Q\alpha Y$, then $P \asymp Q$ and $X \asymp Y$.*

Let $P\alpha X \asymp Q\alpha Y$. Then $P\alpha$ and Q are initial segments of one and the same word, and so $P\alpha$ is an initial segment of Q or Q is an initial segment of P [§18.4.6]. But $P\alpha$ is not an initial segment of Q, since Q, being a word in the alphabet A, does not contain the letter α. Therefore, Q is an initial segment of P. Analogously, we perceive that P is an initial segment of Q. Consequently, $P \asymp Q$. From this it further follows that $X \asymp Y$ [§17.5.2].

Analogously, we can prove

1.2. *If $X\alpha R \asymp Y\alpha S$, then $R \asymp S$ and $X \asymp Y$.*

From the statement 1.1 follows the statement

1.3. *If $P\alpha R \rightleftharpoons Q\alpha S$, then $P \rightleftharpoons Q$ and $R \rightleftharpoons S$.*

2. The word $P\alpha R$ in the alphabet Aα will be called the *pair of words P* and R in the alphabet A. The word P will be called the *first element of the pair* $P\alpha R$, and the word R, the *second element of this pair.* By virtue of 1.3 and the definition of a pair, we have

2.1. *Two pairs of words coincide if and only if their first elements coincide and their second elements coincide.*

3. In our definition of a pair of words, the letter α plays the role of a ≪punctuation mark≫, separating the first element of a pair from the second. Of course, this role can be assigned to any other letter ε, as long as it does not occur in the alphabet A. Acting in this manner, we shall specifically emphasize the preferential nature of ε, and shall speak not merely about pairs, but about ε-pairs of words in the alphabet A.

§23 Occurrences

1. We say that the *word P occurs in the word Q* if there exists a pair of words such that R is the first element of this pair, S is the second and

$$Q \rightleftharpoons RPS.$$

We also say then that the word Q *contains* the word P.

1.1. *The two-place verbal predicate*

(1) $$\ll P \text{ occurs in } Q \gg$$

is semidecidable.

This is obvious from the definition of the given predicate.

In view of 1.1, implications with premises of the form (1) may be regarded as strengthened.

We shall prove the strengthened implication

1.2. *If P occurs in Q, then there exists a U such that P is an initial segment of U and U is a terminal segment of Q.*

Let us hold P and Q fixed. In expanded form, the implication 1.2 looks as follows:

1.2.1. *If there exists a pair of words $R\alpha S$ such that $Q \rightleftharpoons RPS$, then there exists a U such that P is an initial segment of U and U is a terminal segment of Q.*

By virtue of our agreement about understanding strengthened implications, the statement 1.2.1 means the same thing as the statement

1.2.2. *Given any pair of words $R\alpha S$, if $Q \rightleftharpoons RPS$, then there exists a U such that P is an initial segment of U and U is a terminal segment of Q.*

Statement 1.2.2 is just what we must prove. For this we must be able to establish, for any pair of words $R\alpha S$, the truth of the material implication

> ≪if $Q \rightleftharpoons RPS$, then there exists a U such that P is an initial segment of U and U is a terminal segment of Q ≫.

We actually are capable of constructing such a U: it is sufficient to put $U \rightleftharpoons PS$.

Let us prove the converse strengthened implication

1.3. *If there exists a U such that P is an initial segment of U and U is a terminal segment of Q, then P occurs in Q.*

Hold P and Q fixed. It is not hard to see that the implication 1.3 means the same thing as the statement

1.3.1. *Given any U, if P is an initial segment of U and U is a terminal segment of Q, then P occurs in Q.*

For the proof of this statement, it is necessary to be able to prove the following material implication for any U:

1.3.2. *If P is an initial segment of U and U is a terminal segment of Q, then P occurs in Q.*

Assume that the premise of 1.3.2 is true. then there exist words R and S such that

$$U \rightleftharpoons PS$$

and

$$Q \rightleftharpoons RU.$$

Hence $Q \rightleftharpoons RPS$, and the pair $R\alpha S$ can of course be formed from the words R and S. Therefore, P occurs in Q. The material implication 1.3.2 is therefore proven, and together with it, Theorem 1.3 is also proven.

Theorems 1.2 and 1.3 may be reformulated as follows:

1.4. *If P occurs in Q, then P is an initial segment of some terminal segment of Q.*

1.5. *If P is an initial segment of some terminal segment of Q, then P occurs in Q.*

These theorems yield the possibility of composing a list of all the words occurring in a given word Q in the following manner.

We compose a list of all the terminal segment of the word Q [§18.2.7]. For each element of this list (i.e., for each terminal segment of Q) we compose a list of all its initial segments [§18.2.6]. We combine the latter lists, getting the desired list of all the words occurring in Q. In fact, every word occurring in Q is an element of the resulting list [1.4] and every element on the resulting list occurs in Q [1.5].

Having a list of all the words occurring in a given word Q at our disposal, we can determine for every word P, whether it occurs in Q. For this we need only compare P with each element of the list [§11.7.2]. If the comparison yields a positive result for some element of the list, then P occurs in Q. But if the comparison yields a negative result for every element, then P does not occur in Q.

By virtue of this, we have

1.6. *The two-place verbal predicate*

$$\ll P \text{ occurs in } Q \gg$$

is decidable.

In view of this, 1.4 and 1.5 may be regarded as material implications. Since they are mutually converse, they can be combined, yielding

1.7. *The word P occurs in the word Q if and only if P is an initial segment of some terminal segment of Q.*

Analogously, we can prove

1.8. *The word P occurs in the word Q if and only if P is a terminal segment of some initial segment of Q.*

2. The following two statements are easily proven:

2.1. *Every initial and terminal segment of the word P occurs in P. In particular, in every word there occur this word, itself, and the empty word.*

2.2. *If P occurs in Q and Q occurs in R, then P occurs in R.*

Let us prove the statement

2.3. *If P occurs in Q and Q occurs in P then $P \leftrightharpoons Q$.*

Suppose P occurs in Q and Q occurs in P. Then there exist pairs, $R\alpha S$ and $T\alpha U$, such that

$$(1) \qquad Q \leftrightharpoons RPS$$

and

$$(2) \qquad P \leftrightharpoons TQU.$$

We have

$$
\begin{array}{lrll}
(3) & P \eqcirc TRPSU & & [(2),\ (1)], \\
 & [P^\delta \eqcirc [TR^\delta[P^\delta[SU^\delta & & [(3),\ \S 19.1.3], \\
(4) & \eqcirc [P^\delta[TR^\delta[SU^\delta & & [\S 18.10.11], \\
(5) & \Lambda \eqcirc [TR^\delta[SU^\delta & & [\S 17.5.2,\ (4)], \\
(6) & [TR^\delta \eqcirc \Lambda & & [(5),\ \S 17.6.3], \\
(7) & [SU^\delta \eqcirc \Lambda & & [(5),\ \S 17.6.3], \\
(8) & TR \eqcirc \Lambda & & [(6),\ \S 19.1.6], \\
(9) & SU \eqcirc \Lambda & & [(7),\ \S 19.1.6], \\
(10) & R \eqcirc \Lambda & & [(8),\ \S 17.6.3], \\
(11) & S \eqcirc \Lambda & & [(9),\ \S 17.6.3], \\
 & P \eqcirc Q & & [(1),\ (10),\ (11)].
\end{array}
$$

Q.E.D.

3. We shall call words of the form $R * P * S$, where R, P and S are words in an alphabet A and $*$ is a letter not belonging to A, *occurrences* (more precisely, $*$-occurrences in the alphabet A).

3.1. *If the occurrences $R * P * S$ and $T * U * V$ coincide, then $R \eqcirc T$, $P \eqcirc U$ and $S \eqcirc V$.*

This is easily proven with the aid of §22.1.2. Theorem 3.1 asserts that every occurrence can be represented in the form $R * P * S$ in a unique manner. The word R is called the *left wing of the occurrence $R * P * S$*, the word P, the *base of this occurrence*, the word S, the *right wing of this occurrence.*

By virtue of 3.1 and the definition of an occurrence, we have

3.2. *The occurrences X and Y coincide if and only if their left wings coincide, their bases coincide and their right wings coincide.*

Occurrences with the base P will be called *occurrences of the word P*; occurrences $R * P * S$ such that $RPS \eqcirc Q$ will be called *occurrences in the word Q*; Occurrences of the word P which are in the word Q will be called *occurrences of the word P in the word Q.*

We obviously have

3.3. *The three-place predicate*

$$\ll X \text{ is an occurrence of } P \text{ in } Q \gg$$

is decidable.

We have the material implication

3.4. *If P occurs in Q, then there exists an occurrence of the word P in the word Q.*

In fact, then there exists a pair $R\alpha S$ such that $Q \leftrightharpoons RPS$ and, setting $X \rightleftharpoons R * P * S$, we see that X is an occurrence of P in Q.

The converse implication

$\ll$if there exists an occurrence of P in Q, then P occurs in $Q \gg$

can be regarded as a strengthened version of [3.3]. This strengthened implication asserts the same thing as the following universal statement:

$\ll$given any X, if X is an occurrence of the word P in the word Q, then P occurs in $Q \gg$.

The latter is obviously true. Therefore, we have

3.5. *If there exists an occurrence of the word P in the word Q, then P occurs in Q.*

In view of 3.4, 3.5 and also the rule of modus ponens for material implication [§13.4.1] and for a strengthened implication [§14.2.1], we can make use of the available method for recognizing the truth of the predicate

$$\ll P \text{ occurs in } Q \gg$$

in order to construct a method for recognizing the truth of the predicate

$\ll$there exists an occurrence of P in $Q \gg$.

Therefore, we have

3.6. *The two-place predicate*

$\ll$there exists an occurrence of P in $Q \gg$

is decidable.

This makes it possible to regard 3.5 as a material implication. Uniting it with the material implication 3.4, we obtain

3.7. *The word P occurs in the word Q if and only if there exists an occurrence of P in Q.*

It is easy to prove the statement

3.8. *If R is the left wing of an occurrence of the word P in the word Q, then RP is an initial segment of Q.*

We understand this statement to be equivalent to the following strengthened implication:

3.9. *If there exists an X such that X is an occurrence of the word P in the word Q and R is the left wing of X, then RP is an initial segment of Q.*

As such, 3.9 is proven without difficulty.
Also easily proven is

3.10. *If RP is an initial segment of Q, then R is the left wing of an occurrence of the word P in the word Q.*

We may understand this both as a material implication [§18.2.8] and as a strengthened implication. It is true for both understandings [§14.3].
Uniting the material implications 3.8 and 3.10, we obtain

3.11. *The word R is the left wing of an occurrence of the word P in the word Q if and only if RP is an initial segment of Q.*

Analogously, we obtain

3.12. *The word S is the right wing of an occurrence of the word P in the word Q if and only if PS is a terminal segment of Q.*

4. We shall call an occurrence with an empty left wing an *initial occurrence.* We shall call an occurrence with an empty right wing a *terminal occurrence.*
The following statements are easily proven:

4.1. *An initial occurrence of the word P in the word Q exists if and only if P is an initial segment of Q; a terminal occurrence of the word P in the word Q exists if and only if P is a terminal segment of Q. When an initial occurrence of the word P in the word Q does exist, such an occurrence of P in Q is unique and has the form*

$$*P*(P \leftharpoonup Q).$$

When a terminal occurrence of the word P in the word Q does exist, such an occurrence is unique and has the form

$$(P \rightharpoondown Q)*P*.$$

4.2. *$**P$ is an initial occurrence of the empty word in P; $P**$ is a terminal occurrence of the empty word in P.*

5. Various occurrences of a word P in a word Q, where P and Q are held fixed, will now be of some interest to us. For the sake of conciseness, we shall simply call them ≪occurrences≫ in this part of the section, not mentioning P and Q.

5.1. *Occurrences X and Y coincide if and only if their left (right) wings coincide* [§17.5.2, 3.2, §17.5.1].

We shall say that the *occurrence X precedes the occurrence Y* if the left wing of X is a proper initial segment of Y's left wing.

5.2. *One of the following three possibilities holds for the occurrences X and Y: X precedes Y, Y precedes X, $X \asymp Y$* [§18.4.5, §18.4.1].

5.3. *An occurrence does not precede itself* [§18.4.17].

5.4. *If the occurrence X precedes the occurrence Y and Y precedes the occurrence Z, then X precedes Z*[§18.4.18].

Let us prove the statement

5.5. *The three possibilities for occurrences of X and Y stated in* Theorem 5.2 *exclude each other. In other words, it is impossible that X precede Y and Y precede X; it is impossible that X precede Y and coincide with Y; it is impossible that Y precede X and coincide with X.*

In fact, the second and third cases are eliminated at once by 5.3. If the first case held, then the occurrence X would precede itself [5.4], which is impossible by virtue of 5.3.

We shall prove the statement

5.6. *If the occurrence X precedes the occurrence Y, then the right wing of Y is a proper terminal segment of X's right wing.*

Let R, S, T and U denote X's left wing, X's right wing, Y's left wing and Y's right wing, respectively. We have

(1) $$RPS \asymp Q,$$

(2) $$TPU \asymp Q.$$

Since X precedes Y, R is a proper initial segment of T and there exists a word V such that

(3) $$T \asymp RV,$$

(4) $$V \not\asymp \Lambda,$$

since $T \not\asymp R$ [(3)]. We have

(5) $$RPS \asymp RVPU \quad [(1)\text{–}(3)],$$

(6) $$PS \asymp VPU \quad [(5),\ \S 17.5.2].$$

If the word S were a terminal segment of U, then there would be a word W such that

(7) $$U \asymp WS$$

and we would have

(8) $$P \asymp VPW \quad [(5),\ (6),\ \S 17.5.1],$$

whence $V \asymp \Lambda$ [(8), §19.2.3], contrary to (4).

Therefore, S is not a terminal segment of U, and so by virtue of the fact that S and U are terminal segments of one and the same word [(6)], U is a proper terminal segment of S [§18.4.9]. Q.E.D.

Analogously, we prove

5.7. *If the right wing of the occurrence Y is a proper terminal segment of the right wing of the occurrence X, then X precedes Y.*

By virtue of 5.6 and 5.7 we have

5.8. *The occurrence X precedes the occurrence Y if and only if Y's right wing is a proper terminal segment of X's right wing.*

The length of the left wing of an occurrence X will be called the *depth of the occurrence X.*

5.9. *If the occurrence X precedes the occurrence Y, then the depth of X is less than the depth of Y* [§19.1.7].

5.10. *If the depth of the occurrence X is less than the depth of Y, then X precedes Y* [5.2, 5.9, §18.8.7, §18.8.8, §18.8.12].

5.11. *If the depths of the occurrences X and Y coincide, then $X \rightleftharpoons Y$* [5.2, 5.9].

6. We shall call the occurrence of a word P in a word Q, preceding every other occurrence of P in Q, a *first occurrence of the word P in the word Q.*

We obviously have

6.1. *If the word P is an initial segment of the word Q, then the initial occurrence of P in Q* [4.1] *is a first occurrence of P in Q.*

6.2. *The occurrence $**P$ is a first occurrence of the empty word in P* [4.2, 6.1].

We shall prove the following existence theorem for a first occurrence:

6.3. *If the word P occurs in the word Q, then there exists a first occurrence of P in Q.*

Let P occur in Q. Then there is an occurrence X of the word P in the word Q [3.4]. Denote the depth of X by the letter N. Construct the following one-place arithmetical[9] predicate F with the free natural variable M:

$$\ll \text{there exists an occurrence of } P \text{ in } Q \text{ with depth } M \gg$$

or, what is the same thing,

$$\ll M \text{ is the depth of an occurrence of } P \text{ in } Q \gg .$$

The predicate F, as is easily seen, is decidable. The natural number N satisfies it. Hence, the Least Number Theorem [§21.2.1], according to which the smallest number satisfying F can be constructed, is applicable to F and N. Let this number be L. L satisfies F, but every number less than L fails to satisfy F. Since L satisfies F, there is an occurrence Y of P in Q with depth L. We shall show that Y is the desired first occurrence of P in Q.

[9] We call predicates with natural free variables *arithmetical.* Admissible values of such variables are natural numbers. See §10.1.

Let Z be an occurrence of P in Q, distinct from Y. According to the construction of the predicate F, the depth of Z satisfies F. Hence, the depth of Z is not less than L, i.e., the depth of Y. The depth of Z differs from the depth of Y, Since $Z \neq Y$ [5.11]. Consequently, the depth of Y is less than that of Z [§18.6.8]. Therefore, Y precedes Z [5.10].

Thus, Y precedes every other occurrence of P in Q. Q.E.D.

The truth of the following statement is almost obvious:

6.4. *If the word P occurs in the word Q, then the first occurrence of P in Q is unique* [5.5].

7. The following statements are easily proven:

7.1. *If X is an occurrence of the word P in the word Q, then RX is an occurrence of the word P in the word RQ and XR is an occurrence of the word P in the word QR.*

According to 7.1, if X and Y are occurrences of the word P in the word Q, then RX and RY are also occurrences of the word P in one and the same word—RQ. Hence, it is possible to speak of RX's occurrence preceding RY's. One may also speak of XR's occurrence preeceding YR's. The following statements are easily proven:

7.2. *If X and Y are occurrences of the word P in the word Q, then RX precedes RY if and only if X precedes Y.*

7.3. *If X and Y are occurrences of the word P in the word Q, then XR precedes YR if and only if X precedes Y.*

Let us prove the statement

7.4. *If X is the first occurrence of the word R in the word Q, then XR is the first occurrence of P in QR.*

Let X be the first occurrence of P in Q. Denoting X's left wing by S and X's right wing by T, we have

(1) $$X \rightleftharpoons S * P * T,$$

(2) $$XR \rightleftharpoons S * P * TR.$$

XR is an occurrence of the word P in the word QR [7.1]. We shall show that XR precedes every other occurrence of P in QR.

Let Y be an occurrence of P in QR, and let

(3) $$Y \neq XR.$$

It is required to prove that XR precedes Y. Denoting Y's left wing by U and Y's right wing by V, we have

(4) $$Y \rightleftharpoons U * P * V.$$

Since X is an occurrence in Q and Y is an occurrence in QR, we have

$$(5) \qquad SPT \rightleftharpoons Q \qquad [(1)],$$
$$(6) \qquad UPV \rightleftharpoons QR \qquad [(4)],$$
$$(7) \qquad UPV \rightleftharpoons SPTR \quad [(5), (6)].$$

If we had $U \rightleftharpoons S$, we would also have $V \rightleftharpoons TR$ [(7), §17.5.2], and the occurrences XR and Y would coincide [(2), (4)], contrary to (3). Consequently,

$$(8) \qquad U \not\rightleftharpoons S.$$

We shall show that U cannot be a proper initial segment of S.

Suppose that U is a proper initial segment of S. Then there is a word W such that

$$(9) \qquad S \rightleftharpoons UW.$$

By virtue of (8),

$$(10) \qquad W \not\rightleftharpoons \Lambda \quad [(9)].$$

We have

$$(11) \qquad PV \rightleftharpoons WPTR \quad [(7), (9)].$$

By virtue of (10) and §19.2.3, WP is not an initial segment of P. Hence, P is an initial segment of WP [(11), §18.3.4] and there exists a word K such that

$$(12) \qquad WP \rightleftharpoons PK.$$

We have

$$Q \rightleftharpoons UPKT \quad [(5), (9), (12)].$$

Therefore,

$$(13) \qquad U * P * KT$$

is an occurrence of the word P in the word Q. The left wing of this occurrence is a proper initial segment of the left wing of the occurrence X. Consequently, the occurrence (13) precedes the occurrence X. This contradicts the fact that X is the first occurrence of P in Q.

We arrived at this contradiction by assuming that U is a proper initial segment of S. Consequently, U is not a proper initial segment of S. Therefore, S is an initial segment of U [§18.4.5, (7)]. By virtue of (8), S is a proper initial segment of U. Hence, the occurrence XR precedes the occurrence Y. Q.E.D.

Let us prove the statement

7.5. *If X is the first occurrence of the word P in the word Q and P is not an initial segment of the word ξQ, then ξX is the first occurrence of the word P in the word ξQ.*

Denoting X's left wing by U and X's right wing by V, we have

(14) $$X \rightleftharpoons U * P * V.$$

ξX is an occurrence of the word P in the word ξQ [7.1]. It is required to prove that ξX precedes every occurrence distinct from ξX of the word P in the word ξQ.

Let Y be an occurrence of the word P in the word ξX, and let

(15) $$Y \neq \xi X.$$

Denoting Y's left wing by R and Y's right wing by S, we have

(16) $$Y \rightleftharpoons R * P * S$$

and

(17) $$RPS \rightleftharpoons \xi Q.$$

If the word R were empty, the word P would be an initial segment of the word ξQ, contrary to the hypothesis. Therefore, $R \neq \Lambda$ and there exists a word T and a letter η such that

(18) $$R \rightleftharpoons \eta T.$$

We have

(19) $$\eta TPS \rightleftharpoons \xi Q \quad [(17), (18)],$$

and so

(20) $$\eta \rightleftharpoons \xi \quad [(19)],$$

(21) $$TPS \rightleftharpoons Q \quad [(19)],$$

(22) $$R \rightleftharpoons \xi T \quad [(18), (20)].$$

Therefore, $T * P * S$ is an occurrence of the word P in the word Q [(21)]. We have

(23) $$\xi X \rightleftharpoons \xi U * P * V \quad [(14)],$$

(24) $$Y \rightleftharpoons \xi T * P * S \quad [(16), (22)].$$

Hence

(25) $$\xi U \neq \xi T \quad [(23), (24), (15), 5.1],$$

(26) $$U \neq T \quad [(25)],$$

$$T * P * S \neq X \quad [(14), (26), 5.1].$$

Therefore, $T * P * S$ is an occurrence of the word P in the word Q, distinct from the first occurrence X of P in Q. Consequently, X precedes $T * P * S$. Therefore, ξX precedes $\xi T * P * S$ [7.2], i.e., ξX precedes Y [(24)]. Q.E.D.

From the appropriate definitions easily follow the statements:

7.6. *If the word P occurs in the word Q, then the left wing of P's first occurrence in Q is an initial segment of the left wing of every occurrence of P in Q.*

7.7. *If the word P occurs in the word Q, then the right wing of every occurrence of P in Q is an initial segment of the right wing of P's first occurrence in Q.*

With the aid of the statement 3.11, we easily prove

7.8. *The word R is the left wing of the first occurrence of the word P in the word Q if and only if the following conditions are fulfilled:*

a) *RP is an initial segment of Q;*

b) *given any initial segment S of the word Q, such that SP is also an initial segment of Q, R is an initial segment of S.*

Analogously, with the aid of 3.12, we can prove the statement

7.9. *The word T is the right wing of the first occurrence of the word P in the word Q if and only if the following conditions are fulfilled:*

a) *PT is a terminal segment of Q;*

b) *given any terminal segment of U of the word Q, such that PU is also a terminal segment of Q, U is a terminal segment of T.*

It is also easy to prove the statement

7.10. *If $P \neq \Lambda$ and P occurs in Q, then P does not occur in the left wing of its first occurrence in Q.*

8. We shall call the word PSR the *result of substituting the word S for the occurrence $P * Q * R$.*

We obviously have

8.1. *Given any occurrence X and any word S, there is a uniquely defined result of substituting the word S for X. It is a word in the alphabet under consideration.*

We further have

8.2. *If the word Q occurs in the word T, then there is a uniquely defined result of substituting the word S for the first occurrence of the word Q in the word T. It then is a word in the alphabet under consideration* [6.3, 6.4, 8.1].

We shall denote the result of substituting the word S for the first occurrence of the word Q in the word T by the symbol

$$\Sigma(S, Q, T).$$

Therefore, we have

8.3. *The expression $\Sigma(S, Q, T)$ makes sense if and only if the word Q occurs in the word T. In this case, it denotes a certain word in the alphabet under consideration.*

From the definition of the result of substituting a word for an occurrence follows the statement

8.4. *The equality*

$$W \eqcirc \Sigma(S, Q, T)$$

holds if and only if there are an initial segment U of the word W and a terminal segment V of the word W such that:

a) *U is the left wing of Q's first occurrence in T;*

b) *V is the right wing of Q's first occurrence in T;*

c) $W \eqcirc USV$.

We have, as is easy to see,

8.5. $\Sigma(S, \Lambda, T) \eqcirc ST$.

§24 Systems of Words

1. In this section, we shall be dealing with an alphabet A and two different letters, $*$ and γ, not occurring in A. By ≪words≫ we shall understand words in the alphabet A, unless stipulated otherwise; by≪occurrences≫, $*$-occurrences in the alphabet A.

We shall call words in the alphabet Aγ, beginning with the letter γ and ending with this letter, *γ-systems in the alphabet* A or, more briefly, *systems*.

The letters P, Q, R, S will be applied as verbal variables in the alphabet A; the letters X, Y, Z—as verbal variables in the alphabet Aγ.

The following statements are obvious:

1.1. *If X is a system, then $XP\gamma$ and γPX are systems.*

1.2. *If X is a system, then every terminal segment of X, beginning with the letter γ, is a system and every initial segment of X, ending with the letter γ, is a system.*

1.3. *The verbal predicate in the alphabet $A\gamma$* ≪ X is a system≫ *is decidable.*

It is easy to prove the statement

1.4. *The one-place verbal predicate in the alphabet* Aγ *with the free variable* Y, ≪there exists a system X and a word P in the alphabet A such that $Y \eqcirc XP\gamma$ ≫, *is decidable.*

We shall prove the statement

1.5. *Every system distinct from γ can be represented as γPX, where P is a word in the alphabet* A *and X is a system.*

Let Y be a system and let $Y \not\eqcirc \gamma$. We show that Y can be represented in the form γPX. By definition, γ is an initial segment of the system Y, and so

(1) $$Y \eqcirc \gamma(\gamma \leftharpoondown Y) \quad [§18.8.4],$$

where $(\gamma \leftharpoondown Y) \not\eqcirc \Lambda$, since $Y \not\eqcirc \gamma$ [(1)]. Hence,$(\gamma \leftharpoondown Y)$ has a last letter [§17.4.3]. In view of (1), it is also the last letter of the system Y, i.e., the letter γ. Therefore, γ occurs in $(\gamma \leftharpoondown Y)$, and so the first occurrence of the letter γ in the word $(\gamma \leftharpoondown Y)$ exists [§23.6.3]. This occurrence has the form

$$P * \gamma * W,$$

where P and W are words in the alphabet $A\gamma$. In view of the fact that the occurrence (2) is the first occurrence of γ in $(\gamma \leftharpoonup Y)$, γ does not occur in P [§23.7.10]. Therefore, P is a word in the alphabet A. In view of the fact that (2) is an occurrence in $(\gamma \leftharpoonup Y)$, we have

$$(3) \qquad (\gamma \leftharpoonup Y) \doteq P\gamma W,$$

and hence

$$\begin{aligned} Y &\doteq \gamma P\gamma W \quad [(1), (3)] \\ &\doteq \gamma PX, \end{aligned}$$

where $X \rightleftharpoons \gamma W$. Being a terminal segment of the system Y beginning with the letter γ [(3)], X is a system [1.2], which is just what remained to be proven.

Let us prove the following theorem:

1.6. *Let F be a one-place verbal predicate in the alphabet $A\gamma$. Let γ satisfy F and let the following condition be fulfilled: ≪whatever be the system X and the word P in the alphabet* A , *the implication*

If X satisfies F, then γPX satisfies F ≫

holds. Then every system satisfies F.

Let G denote the following one-place arithmetical predicate with the free natural variable N: ≪every system whose length does not exceed N satisfies F≫. The number Λ satisfies G, since there exists no system of length not exceeding Λ [1.5]. Let the natural number N satisfy G. We shall show that $N\mathsf{I}$ also satisfies G.

Suppose that X is a system and $[X^\delta \leqslant N\mathsf{I}$. Let us show that X satisfies F. If $X \doteq \gamma$, then X satisfies F by hypothesis. Let $X \not\doteq \gamma$. Then there exist a word P in the alphabet A and a system Y such that

$$(4) \qquad X \doteq \gamma PY \quad [(1.5)].$$

In view of (4), $[Y^\delta < [X^\delta$ [§19.1.2], and so $[Y^\delta \leqslant N$. Since N satisfies the predicate G, Y satisfies F. Therefore, according to our hypothesis and in view of (4), X also satisfies F. We proved this for every system X whose length does not exceed $N\mathsf{I}$. Consequently, $N\mathsf{I}$ satisfies the predicate G. Thus, whatever be the natural number N, the implication ≪if N satisfies G, then $N\mathsf{I}$ satisfies G ≫ holds. Therefore, every natural number satisfies G. It easily follows from this that every system satisfies the predicate F. Q.E.D.

With the aid of Theorem 1.6, we can easily prove the statement

1.7. *Every system distinct from γ is representable in the form $XP\gamma$, where X is a system and P is a word in the alphabet* A.

After this, we can prove, analogously to the way in which 1.6 was proven, the theorem

1.8. *Let F be a one-place verbal predicate in the alphabet $\text{A}\gamma$. Assume that γ satisfies F and the following condition is fulfilled: given any system X and word P in the alphabet* A, *the implication*

$$\ll \text{if } X \text{ satisfies } F, \text{ then } XP\gamma \text{ satisfies } F \gg$$

holds. Then every system satisfies F.

2. Words of the form $\gamma P\gamma$, where P is a word in the alphabet A, will be called *atoms*. Occurrences of atoms in systems will be called *elements*. Occurrences of atoms in a system X will be called *elements of the system X*.

We have the statement

2.1. *Every element is the occurrence of a uniquely determined atom in a uniquely determined system* [§23.5.1].

We shall prove the following theorem:

2.2. *Every element of the system γPX, where P is a word in the alphabet* A *and X is a system, has one of the following forms:*

a) $*\gamma P\gamma * (\gamma \leftharpoondown X)$;

b) γPW, *where W is an element of the system X.*

Let V be an element of the system γPX, i.e., an occurrence in γPX of the atom $\gamma Q\gamma$, where Q is a word in the alphabet A. Denoting the wings of this occurrence by Y and Z, we have

$$V \leftrightharpoons Y * \gamma Q\gamma * Z, \tag{1}$$

$$Y\gamma Q\gamma Z \leftrightharpoons \gamma PX. \tag{2}$$

Since γ is an initial segment of the system X, we have

$$X \leftrightharpoons \gamma(\gamma \leftharpoondown X). \tag{3}$$

We shall consider the two possible cases: Case 1), when $Y \leftrightharpoons \Lambda$, and Case 2), when $Y \not\leftrightharpoons \Lambda$.

In Case 1) we have

$$Q\gamma Z \leftrightharpoons P\gamma(\gamma \leftharpoondown X) \quad [(2), \S17.5.2, (3)], \tag{4}$$

where P and Q are words in the alphabet A, not containing the letter γ. Hence

$$Q \leftrightharpoons P \tag{5}$$

and

$$Z \leftrightharpoons (\gamma \leftharpoondown X) \tag{6}$$

[(4), §22.1.1] and, consequently,

$$V \leftrightharpoons *\gamma P\gamma * (\gamma \leftharpoondown X) \quad [(1), (5), (6)].$$

Therefore, in Case 1) V has the form a).

Now consider Case 2). In this case Y has the initial letter γ, in view of which

(7) $$Y \rightleftharpoons \gamma(\gamma \leftharpoondown Y).$$

We have

(8) $$(\gamma \leftharpoondown Y)\gamma Q\gamma Z \rightleftharpoons PX \quad [(2), (7), §17.5.2].$$

$(\gamma \leftharpoondown Y)\gamma$ is not an initial segment of the word P in the alphabet A. Since $(\gamma \leftharpoondown Y)\gamma$ and P are initial segments of one and the same word [(8)], P is an initial segment of $(\gamma \leftharpoondown Y)$ [§18.4.6], and we have

(9) $$P(P \leftharpoondown (\gamma \leftharpoondown Y)) \rightleftharpoons (\gamma \leftharpoondown Y) \quad [§18.8.4].$$

Hence

(10) $$(P \leftharpoondown (\gamma \leftharpoondown Y))\gamma Q\gamma Z \rightleftharpoons X \quad [(8), (9), §17.5.2].$$

Setting

(11) $$W \rightleftharpoons (P \leftharpoondown (\gamma \leftharpoondown Y)) * \gamma Q\gamma * Z,$$

we see that W is an occurrence of the atom $\gamma Q\gamma$ in the system X [(10),(11)] and, consequently, an element of the system X. Furthermore,

$$\begin{aligned} V &\rightleftharpoons \gamma(\gamma \leftharpoondown Y) * \gamma Q\gamma * Z && [(1), (7)] \\ &\rightleftharpoons \gamma P(P \leftharpoondown (\gamma \leftharpoondown Y)) * \gamma Q\gamma * Z && [(9)] \\ &\rightleftharpoons \gamma PW && [(11)]. \end{aligned}$$

Therefore, in Case 2) V has the form b) and the proof of Theorem 2.2 is completed.

Analogously, we have

2.3. *Every element of the system $XP\gamma$, where X is a system and P is a word in the alphabet A has one of the following forms:*

a) $(\gamma \rightharpoondown X) * \gamma P\gamma *$;

b) *$WP\gamma$, where W is an element of the system X.*

Let us prove the theorem

2.4. *Every system distinct from γ begins with an atom.*

Suppose that X is a system and $X \not\rightleftharpoons \gamma$. Then we can single out a word P in the alphabet A and a system Z such that

$$X \rightleftharpoons \gamma PZ \quad [1.5].$$

The system Z has the initial letter γ, and so

$$Z \rightleftharpoons \gamma(\gamma \leftharpoondown Z) \quad [§18.8.4].$$

Consequently,

$$X \rightleftharpoons \gamma P\gamma(\gamma \leftharpoondown Z),$$

where $\gamma P\gamma$ is an atom. By the same token, the theorem is proven.

Analogously, we can prove the statement

2.5. *Every system distinct from γ ends with an atom* [1.7].

3. We shall call the words P (in the alphabet A), such that the atom $\gamma P\gamma$ occurs in X, terms of the system X.

We shall say that P is a *first term of the system* X if $\gamma P\gamma$ is an initial segment of X; we shall say that P is a *last term of the system* X if $\gamma P\gamma$ is a terminal segment of X. We shall say that the system X connects the word P to the word Q, if P is a first term of X and Q is a last term of X.

We have the following true statements:

3.1. *Every system distinct from γ has a unique first term and a unique last term* [2.4, 2.5, §24.1.1, §17.5.1, §17.5.2].

3.2. *The first and last terms of a system distinct from γ are its terms.*

3.3. *If X is a system, then P is the first term of the system γPX and the last term of the system $XP\gamma$.*

3.4. *If X is a system, then each of its terms is a term of the system γPX as well as of the system $XP\gamma$.*

3.5. *If X is a system, then every term of the system γPX either coincides with P or else is a term of the system X* [2.2, §23.3.2, §23.7.1, §17.5.1, §17.5.2].

3.6. *If X is a system, then every term of the system $XP\gamma$ either coincides with P or else is a term of the system X* [2.3, §23.3.2, §23.7.1, §17.5.1, §17.5.2].

It is natural to call the system γ *empty*: no word is its term. Of course, the empty system should be distinguished from the empty word.

4. Whatever be the system X, the inequality $[[X^{\gamma\delta} \geqslant \mathsf{I}$[10] holds, and so the expression $(([[X^{\gamma\delta} - \mathsf{I})$ makes sense. The number which is its value will be called the *volume of the system* X. We shall denote the volume of X by the symbol

$$[X^{\circ}.$$

According to the definition, we have

$$(1) \qquad [X^{\circ} \rightleftharpoons (([[X^{\gamma\delta} - \mathsf{I})$$

and, consequently,

$$[X^{\circ}\mathsf{I} \rightleftharpoons [[X^{\gamma\delta} \quad [(1)].$$

Here X is an arbitrary system.

4.1. *If the system X is distinct from γ, then* $\Lambda < [X^{\circ}$ [2.4, §19.4.1, §19.4.3].

4.2. $[\gamma^{\circ} \rightleftharpoons \mathsf{I}$.

5. To each element Y, we ascribe the natural number $[[Z^{\gamma\delta}\mathsf{I}$ as *its number*, where Z is the left wing of Y. We shall denote the number of the element Y by the symbol

$$[Y^{\triangle}.$$

5.1. *The number of every element X is a positive natural number.*

This follows directly from the definition of the number of an element.

---

[10]Here $[X^{\gamma}$ denotes the projection of X onto the one-letter alphabet γ.

5.2. *The number of every element of a system X is majorized by the volume of the system X.*

Let Y be an element of the system X. We shall prove that

$$(1) \qquad [Y^{\triangle} \Bumpeq\leqslant [X^{\circ}.$$

Y is an occurrence in X of a uniquely defined atom $\gamma P\gamma$, where P is a word in the alphabet A [2.1]. Denote Y's left wing by Z. By the definition of an element's number, we have

$$[Y^{\triangle} \Bumpeq [[Z^{\gamma\delta}\mathbf{I}.$$

The word $Z\gamma P\gamma$ is an initial segment of X [§23.3.8]. Hence

$$[[Z\gamma P\gamma^{\gamma\delta} \leqslant [[X^{\gamma\delta} \quad [§19.4.2],$$

from which

$$(3) \qquad [[Z^{\gamma\delta}\mathbf{II} \leqslant [[X^{\gamma\delta} \quad [§19.4.1, §19.4.3].$$

Consequently,

$$(4) \qquad [[Z^{\gamma\delta}\mathbf{I} < [X^{\circ} \quad [(3), 4(1), §17.5.1],$$

i.e., we have (1) [(4), (2)]. Q.E.D.

Let us prove the following theorem:

5.3. *Given any system X and any natural number N such that*

$$(5) \qquad \mathbf{I} \leqslant N \leqslant [X^{\circ},$$

one can find an element Y of X such that

$$(6) \qquad [Y^{\triangle} \Bumpeq N.$$

We shall carry out the proof by means of the method of restricted arithmetical induction. We shall hold the system X fixed and consider the following one-place arithmetical predicate F with the free variable M:

(7) $\ll$ one can find an element Y of the system X such that $[Y^{\triangle} \Bumpeq M\mathbf{I} \gg$.

If $X \Bumpeq \gamma$, then $[[X^{\gamma\delta} \Bumpeq \mathbf{I}$ and $[X^{\circ} \Bumpeq \Lambda$[4(1)]. Then there exist no natural numbers N satisfying the condition (5), and it is trivially true that for every such N, one can find an element Y of the system X, satisfying the condition (6) [§9.3]. In what follows, we shall assume that $X \not\Bumpeq \gamma$. Then X begins with an atom [2.4], i.e., there is a word P in the alphabet A such that the atom $\gamma P\gamma$ is an initial segment of X. There is a word Z in the alphabet Aγ, such that

$$(8) \qquad X \Bumpeq \gamma P\gamma Z.$$

Put

$$(9) \qquad Y \rightleftharpoons *\gamma P\gamma * Z.$$

In view of (8) and (9), Y is an occurrence of an atom in the system X, i.e., this system's element. This element's left wing is empty and so, according to the definition of an element's number,

$$\begin{aligned}[Y^{\triangle} &\asymp [[\Lambda^{\gamma\delta}\mathbf{I} \\ &\asymp \Lambda\mathbf{I} \qquad [\S 19.4.4].\end{aligned}$$

Therefore, the natural number Λ satisfies the predicate F.

In view of (8), we have $[[X^{\gamma\delta} \geqslant \mathbf{II}$, and so $[X^{\circ} \geqslant \mathbf{I}$ [4(1)] and $([X^{\circ} - \mathbf{I})$ is a natural number.

Let M be a number less than $([X^{\circ} - \mathbf{I})$. We shall prove the implication

$$(10) \qquad \ll \text{if } M \text{ satisfies } F, \text{ then} M\mathbf{I} \text{ satisfies } F \gg.$$

Let M satisfy F. One can find an element U of the system X, such that

$$(11) \qquad [U^{\triangle} \asymp M\mathbf{I}.$$

U is an occurrence of a uniquely determined atom $\gamma Q\gamma$ in X. Here Q is a word in the alphabet A. Denoting the left and right wings of the occurrence U by V and W, we have

$$(12) \qquad X \asymp V\gamma Q\gamma W.$$

Since V is U's left wing, by the definition of an element's number we have

$$(13) \qquad [U^{\triangle} \asymp [[V^{\gamma\delta}\mathbf{I}.$$

We further have

$$\begin{aligned}
&(14) & M &\asymp [[V^{\gamma\delta} && [(13), (11), \S 17.5.1], \\
&(15) & [[X^{\gamma\delta} &\asymp M\mathbf{II}[[W^{\gamma\delta} && [(12), \S 19.4.1, (14), \S 19.4.3, \S 19.4.5], \\
&(16) & [X^{\circ} &\asymp M\mathbf{I}[[W^{\gamma\delta} && [(15), \S 18.10.11, 4(1), \S 17.5.1].
\end{aligned}$$

Since $M < ([X^{\circ} - \mathbf{I})$ by assumption, we have $M\mathbf{I} < [X^{\circ}$. Consequently,

$$[[W^{\gamma\delta} \not\asymp \Lambda \quad [(16), \S 18.6.12],$$

and so

$$W \not\asymp \Lambda \quad [\S 19.4.4]$$

and

$$(17) \qquad \gamma W \not\asymp \gamma \quad [\S 17.5.2].$$

γW is a terminal segment of the system X [(12)] and, consequently, of the system [1.2]. By virtue of (17), γW begins with an atom, i.e., there is a word R in the alphabet A such that $\gamma R\gamma$ is an initial segment of γW [2.4]. We have

(18) $$\gamma W \coloneqq \gamma R\gamma(\gamma R\gamma \leftharpoondown \gamma W) \qquad [\S 18.8.4],$$

(19) $$X \coloneqq V\gamma Q\gamma R\gamma(\gamma R\gamma \leftharpoondown \gamma W) \quad [(12), (18)].$$

Set

(20) $$T \rightleftharpoons V\gamma Q * \gamma R\gamma * (\gamma R\gamma \leftharpoondown \gamma W)$$

by definition. T is an occurrence of the atom $\gamma R\gamma$ in the system X [(19), (20)], i.e., an element of X. T's left wing is the word $V\gamma Q$, for which we have

$$\begin{aligned}[][[V\gamma Q^{\gamma\delta} &\coloneqq V^{\gamma\delta}\mathsf{I} \quad [\S 19.4.1, \S 19.4.3, \S 18.4.5]\\ &\coloneqq M\mathsf{I} \quad [(14)].\end{aligned}$$

From this, by the definition of an element's number, we get

$$\begin{aligned}[][T^{\triangle} &\coloneqq [[V\gamma Q^{\gamma\delta}\mathsf{I}\\ &\coloneqq M\mathsf{II}.\end{aligned}$$

Therefore, the number $M\mathsf{I}$ satisfies the predicate F. By the same token, the implication (10) is proven [§13.6.2].

Applying the method of restricted arithmetical induction, we see that every natural number M, such that $M \leqslant ([X^{\circ} - \mathsf{I})$, satisfies the predicate F.

But if the number N satisfies the condition (5), then

$$(N - \mathsf{I}) \leqslant ([X^{\circ} - \mathsf{I})$$

and so $(N - \mathsf{I})$ satisfies F. This means that there exists an element of the system X with the number N. Q.E.D.

5.4. *If Y and U are elements of one and the same system and $[Y^{\triangle} \coloneqq [U^{\triangle}$, then $Y \coloneqq U$.*

In fact, let Y and U be elements of the system X, and let

(21) $$[Y^{\triangle} \coloneqq [U^{\triangle}.$$

Y and U are occurrences of certain atoms, $\gamma P\gamma$ and $\gamma Q\gamma$, respectively, in X. Here P and Q are words in the alphabet A. Denoting U's left and right wings by V and T, Y's left and right wings by Z and W, we have

(22) $$Y \coloneqq Z * \gamma P\gamma * W,$$

(23) $$U \coloneqq V * \gamma Q\gamma * T,$$

(24) $$X \coloneqq Z\gamma P\gamma W,$$

(25) $$X \coloneqq V\gamma Q\gamma T,$$

(26) $$[Y^{\triangle} \coloneqq [[Z^{\gamma\delta}\mathsf{I},$$

(27) $$[U^{\triangle} \coloneqq [[V^{\gamma\delta}\mathsf{I},$$

(28) $$[[Z^{\gamma\delta} \coloneqq [[V^{\gamma\delta} \qquad [(21)\text{-}(27)].$$

If the word $V\gamma$ were an initial segment of the word Z, then we would have

$$[[V\gamma^{\gamma\delta} \leqslant [[Z^{\gamma\delta} \quad [\S 19.4.2],$$

i.e.,

(29) $$[[V^{\gamma\delta}\mathsf{I} \leqslant [[Z^{\gamma\delta} \quad [\S 19.4.1, \S 19.4.3],$$

whence

$$[[V^{\gamma\delta}\mathsf{I} \leqslant [[V^{\gamma\delta} \quad [(28), (29)],$$

and, consequently,

$$\mathsf{I} \leqslant \Lambda \quad [\S 17.5.2],$$

which is absurd. Therefore, the word $V\gamma$ is not an initial segment of the word Z. But since $V\gamma$ and Z are initial segments of one and the same word X, Z is an initial segment of V [§18.4.6]. It is proven in an analogous manner that V is an initial segment of Z. Consequently,

(30) $$Z \leftrightharpoons V \quad [\S 18.3.3].$$

We further have

(31) $$Z\gamma P\gamma W \leftrightharpoons V\gamma Q\gamma T \quad [(24), (25)],$$

(32) $$P\gamma W \leftrightharpoons Q\gamma T \quad [(31), (30), \S 17.5.2],$$

(33) $$P \leftrightharpoons Q \quad [(32), \S 22.1.1],$$

(34) $$W \leftrightharpoons T \quad [(32), \S 22.1.1],$$

$$Y \leftrightharpoons U \quad [(22), (23), (30), (33), (34)].$$

Q.E.D.

6. We shall say that the *element* Y *precedes the element* Z if Y and Z are elements of one and the same system and Y's left wing is proper initial segment of Z's left wing.

6.1. *No element precedes itself* [§18.4.17].

6.2. *If the element* Y *precedes the element* Z, *then* $[Y^{\triangle} < [Z^{\triangle}$.

In fact, let the element Y precede the element Z (Y and Z are elements of one and the same system). Denote Y's left wing by the letter U, and Z's left wing by the letter V. U is a proper initial segment of V, since Y precedes Z. Hence $U \not\leftrightharpoons V$ and, consequently, $Y \not\leftrightharpoons Z$. Since U is an initial segment of V, we have

$$[[U^{\gamma\delta} \leqslant [[V^{\gamma\delta} \quad [\S 19.4.2],$$

and so, by the definition of an element's number,

(1) $$[Y^{\triangle} \leqslant [Z^{\triangle} \quad [\S 18.6.10, \S 18.10.11].$$

If the equality $[Y^{\triangle} \leftrightharpoons [Z^{\triangle}$ held, we would have the equality $Y \leftrightharpoons Z$ [5.4], but we have seen that $Y \not\leftrightharpoons Z$. Therefore,

$$[Y^{\triangle} \not\leftrightharpoons [Z^{\triangle}$$

and, consequently,

(2) $$[Y^{\triangle} < [Z^{\triangle}.$$

Q.E.D.

6.3. *If Y and Z are elements of one and the same system and $[Y^\triangle < [Z^\triangle$, then Y precedes Z.*

In fact, let Y and Z be elements of the system X, and let (2) hold. Y and Z are then occurrences in X. Denote the left wing of the occurrence Y by U, and the left wing of the occurrence Z by V. U and V are initial segments of the word X [§23.3.8], by virtue of which V is an initial segment of U or U is a proper initial segment of V [§18.4.5]. But if V were an initial segment of U, we would have

$$[[V^{\gamma\delta}\mathsf{I} \leqslant [[U^{\gamma\delta}\mathsf{I} \quad [\S 19.4.2, \S 18.6.10, \S 18.10.11],$$

i.e.,

$$[Z^\triangle \leqslant [Y^\triangle,$$

which is incompatible with (2) [§18.6.13]. Consequently, V is not an initial segment of U, and so U is a proper initial segment of V. This means that Y precedes Z. Q.E.D.

6.4. *The element Y precedes the element Z if and only if Y and Z are elements of one and the same system and $[Y^\triangle < [Z^\triangle$* [6.2, 6.3].

6.5. *If the element Y precedes the element Z, then Z does not precede Y* [6.4].

7. The following statements are immediate consequences of the definition of an element:

7.1. *Every element is a word of the form*

$$(1) \qquad Y * \gamma P \gamma * Z.$$

We also have

7.2. *Every element is representable in the form* (1) *in a unique manner* [§22.1.1, §22.1.2].

By virtue of 7.1 and 7.2, the following definition is legitimate:

We call the word P in the representation (1) of the element X the *kernel* of the element X.

The next statement easily follows from the definitions.

7.3. *The word P is the kernel of an element of the system X if and only if P is a term of X.*

8. Let X be a system and let N be a natural number for which $\mathsf{I} \leqslant N \leqslant [X^\circ$. The kernel of the element Y of the system X, such that $[Y^\triangle \doteq N$, will be called the *Nth term of the system X*.

If $\mathsf{I} < N < [X^\circ$, then we shall call the Nth term of the system X an *interior term of X*.

§25 Schemes

1. In this section, A denotes an alphabet and α, β, γ and $*$, certain pairwise distinct letters not occurring in A. We introduce the alphabet Б into our presentation, setting

$$Б \rightleftharpoons A\alpha\beta$$

by definition. The letters P and Q will be employed as verbal variables in the alphabet A; the letters X, Y and Z—as verbal variables in the alphabet Бγ; the letters T, U, V, W—as verbal variables in the alphabet Б.

We shall use the letter γ for the construction of γ-systems in the alphabet Б. As in §24, we shall call them simply *systems.* We shall use the letter $*$ for the construction of $*$-occurrences in the alphabet Бγ, which we shall call simply *occurrences.*

2. We shall call words of the form $P\xi Q$, where ξ is a letter of the alphabet $\alpha\beta$, *substitution formulas* (in the alphabet A); words of the form $P\alpha Q$—*simple substitution formulas,* words of the form $P\beta Q$—*final substitution formulas.*

In the obvious way, we have

2.1. *Every substitution formula is simple or final.*

2.2. *No simple substitution formula is final.*

2.3. *Every substitution formula can be represented in the form $P\xi Q$, where ξ is a letter of the alphabet $\alpha\beta$, in a unique fashion* [§22.1.3].

The word P will be called the *left side* of the substitution formula $P\xi Q$; the word Q, the *right side* of this substitution formula.

We have

2.4. *The left and right sides of every substitution formula are words in the alphabet* A.

It is not hard to see that the following verbal predicates are decidable: $\ll X$ is a substitution formula$\gg$, $\ll X$ is a simple substitution formula$\gg$, $\ll X$ is a final substitution formula$\gg$, $\ll X$ is a substitution formula and P is the left side of $X \gg$, $\ll X$ is a substitution formula and Q is the right side of $X \gg$.

3. We shall say that the substitution formula F *acts on the word* P if the left side of F occurs in P. In this case, by the *result of the action of F on P* we shall mean the result of substituting the right side of the substitution formula F for the first occurrence of its left side in the word P.

We have

3.1. *The result of the action of the substitution formula F on the word P is defined if and only if F acts on P. It is then a word in the alphabet* A [§23.8.2].

4. We shall call a γ-system X in the alphabet Б a *scheme* (more precisely—a *γ-scheme in the alphabet* Б) if all its terms are substitution formulas.

It is not difficult to see that the verbal predicate $\ll X$ is a scheme$\gg$ is decidable.

We shall say that the *scheme* X *acts on the word* P if at least one term of X acts on P. This definition makes sense in view of the fact that the terms of X are substitution formulas.

We shall say that the *element* Y *acts on the word* P if this element's kernel acts on P. In this case, the result of the action of Y's kernel on P will be regarded as the *result of the action of* Y *on* P.

4.1. *If the scheme X acts on the word P, then there is a unique element of X acting on P and preceding every other element of X acting on P.*

In fact, suppose that the scheme X acts on the word P. Construct the one-place arithmetical predicate G

$$\ll N \text{ is the number of an element of the scheme } X, \text{ acting on } P \gg,$$

where X and P are held fixed and N is a free natural variable. This predicate is obviously decidable. Since X acts on P, there is a term Q of the scheme X acting on P. Q is the kernel of an element Y of the scheme X [§24.7.3]. $[Y^{\triangle}$ is a positive natural number [§24.5.1]. Y acts on P, since the kernel Q of the element Y acts on P. Therefore, $[Y^{\triangle}$ satisfies the predicate G.

According to the Least Number Theorem, one can find the smallest number satisfying G. Let this be M. Since M satisfies G, M is the number of an element Z of the scheme X, acting on P.

In fact, let W be an element of the scheme X, acting on P, and let $W \not\neq Z$. We have $[W^{\triangle} \not\neq [Z^{\triangle}$ [§24.5.4], i.e., $[W^{\triangle} \not\neq M$. $[W^{\triangle}$ is the number of an element of the scheme X, acting on P, i.e., $[W^{\triangle}$ satisfies the predicate G. But since M is the smallest number satisfying G, and $[W^{\triangle} \not\neq M$, we have $M < [W^{\triangle}$, i.e., $[Z^{\triangle} < [W^{\triangle}$. Consequently, Z precedes W [§24.6.3]. Therefore, Z precedes every other element of the scheme X, acting on P.

The uniqueness of the element of the scheme X, acting on P and preceding every other element of the scheme X, acting on P, follows from the statement §24.6.5.

We shall say that the *element* Y *is active on the word* P *in the scheme* X if Y is an element of the scheme X, Y acts on P and precedes every other element of the scheme X, acting on P.

The following statement is a rephrasing of the statement 4.1:

4.2. *If the scheme X acts on the word P, then there is a unique element active on P in X.*

We shall regard the result of the action on P of the element active on P in X as the *result of the action of the scheme* X *on the word* P.

It is easy to prove the statement

4.3. *The result of the action of the scheme X on the word P is defined if and only if X acts on P. It then is a word in the alphabet* A.

We shall say that a *scheme* X *converts a word* P *into a word* Q if Q is the result of the action of X on P. In what follows, the notation

$$X : P \Vdash Q \tag{1}$$

will mean that X is a scheme and X converts the word P into the word Q. As can be seen without difficulty, the three-place verbal predicate (1) is decidable.

The following statement is obvious:

4.4. *If* $X : P \Vdash Q$ *and* $X : P \Vdash R$, *then* $Q \rightleftharpoons R$.

We shall say that a *scheme X simply converts a word P into a word Q* if X converts P into Q and the kernel of the element active on P in X is a simple substitution formula; we shall say that a *scheme X finally converts a word P into a word Q* if X converts P into Q and the kernel of the element active on P in X is a final substitution formula.

In the sequel,

$$X : P \vdash Q$$

will mean that X simply converts P into Q;

$$X : P \vdash\cdot Q$$

will mean that X finally converts P into Q.

In view of the fact that the kernel of every element of a scheme X is its term and, consequently, is a substitution formula, we have

4.5. *If* $X : P \Vdash Q$, *then* $X : P \vdash Q$ *or* $X : P \vdash\cdot Q$ [2.1].

We also have

4.6. *If* $X : P \vdash Q$, *then it is not the case that* $X : P \vdash\cdot Q$ [2.2].

In the future, the notation

(2) $$X : P\rceil$$

will mean that the scheme X does not act on the word P.

4.7. *Given any scheme X and word P, one and only one of the following three assertions holds:*

a) *there exists a unique word Q such that* $X : P \vdash Q$;
b) *there exists a unique word Q such that* $X : P \vdash\cdot Q$;
c) $X : P\rceil$.

Further, we have

4.8. *If Y is the only term of a scheme X acting on P, then Y is active on P in X.*

5. We shall say of a γ-system X in an alphabet A that it is a *converting system of the scheme Z* if for any P and Q we have the material implication

$$\ll \text{if } \gamma P \gamma Q \gamma \text{ occurs in } X \text{ then } Z : P \vdash Q \gg .$$

It is not difficult to see that the predicate $\ll X$ is a converting system of the scheme $Z \gg$ is decidable.

The following statement is trivially true:

5.1. *The converting system $\gamma P\gamma$ of the scheme Z connects P to P* [§24.3].

It is easy to prove

5.2. *If $Z : P \vdash Q$, then $\gamma P\gamma Q\gamma$ is a converting system of the scheme Z, which connects P to Q.*

We shall prove the theorem

5.3. *If a converting system X of the scheme Z connects P to Q and a converting system Y of the scheme Z connects Q to R, then the converting system*

$$(1) \qquad (\gamma Q\gamma \rightharpoondown X)\gamma Q\gamma(\gamma Q\gamma \leftharpoondown Y)$$

of the scheme Z connects P to R.

Indeed, let the theorem's hypothesis be satisfied. Then P is the first term of X, Q is the last term of X and the first term of Y, R is the last term of Y. This means that $\gamma P\gamma$ is an initial segment of X, $\gamma Q\gamma$ is a terminal segment of X and an initial segment of Y, $\gamma R\gamma$ is a terminal segment of Y. In view of this, the initial complement $(\gamma Q\gamma \rightharpoondown X)$ and the terminal complement $(\gamma Q\gamma \leftharpoondown X)$ make sense. They, like the systems X and Y, are words in the alphabet $A\gamma$. Moreover, we have

$$(2) \qquad (\gamma Q\gamma \rightharpoondown X)\gamma Q\gamma \leftrightharpoons X \quad [\S 18.9.4],$$
$$(3) \qquad \gamma Q\gamma(\gamma Q\gamma \leftharpoondown Y) \leftrightharpoons Y \quad [\S 18.8.4],$$

by virtue of which X is an initial segment and Y is a terminal segment of the word (1). Hence the word $\gamma P\gamma$ is an initial segment of (1) and the word $\gamma R\gamma$ is a terminal segment of (1). It follows from this that the letter γ is an initial, as well as a terminal, segment of (1). Consequently, (1) is a γ-system in the alphabet A, P is the first term of this γ-system and R is its last term. In other words, the γ- system (1) connects P to R. It remains to prove that this system is a converting system of the scheme, i.e., that for all words S and T in A such that $\gamma S\gamma T\gamma$ occurs in (1), we have $Z : S \vdash T$.

Thus, let $\gamma S\gamma T\gamma$ occur in the word (1), where S and T are words in the alphabet A. We have

$$(\gamma Q\gamma \rightharpoondown X)\gamma Q\gamma(\gamma Q\gamma \leftharpoondown Y) \leftrightharpoons U\gamma S\gamma T\gamma V,$$

where U and V are words in the alphabet $A\gamma$.

The words $U\gamma S\gamma T\gamma$ and X are initial segments of one and the same word [(2)]. Therefore, $U\gamma S\gamma T\gamma$ is an initial segment of X or X is a proper initial segment of $U\gamma S\gamma T\gamma$ [§17.4.5].

In the first case, $X \leftrightharpoons U\gamma S\gamma T\gamma V'$, where V' is a word in $A\gamma$. But then $\gamma S\gamma T\gamma$ occurs in the converting system X of the scheme Z, and so $Z : S \vdash T$.

In the second case, $U\gamma S\gamma T\gamma \leftrightharpoons XV''$, where V'' is a non-empty word in $A\gamma$. Consequently, the words $U\gamma S\gamma$ and X are initial segments of one and the same word, and so X is an initial segment of $U\gamma S\gamma$ or $U\gamma S\gamma$ is a proper initial segment of X. Furthermore, $V'' \rightleftharpoons V'''\gamma$ in this case, where V''' is a word in $A\gamma$, and so $U\gamma S\gamma T \rightleftharpoons XV'''$.

Suppose that $U\gamma S\gamma$ is a proper initial segment of X. Then $X \rightleftharpoons U\gamma S\gamma U'$, where U' is a non-empty word in $A\gamma$. Since the last letter of the system X is γ, we have $U' \rightleftharpoons U''\gamma$, where U'' is a word in $A\gamma$. Therefore, $X \rightleftharpoons U\gamma S\gamma U''\gamma$, and so $U\gamma S\gamma T \rightleftharpoons U\gamma S\gamma U''\gamma V'''$. Hence $T \rightleftharpoons U''\gamma V'''$, which is impossible, since T is a word in A.

Consequently, X is an initial segment of $U\gamma S\gamma$, and so

$$U\gamma S\gamma \rightleftharpoons XW \tag{4}$$

for some W. Consequently, $U\gamma S\gamma T\gamma V \rightleftharpoons XWT\gamma V$. But $U\gamma S\gamma T\gamma V$ is our system (1). Hence,

$$(\gamma Q\gamma \rightharpoonup X)\gamma Q\gamma(\gamma Q\gamma \leftharpoonup Y) \rightleftharpoons XWT\gamma V.$$

In view of (2), this equality can be written out in the following way:

$$X(\gamma Q\gamma \leftharpoonup Y) \rightleftharpoons XWT\gamma V,$$

and therefore

$$(\gamma Q\gamma \leftharpoonup Y) \rightleftharpoons WT\gamma V \quad [\S 17.5.2]. \tag{5}$$

$\gamma S\gamma$ and W are terminal segments of one and the same word [(4)]. Therefore, $\gamma S\gamma$ is a terminal segment of W or W is a terminal segment of $S\gamma$.

In the first case, there is a word Ω in the alphabet $A\gamma$ such that

$$W \rightleftharpoons \Omega\gamma S\gamma. \tag{6}$$

We then have

$$\begin{aligned} Y &\rightleftharpoons \gamma Q\gamma WT\gamma V && [(3), (5)] \\ &\rightleftharpoons \gamma Q\gamma\Omega\gamma S\gamma T\gamma V && [(6)] \end{aligned}$$

and, therefore, $\gamma S\gamma T\gamma$ occurs in the converting system Y of the scheme Z. Hence, we have $Z : S \vdash T$ in the given case.

Let us consider the second case, when W is a terminal segment of $S\gamma$.

If $W \rightleftharpoons \Lambda$, then

$$X \rightleftharpoons U\gamma S\gamma \quad [(4)] \tag{7}$$

and

$$(\gamma Q\gamma \leftharpoonup Y) \rightleftharpoons T\gamma V \quad [(5)]. \tag{8}$$

Therefore, we have

$$U\gamma S\gamma \rightleftharpoons (\gamma Q\gamma \rightharpoonup X)\gamma Q\gamma \quad [(7), (2)],$$

and hence

$$(10) \qquad S \mathrel{\overline{\overline{\ }}} Q \quad [(9), §17.5.1, §22.1.2].$$

We have, therefore,

$$\begin{aligned} Y &\mathrel{\overline{\overline{\ }}} \gamma S \gamma (\gamma Q \gamma \leftharpoondown Y) && [(3), (10)] \\ &\mathrel{\overline{\overline{\ }}} \gamma S \gamma T \gamma V && [(8)]. \end{aligned}$$

Consequently, in the given case the word $\gamma S \gamma T \gamma$ occurs in the converting system Y of the scheme Z, by virtue of which $Z : S \vdash T$.

Now let $W \not\mathrel{\overline{\overline{\ }}} \Lambda$. Then in view of (4), γ is the last letter of W, i.e., there is a word θ in the alphabet $\mathrm{A}\gamma$ such that

$$(11) \qquad W \mathrel{\overline{\overline{\ }}} \theta \gamma.$$

We have

$$\begin{aligned} (12) \qquad U \gamma S &\mathrel{\overline{\overline{\ }}} X \theta && [(4), (11)] \\ &\mathrel{\overline{\overline{\ }}} (\gamma \rightharpoondown X) \gamma \theta && [§18.9.4], \end{aligned}$$

in view of which $\gamma\theta$ is a terminal segment of S or S is a terminal segment of θ [§18.4.10]. But $\gamma\theta$ is not a terminal segment of a word S in the alphabet A. Therefore, S is a terminal segment of θ, i.e., there is a word Γ in the alphabet $\mathrm{A}\gamma$ such that

$$(13) \qquad \theta \mathrel{\overline{\overline{\ }}} \Gamma S.$$

We have

$$U \gamma \mathrel{\overline{\overline{\ }}} X \Gamma \quad [(12), (13), §17.5.1],$$

by virtue of which the word Γ is empty or has the last letter γ. Further, we have

$$\begin{aligned} (\gamma Q \gamma \leftharpoondown Y) &\mathrel{\overline{\overline{\ }}} \theta \gamma T \gamma V && [(5), (11)] \\ &\mathrel{\overline{\overline{\ }}} \Gamma S \gamma T \gamma V && [(13)], \end{aligned}$$

and, consequently,

$$(14) \qquad Y \mathrel{\overline{\overline{\ }}} \gamma Q \gamma \Gamma S \gamma T \gamma V \quad [(3)].$$

By virtue of (14), $\gamma S \gamma T \gamma$ occurs in Y in both of the possible cases here: when $\Gamma \mathrel{\overline{\overline{\ }}} \Lambda$ and when γ is the last letter of Γ. Consequently, $Z : S \vdash T$ in the given case, too. Theorem 5.3 is proven.

We shall prove the theorem

5.4. *If the first terms of converting systems X and Y of a scheme Z exist and coincide, then X is an initial segment of Y or Y is an initial segment of X.*

Let the theorem's hypotheses be satisfied, and let P be the common first term of converting systems X and Y of a scheme Z. Then $\gamma P \gamma$ is a common initial segment of the systems X and Y. Let W be the unique greatest common initial segment of the systems X and Y, which exists by §18.5.9. Let U and V be words in the alphabet $A\gamma$, which are relatively prime on the left and such that

(15) $$X \leftrightharpoons WU$$

and

(16) $$Y \leftrightharpoons WV.$$

They exist according to §18.5.8 and the definition of a greatest common initial segment. $\gamma P \gamma$ is an initial segment of W [§18.5.10]. γ occurs in W, and so the last occurrence of γ in W exists. Let this be the occurrence $\Sigma * \gamma * \triangle$, where Σ and $\triangle$ are words in the alphabet $A\gamma$. We have

(17) $$\Sigma\gamma \triangle \leftrightharpoons W.$$

Moreover, γ does not occur in $\triangle$, since $\Sigma * \gamma * \triangle$ is the last occurrence of γ in W. Consequently, $\triangle$ is a word in the alphabet A.

If $U \leftrightharpoons \Lambda$, then X is an initial segment of Y [(15), (16)] and there is nothing more to be proven; if $V \leftrightharpoons \Lambda$, then Y is an initial segment of X [(15), (16)]. It remains to assume that

(18) $$U \not\leftrightharpoons \Lambda$$

and

(19) $$V \not\leftrightharpoons \Lambda.$$

However, we shall reduce this assumption to an absurdity.

Since the systems X and Y end with the letter γ, the words U and V also end with it [(15), (16), (18), (19)]. Therefore, γ occurs in U as well as in V. Hence, there are first occurrences of γ in U and V [§23.6.3]. Let these be the occurrences $\Omega * \gamma * \Gamma$, the first occurrence of γ in U, and $\theta * \gamma * \Xi$, the first occurrence of γ in V. We have

(20) $$\Omega\gamma\Gamma \leftrightharpoons U$$

and

(21) $$\theta\gamma\Xi \leftrightharpoons V.$$

The letter γ occurs neither in Ω nor in θ, which, therefore, are words in the alphabet A. We have

(22) $$X \leftrightharpoons \Sigma\gamma \triangle \Omega\gamma\Gamma \quad [(15), (17), (20)],$$

(23) $$Y \leftrightharpoons \Sigma\gamma \triangle \theta\gamma\Xi \quad [(16), (17), (21)].$$

Here $\triangle\,\Omega$ and $\triangle\,\theta$ are words in A.

If $\Sigma \doteq \Lambda$, then we have

$$(24)\qquad \begin{aligned} \gamma\,\triangle &\doteq W && [(17)],\\ X &\doteq \gamma\,\triangle\,\Omega\gamma\Gamma && [(22)],\\ Y &\doteq \gamma\,\triangle\,\theta\Xi && [(23)]. \end{aligned}$$

It follows from this, in view of the fact that $\triangle\,\Omega$ and $\triangle\,\theta$ are words in A, that $\triangle\,\Omega$ is the first term of the system X, while $\triangle\theta$ is the first term of the system Y. But the first terms of these systems coincide, by virtue of which $\triangle\,\Omega \doteq\triangle\,\theta$ and, consequently, $\gamma\,\triangle\,\Omega\gamma$ is a common initial segment of the systems X and Y. Therefore, $\gamma\,\triangle\,\Omega\gamma$ is an initial segment of the greatest common initial segment W of these systems. Hence, in the case when $\Sigma \doteq \Lambda$, $\gamma\triangle\Omega\gamma$ is an initial segment of $\gamma\,\triangle$ [(24)] and, consequently, $\Omega\gamma$ is an initial segment of Λ, which is absurd.

Suppose, finally, that $\Sigma \not\doteq \Lambda$. Since γ is the first letter of the system X, this same letter is the first letter of a non-empty word Σ [(22)], and hence the first letter of the word $\Sigma\gamma$. Moreover, as we know, Σ is a word in the alphabet $\mathrm{A}\gamma$. It follows from this that the word $\Sigma\gamma$ is a system. Furthermore, $\Sigma\gamma \not\doteq \gamma$ since $\Sigma \not\doteq \Lambda$. The system $\Sigma\gamma$ ends with an atom, i.e., there exists a word K in the alphabet $\mathrm{A}\gamma$ and a word Q in the alphabet A, such that

$$(25)\qquad \Sigma\gamma \doteq K\gamma Q\gamma.$$

We have

$$(26)\qquad X \doteq K\gamma Q\gamma\,\triangle\,\Omega\gamma\Gamma \quad [(22),\,(25)],$$

$$(27)\qquad Y \doteq K\gamma Q\gamma\,\triangle\,\theta\gamma\Xi \quad [(23),\,(25)].$$

Therefore, the word $\gamma Q\gamma\triangle\Omega\gamma$ occurs in the converting system X of the scheme Z, while the word $\gamma Q\gamma\triangle\theta\gamma$ occurs in the converting system Y of the scheme Z. Here Q, $\triangle\Omega$ and $\triangle\theta$ are words in the alphabet A. We therefore have

$$Z : Q \vdash \triangle\Omega$$

and

$$Z : Q \vdash \triangle\theta,$$

whence $\triangle\Omega \doteq\triangle\theta$ [4.7]. Consequently, the word $K\gamma Q\gamma\triangle\Omega\gamma$ is a common initial segment of the systems X and Y. It is therefore an initial segment of their greatest common initial segment of W, i.e., of the word $K\gamma Q\gamma\triangle$ [(17), (25)]. It follows from this that $\Omega\gamma$ is an initial segment of Λ, which is absurd.

The theorem is proven.

We have

5.5. *Every γ-system in an alphabet* A*, occurring in a converting system of a scheme Z is a converting system of the scheme Z* [§23.2.2].

Let us prove the statement

5.6. *If X is a converting system of a scheme Z, connecting Q to R, and $Z : Q\rceil$, then $X \rightleftharpoons \gamma Q \gamma$ and $Q \rightleftharpoons R$.*

Assume that the hypothesis of the material implication 5.6 holds. Then Q is the first term of the system X, while R is its last term. $\gamma Q \gamma$ is an initial segment of X, i.e., there is a word U in the alphabet $A\gamma$ such that

$$X \rightleftharpoons \gamma Q \gamma U. \tag{28}$$

γU is a terminal segment of X [(28)], beginning with γ, and so is a system [§24.1.2]. If $U \not\rightleftharpoons \Lambda$, then $\gamma U \not\rightleftharpoons \gamma$ and the system γU has a first term [§24.3.1]. Let this be the word S in the alphabet A. We have

$$\gamma U \rightleftharpoons \gamma S \gamma V, \tag{29}$$

where V is a word in the alphabet $A\gamma$. Then

$$X \rightleftharpoons \gamma Q \gamma S \gamma V \quad [(28), (29)]$$

and, therefore, the word $\gamma Q \gamma S \gamma$, where Q and S are words in the alphabet A, occur in the converting system X of the scheme Z. Consequently,

$$Z : Q \vdash S,$$

contrary to our assumption that $Z : Q\rceil$ [4.7]. We arrived at this contradiction assuming that $U \not\rightleftharpoons \Lambda$. Consequently, $U \rightleftharpoons \Lambda$, and hence $X \rightleftharpoons \gamma Q \gamma$ [(28)]. It follows from this that Q is the last term of the system X. But since R is the unique last term of this system [§24.3.1], we have $Q \rightleftharpoons R$, which is just what remained to be proven.

The next statement is proven analogously.

5.7. *If X is a converting system of a scheme Z, connecting Q to R, and $Z : Q \vdash\cdot P$, then $X \rightleftharpoons \gamma Q \gamma$ and $Q \rightleftharpoons R$.*

6. We shall say that a *scheme Z simply transforms a word P into a word Q* if there exists a converting system of the scheme Z, which connects P to Q. The notation

$$Z : P \models Q \tag{1}$$

will mean that the scheme Z simply transforms the word P into the word Q.

As can be seen without difficulty, for any given Z, P and Q, the statement (1) is semidecidable.

The following statements, in which Z is a scheme and P and Q are words in an alphabet A, hold:

6.1. $Z : P \models P$ [5.1].

6.2. *If* $Z : P \vdash Q$, *then* $Z : P \models Q$ [5.2].

6.3. *If* $Z : P \models Q$ *and* $Z : Q \models R$, *then* $Z : P \models R$ [5.3, §14.3].

6.4. *If* $Z : P \models Q$, *then if* $Z : Q \vdash R$, *then* $Z : P \models R$ [6.2, 6.3].

6.5. *If X is a converting system of a scheme Z, connecting a word P to a word Q, and Y is a converting system of the scheme Z, connecting the word P to a word R, then* $Z : Q \models R$ *or* $Z : R \models Q$.

Indeed, let X be a converting system of a scheme Z, connecting a word P to a word Q, and let Y be a converting system connecting the word P to a word R. Then P is the first term of X as well as of Y, Q is the last term of X and R is the last term of Y. In view of the fact that the first terms of the converting systems X and Y coincide, X is an initial segment of Y or Y is an initial segment of X [5.4]. Suppose that X is an initial segment of Y. Then

(2) $$Y \rightleftharpoons X(X \leftharpoonup Y) \quad [§18.8.4].$$

$\gamma Q\gamma$ is a terminal segment of X, and so

(3) $$X \rightleftharpoons (\gamma Q\gamma \rightharpoonup X)\gamma Q\gamma \quad [§18.9.4].$$

We therefore have

(4) $$Y \rightleftharpoons (\gamma Q\gamma \rightharpoonup X)\gamma Q\gamma(X \leftharpoonup Y) \quad [(2), (3)].$$

Therefore, $\gamma Q\gamma(X \leftharpoonup Y)$ is a terminal segment of Y. $\gamma R\gamma$ also is a terminal segment of Y, since R is the last term of Y. Hence, $\gamma R\gamma$ is a terminal segment of $\gamma Q\gamma(X \leftharpoonup Y)$ or $\gamma Q\gamma(X \leftharpoonup Y)$ is a terminal segment of $R\gamma$ [§17.4.10]. The latter, however is impossible, since $[[\gamma Q\gamma(X \leftharpoonup Y)^{\gamma\delta} \geqslant \mathsf{II}$, but $[[R\gamma^{\gamma\delta} \rightleftharpoons \mathsf{I}$. Consequently, $\gamma R\gamma$ is a terminal segment of $\gamma Q\gamma(X \leftharpoonup Y)$. γ is a terminal segment of the word $\gamma(X \leftharpoonup Y)$ when $X \rightleftharpoons Y$, just as when $X \not\rightleftharpoons Y$. Hence the word $\gamma Q\gamma(X \leftharpoonup Y)$ in the alphabet $\mathrm{A}\gamma$ is a γ-system in the alphabet A. But since this system occurs in the converting system Y of the scheme Z [(4)], it is itself a converting system of the scheme Z. Its first term is Q and its last term is R, i.e., it connects Q to R. Therefore, we have $Z : Q \models R$ in the given case.

Analogously, we discover that when Y is an initial segment of X, we will have $Z : R \models Q$. Therefore, we have $Z : Q \models R$ or $Z : R \models Q$. Q.E.D.

Using the notation (1), the statement 6.5 can be formulated in the following manner:

6.6. *If* $Z : P \models Q$ *and* $Z : P \models R$, *then* $Z : Q \models R$ *or* $Z : R \models Q$.

We have

6.7. *If* $Z : Q \models R$, *then if* $Z : Q]$, *then* $Q \rightleftharpoons R$.

We regard this statement as a strengthened implication with the semidecidable premise $\ll Z : Q \models R \gg$ and a decidable conclusion. Laying bare the meaning of the strengthened implication according to §14.1, we convince ourselves of its truth [5.6].

Analogously, we have

6.8. *If $Z : Q \models R$, then if $Z : Q \vdash\cdot R$, then $Q \leftrightharpoons R$* [§14.1, 5.7].

7. We say that Z *naturally transforms* P *into* Q and write

$$Z : P \models Q\rceil,$$

if $Z : P \models Q$ and $Z : Q\rceil$.

We say that Z *finally transforms* P *into* Q, and write

$$Z : P \models\cdot Q,$$

if there exists a word R such that $Z : P \models R$ and $Z : R \vdash\cdot Q$.

7.1. *If $Z : P \models\cdot Q$, then there exists a unique word R such that $Z : P \models R$ and $Z : R \vdash\cdot Q$.*

In fact, let the words R and S be such that

$$Z : P \models R, \quad Z : R \vdash\cdot Q$$

and, on the other hand,

$$Z : P \models S, \quad Z : S \vdash\cdot Q.$$

Then $Z : R \models S$ or $Z : S \models R$ [6.6]. Suppose, for example, that $Z : R \models S$. But by assumption, we also have $Z : R \vdash\cdot Q$, so that $R \leftrightharpoons S$ [6.8]. The case $X : S \models R$ is treated analogously.

We shall say that R is the *next to the last word for* P *and* Q if $Z : P \models R$ and $Z : R \vdash\cdot Q$.

We shall say that Z *changes* P into Q and write

$$Z : P \Rightarrow Q$$

if $X : P \models Q\rceil$ or $Z : P \models\cdot Q$.

7.2. *If $Z : P \models Q\rceil$ and $Z : P \models R$, then $Z : R \models Q\rceil$* [6.6, 6.7, 6.1].

Let us agree to write

$$Z : P \models Q \vdash\cdot R$$

instead of

$$Z : P \models Q \quad Z : Q \vdash\cdot R.$$

We shall prove

7.3. *It is not the case that*

(1) $$Z : P \models Q\rceil \text{ and } Z : P \models\cdot R.$$

Let us assume that both of the statements in (1) are true. Then the following statements are true:

(2) $$Z : P \models Q,$$

(3) $$Z : Q\rceil$$

and there exists an S such that

(4) $$Z : P \models S$$

and that

$$Z : S \vdash\cdot R.$$

Let us hold such an S fixed.

We have $Z : S \models Q$ or $Z : Q \models S$ [(2), (4), 6.6]. If $Z : S \models Q$, then in view of (5), $S \eqcirc Q$ [6.8]. If $Z : Q \models S$, then in view of (3), we again have $S \eqcirc Q$ [6.7]. Thus, $S \eqcirc Q$. But then $Z : Q\rceil$ and $Z : Q \vdash\cdot R$, which is impossible [4.7]. The theorem is proven.

We shall say that the *scheme* Z *is applicable to the word* P, and shall write

$$!\,Z(P),$$

if there exists a word Q such that $Z : P \Rightarrow Q$.

Let us prove

7.4. *If* $!\,Z(P)$, *then there exists a unique word* Q *such that* $Z : P \Rightarrow Q$.

The existence of a Q such that $Z : P \Rightarrow Q$ follows from the definition of $!\,Z(P)$. We shall show that such a Q is unique. Indeed, let $Z : P \Rightarrow Q$ and $Z : P \Rightarrow R$. Then in view of 7.3,

$$Z : P \Rightarrow Q\rceil \text{ and } Z : P \models R\rceil$$

or

$$Z : P \models\cdot Q \text{ and } Z : P \models\cdot R.$$

In the first case, $Z : P \models Q$ and $Z : P \models R$. Consequently, $Z : Q \models R$ or $Z : R \models Q$ [6.6]. But $Z : Q\rceil$ and $Z : R\rceil$, so that $Q \eqcirc R$ [6.7].

In the second case, $Z : P \models S \vdash\cdot Q$ and $Z : P \models T \vdash\cdot R$ for some S and T. But then $Z : S \models T$ or $Z : T \models S$ [6.6]. If $Z : S \models T$, then, since $Z : S \vdash\cdot Q$, we have $S \eqcirc T$ [6.8]. The same equality is also obtained when $Z : T \models S$. Consequently, $Z : S \vdash\cdot Q$ and $Z : S \vdash\cdot R$. But it follows from this that $Q \eqcirc R$ [4.4]. The theorem is proven.

8. We shall prove the following statements:

8.1. *If* $Z : P \models Q$, *then there exists a unique converting system* X *of the scheme* Z, *connecting* P *to* Q *and such that* X *is an initial segment of each converting system of the scheme* Z, *connecting* P *to* Q.

Indeed, let $Z : P \models Q$. Then there exists a unique converting system of the scheme Z, connecting P to Q [6]. The predicate $\ll X$ is a converting system of the scheme Z, connecting P to $Q \gg$ is decidable. Hence, by virtue of the Least Number Theorem [§21.2.1], there exists a converting system X of the scheme Z, connecting P to Q, with minimal length. Now let U be an arbitrary converting system of the scheme Z, connecting P to Q. The first terms of the systems X and U coincide. Therefore, X is an initial segment of U or U is an initial segment of X [5.4]. By virtue of our choice of X, U cannot be a proper initial segment of X. Consequently, X is an initial segment of U. Q.E.D.

Let $Z : P \models Q$. We shall call the converting system X singled out in 8.1 the *natural converting system connecting* P *to* Q.

Let us prove

8.2. *If $Z : P \models Q]$ and X is a converting system of the scheme Z, connecting P to Q, then X is the natural converting system of the scheme Z, connecting P to Q.*

Indeed, let $Z : P \models Q]$ and let X be a converting system of the scheme Z, connecting P to Q. Then, in particular, $Z : P \models Q$ and, therefore, one can construct the natural converting system Y of the scheme Z, connecting P to Q [8.1]. We shall show that $X \rightleftharpoons Y$.

In fact, Y connects P to Q, and so $Y \rightleftharpoons U\gamma Q\gamma$ for some U. Besides, Y is an initial segment of X, and so $X \rightleftharpoons YV$ for some V in the alphabet $A\gamma$. Assume that $V \not\rightleftharpoons \Lambda$. Then, since X is a system, the last letter of V is γ, and so γ occurs in V. Let $R * \gamma * W$ be the first occurrence of γ in V. Then R is a word in A and $V \rightleftharpoons R\gamma W$. Thus $X \rightleftharpoons U\gamma Q\gamma R\gamma W$, and so the word $\gamma Q\gamma R\gamma$ occurs in X, where Q and R are words in A. Since X is a converting system of the scheme Z, $Z : Q \vdash R$ [(5)], which contradicts $Z : Q]$ [4.7]. Consequently, $V \rightleftharpoons \Lambda$ and $X \rightleftharpoons Y$. Q.E.D.

Analogously to the way in which 8.2 was proven, we can prove the statement

8.3. *If $Z : P \models R \vdash\cdot Q$ and X is a converting system of the scheme Z, connecting P to R, then X is the natural converting system of the scheme Z, connecting P to R.*

We shall say that the *scheme Z simply transforms P into Q in N steps* if $Z : P \models Q$ and the volume of the natural converting system connecting P to Q equals $N\mathsf{I}$.

We shall say that the *scheme Z naturally transforms P into Q in N steps* if $Z : P \models Q]$ and Z simply transforms P into Q in N steps.

We shall say that the *scheme Z finally transforms P into Q in N steps* if $Z : P \models\cdot Q$ and in $(N - 1)$ steps Z simply transforms P into the unique [7.1] word R for which $Z : P \models R \vdash\cdot Q$.

We shall say that the *scheme Z changes P into Q in N steps* if Z naturally transforms P into Q in N steps or finally transforms P into Q in N steps.

The following four statements can easily be proven:

8.4. *If $Z : P \models Q$, then the number N, such that Z simply transforms P into Q in N steps, is unique* [8.1].

8.5. *If $Z : P \models Q]$, then the number N, such that Z naturally transforms P into Q in N steps, is unique* [8.4].

8.6. *If $Z : P \models\cdot Q$, then the number N, such that Z finally transforms P into Q in N steps, is unique* [7.1, 8.4].

8.7. *If $Z : P \Rightarrow Q$, then the number N, such that Z changes P into Q in N steps, is unique* [7.1, 7.3, 8.5, 8.6].

Let $!Z(P)$. We shall denote the uniquely determined natural number N, such that Z changes P into some word Q in N steps, by the symbol $(Z : P)$. The number $(Z : P)$ will be called the *number of steps in the scheme Z's performance on the word P*.

Let us prove the following two statements:

8.8. *If $Z : P]$, then Z changes P into P in zero steps.*

Indeed, let X be a converting system of the scheme Z, connecting P to some word R. Then $R \leftrightharpoons P$ and $X \leftrightharpoons \gamma P \gamma$ [5.6]. X is the natural converting system of the scheme Z, connecting P to P, and the volume of X is equal to one. Since $Z : P]$, Z naturally transforms and, therefore, changes P into P in zero steps. Q.E.D.

8.9. *If $Z : P \vdash Q$ and $Z : Q \Rightarrow R$, then $Z : P \Rightarrow R$ and $(Z : P) \leftrightharpoons (Z : Q)\mathsf{I}$.*

Indeed, let this implication's premise be fulfilled. Then, in particular, $Z : P \models Q$. Furthermore, $Z : Q \Rightarrow R$. Consequently, $Z : Q \Rightarrow R]$ or $Z : Q \models\cdot R$. In the first case, $Z : P \models R]$; moreover, if X is a converting system of the scheme Z, connecting Q to R, then $\gamma P X$ is a converting system of Z, connecting P to R. The systems X and $\gamma P X$ are the natural converting systems connecting Q and P, respectively, to R [8.2]. Hence

$$(Z : P) \leftrightharpoons [\gamma P X^\circ \leftrightharpoons \mathsf{I}[X^\circ \leftrightharpoons \mathsf{I}(Z : Q) \leftrightharpoons (Z : Q)\mathsf{I}.$$

The case $Z : Q \models\cdot R$ is treated analogously, with a reference to 8.3. The assertion 8.9 is proven.

9. We shall sometimes apply the following *method of induction on a converting system*. Let Z be a scheme and let F be a one-place verbal predicate in the alphabet $A\gamma$. In order to prove that every converting system X of the scheme Z satisfies the predicate F, we prove that:

1) every converting system of the scheme Z, having the form $\gamma P \gamma$, satisfies F;

2) for every converting system X of the scheme Z and every word Q, the following implication holds:

> ≪if X satisfies F, P is the last term of X and $Z : P \vdash Q$, then the system $XQ\gamma$ satisfies F≫.

Then every converting system X of the scheme Z, distinct from γ, satisfies F.

Indeed, let there be given a scheme Z and a predicate F. Assume that the conditions 1) and 2) are fulfilled. Construct the following one-place verbal predicate G in the alphabet $A\gamma$:

> ≪every initial segment X of the word Y, which is a converting system distinct from γ of the scheme Z, satisfies F≫.

We shall prove that every word Y in the alphabet $A\gamma$ satisfies G.

For this we shall make use of the method of right induction on the construction of a word Y.

First of all, the empty word Λ satisfies G, since every initial segment X of the empty word will fail to be a converting system. Now let the word Y satisfy G and let ξ be a letter of the alphabet $A\gamma$. We shall show that $Y\xi$ also satisfies G. Let X be an initial segment of $Y\xi$ which is a converting system of Z, and let

$X \neq \gamma$. Then$X \rightleftharpoons X_1 Q\gamma$, where X_1 is a converting system and Q is a word in A. Let us show that X satisfies F. If $X_1 \rightleftharpoons \gamma$, then this follows from the premise 1) of the method of induction on a converting system. Let $X_1 \neq \gamma$. Then $X_1 \rightleftharpoons X_2 P\gamma$, where X_2 is a converting system, itself. Further, X is an initial segment of $Y\xi$. Therefore X is an initial segment of Y or $X \rightleftharpoons Y\xi$. If X is an initial segment of Y, then X satisfies F, since Y satisfies G. Consider the case when $X \rightleftharpoons Y\xi$. Then $X_1 Q\gamma \rightleftharpoons Y\xi$. Hence $\xi \rightleftharpoons \gamma$ and $X_1 Q$ is an initial segment of Y. Therefore X_1 is an initial segment of Y. X_1 is a converting system distinct from γ. Since Y satisfies G, X_1 satisfies F. $X \rightleftharpoons X_2 P\gamma Q\gamma$ and X is a converting system. Hence, $Z : P \vdash Q$. By virtue of the premise 2), the word $X_1 Q\gamma$ also satisfies F, i.e., X satisfies F. By the same token, $Y\xi$ satisfies G. Consequently, any Y satisfies G.

It is already not hard to show that the assertion formulated in the conclusion of the method of induction on a converting system, described by us, holds. In fact, let X be a converting system of the scheme Z, distinct from γ. Then X satisfies G and, consequently, every initial segment of X, distinct from γ, is a converting system of Z and satisfies F. In particular, we may take the word X, itself, as this initial segment. Therefore, the method presented by us is justified.

10. Another variant of induction, which will be useful to us in the sequel, is the *method of induction on the steps in the performance of a scheme Z*. It consists in the following. Let Z be a scheme, let F be a one-place verbal predicate in an alphabet A and let P be a word in A. We prove that:

1′) P satisfies F;

2′) for any two words R and S in A, such that $Z : P \models R$ and $Z : R \vdash S$, the following implication holds:

$$\ll \text{if } R \text{ satisfies } F, \text{ then } S \text{ also satisfies } F \gg .$$

Then every word R, such that $Z : P \models R$, satisfies F.

Indeed, let there be given a Z, F and P for which the premises 1′) and 2′) are fulfilled. Construct the following one-place verbal predicate G with the free variable X for words in the alphabet Aγ:

> $\ll$if X is a converting system of the scheme Z, beginning with the word $\gamma P\gamma$ and ending with the word $\gamma S\gamma$, where S is a word in the alphabet A, then S satisfies $F \gg$.

We shall prove that every converting system X of the scheme Z, distinct from γ, satisfies G. To this end, let us make use of the method of induction presented in Part 9. If X has the form $\gamma P\gamma$, then X satisfies G, since P satisfies F by virtue of the condition 1′). Further, if $X \rightleftharpoons X_1\gamma R\gamma$, X satisfies G and $Z : P \vdash S$, then $XS\gamma$ satisfies G. Indeed, if X begins with the word $\gamma P\gamma$, then R satisfies F, since X satisfies G. But then S satisfies F by virtue of 2′). This means that $XS\gamma$ satisfies G.

Now let us prove the assertion formulated in the conclusion of the method of induction on the steps in the performance of a scheme, being described by us. Let $Z : P \models R$. Then there exists a converting system X of the scheme Z, connecting P to R. X begins with the word $\gamma P\gamma$ and ends with the word $\gamma R\gamma$. Since X satisfies G, R satisfies F.

11. Let us consider a simple example.

Let Q be any word in an alphabet A. The word βQ is then a final substitution formula with empty left side and with right side Q. $\gamma\beta Q\gamma$ is a system with the unique term βQ, i.e., a scheme. Let P be any word in the alphabet A. The formula βQ acts on P, since the empty word occurs in P. The result of the action of βQ on P is the result of substituting the word Q for the first occurrence of the empty word in P, i.e., the word QP. Hence

$$\gamma\beta Q\gamma : P \vdash\cdot QP$$

and, consequently,

11.1. $\gamma\beta Q\gamma : P \Rightarrow QP.$

(Here P and Q are any words in the alphabet A.)

Analogously, the word αQ is a simple substitution formula with empty left side and with right side Q. $\gamma\alpha Q\gamma$ is a scheme. Again, let P be a word in the alphabet A. The formula αQ acts on P, since the empty word occurs in P. The result of the action this time is also the word QP. But now

11.2. $\gamma\alpha Q\gamma : P \vdash QP,$

since the formula αQ is, unlike βQ, simple.

We shall prove

11.3. *If* $\gamma\alpha Q\gamma : P \models R$, *then there exists a natural number* N *such that* $R \rightleftharpoons (Q \times N)P$.

Let us make use of the method of induction on the steps in the performance of the scheme [10]. As F, we take the decidable predicate

$$\ll \text{there exists a natural number } N \text{ such that } R \rightleftharpoons (Q \times N)P \gg .$$

It is easy to see that P satisfies F (for $N \rightleftharpoons \Lambda$). Further, let $\gamma\alpha Q\gamma : P \models R$, $\gamma\alpha Q\gamma : R \vdash S$ and $R \rightleftharpoons (Q \times M)P$. Then $S \rightleftharpoons QR$ [11.2] and, therefore, $S \rightleftharpoons Q(Q \times M)P \rightleftharpoons (Q \times M\mathsf{I})P$ [§20], i.e., S also satisfies F. Q.E.D.

11.4. *The scheme* $\gamma\alpha Q\gamma$ *is not applicable to any word* P *in the alphabet* A.

Indeed, let P be an arbitrary word in A and let $!\gamma\alpha Q\gamma(P)$. Then for some R, $\gamma\alpha Q\gamma : P \Rightarrow R$, i.e., $\gamma\alpha Q\gamma : P \models R]$ or $\gamma\alpha Q\gamma : P \models\cdot R$. In the first case, $\gamma\alpha Q\gamma : P \models R$ and $\gamma\alpha Q\gamma : R]$, which is impossible, since the scheme $\gamma\alpha Q\gamma$ acts on any word P in A [11.2]. In the second case, there exists a word S such that $\gamma\alpha Q\gamma : P \models S \vdash\cdot R$. But $\gamma\alpha Q\gamma : S \vdash\cdot R$ is impossible, since the scheme $\gamma\alpha Q\gamma$ does not contain any final substitution formulas. Thus, the assumption that $!\gamma\alpha Q\gamma(P)$ leads to a contradiction. Q.E.D.

REMARK: We have proven this assertion by the method of ≪reductio ad absurdum≫, and so we should understand the negation appearing in it as a *reductional* negation (see §15.4). However, we might have understood it as a *strengthened* negation (see §12.1). We suggest that the reader, herself or himself, give a proof which would correspond to such an approach to understanding the formulation of our theorem.

CHAPTER III

NORMAL ALGORITHMS: DEFINITION AND EXAMPLES

§26 Algorithms

1. Algorithims play an essential role in the sequel. The concept of an algorithm can be given the following preliminary definition, which will be made precise in the sequel:

An *algorithm* is a prescription uniquely determining the course of certain constructive processes.

2. The word ≪prescription≫ in the definition of an algorithm indicates the presence of an addressee—a subject liable to fulfill this prescription. In particular, this can be a person or some good model of one. Fulfilling a prescription, its addressee will be carrying out a certain process, uniquely determined by this prescription. This process will be going on within a certain system (for example, within the system: ≪blackboard, chalk, eraser≫), whose state will be changing during the course of the process. All these changes should be completely determined by the prescription and carried out by the addressee-performer, who must, of course, understand the prescription. Sometimes the prescription may demand that the process end.

3. In our definition of an algorithm, processes are spoken of in the plural. An algorithm can, generally speaking, determine many processes, differing from each other in the system's initial state. (In a particular case, however, an algorithm can hold an initial state fixed, and then we shall have only a single process.)

4. The word ≪uniquely≫ in the definition of an algorithm should not be understood too literally. One or another relation of identity can be established for a system's states. For example, words written with chalk on a blackboard may be regarded as the states of the system ≪blackboard, chalk, eraser≫. In this case, it will be natural to regard states as identical if the corresponding words are identical.

The process determined by an algorithm for a given initial state of the system should be uniquely determined only "up to the identity" of the states obtained during its run. Only by agreeing to apply the abstraction of identification [see §5] in a natural way to the states of our system, and to the changes taking place in it, can we require that this process be unique.

5. In order to formulate and understand an algorithm, it is necessary to have one or another language—the language of the given algorithm. It is natural to require that it be possible to name in this language any of the system's states with which the algorithm deals. Moreover, a state should be uniquely determined by its name.

There will be no essential loss in generality if we agree to assume that an algorithm's language is written and linear. This means that the language has an alphabet consisting of elementary symbols—the letters; moreover, expressions of the given language (statements, names, prescriptions, etc.) are ≪words≫, i.e., linear rows of letters. The process of applying an algorithm can then be simulated by some other process, dealing with words—with names of states.

Thus, we reduce an analysis of algorithms to an analysis of *verbal algorithms*. This is what we shall call algorithms acting on words in any given alphabet.

The process of a performance of a verbal algorithm *in an alphabet* A consists of the successive generation of words in the alphabet A according to the formulated prescription. A certain *initial word* in A is chosen before the beginning of this process. The process can end with the generation of a certain word, which we then declare to be the *result of our algorithm's performance on the initial word*. We shall also say that our algorithm *changes the initial word into the result of its performance on it*. We shall say that an algorithm is *applicable to a given word* if the process of its performance on this word ends, i.e., if there is a result of its performance on the given word as the initial word.

In what follows,

$$!\mathfrak{A}(P) \tag{1}$$

will mean that the algorithm $\mathfrak{A}$ is applicable to the word P;

$$\mathfrak{A} : P \Rightarrow Q \tag{2}$$

will mean that the algorithm $\mathfrak{A}$ changes the word P into the word Q.

The following uniqueness law for the result of an algorithm's performance is obvious:

5.1. *If* $\mathfrak{A} : P \Rightarrow Q$, *then if* $\mathfrak{A} : P \Rightarrow R$, *then* $Q = R$.

5.2. $!\mathfrak{A}(P)$ *if and only if there exists a* Q *such that* $\mathfrak{A} : P \Rightarrow Q$.

By definition, starting with a given initial word P in an alphabet A, an algorithm $\mathfrak{A}$ in this alphabet successively generates words in A. In defining the predicates $!\mathfrak{A}(P)$ and $\mathfrak{A} : P \Rightarrow Q$, we tacitly assumed that P is a word in A. Other alphabets and words in them will be encountered in our future analysis. Let us agree in such cases to regard the algorithm $\mathfrak{A}$ in the alphabet A as inapplicable to a word P, if P is not a word in A, i.e., if a letter occurs in P, but not in A. Therefore, we shall regard the statement $!\mathfrak{A}(P)$ as false in this case.

The statement $\mathfrak{A} : P \Rightarrow Q$ will then also be regarded as false, whatever be the word Q.

6. Considering some algorithm $\mathfrak{A}$ in an alphabet A, we shall often be interested only in the results of its performance on words in a narrower alphabet, Б, making up part of the alphabet A.

We shall say that a verbal algorithm $\mathfrak{A}$ *in an alphabet* A is an algorithm *over an alphabet* Б if A is an extension of Б, i.e., if every letter of the alphabet Б is also a letter of the alphabet A.

Let Б be an alphabet, and let $\mathfrak{B}$ and $\mathfrak{C}$ be verbal algorithms over Б.

We shall say that $\mathfrak{B}$ and $\mathfrak{C}$ are *equivalent with respect to* Б if for any words P and Q in Б, we have the implications:

a) if $\mathfrak{B} : P \Rightarrow Q$, then $\mathfrak{C} : P \Rightarrow Q$,

b) if $\mathfrak{C} : P \Rightarrow Q$, then $\mathfrak{B} : P \Rightarrow Q$.

We shall say that $\mathfrak{B}$ and $\mathfrak{C}$ *are strongly equivalent with respect to* Б if they are equivalent with respect to Б and, moreover, for any word P in Б, we have the implications:

c) if $!\,\mathfrak{B}(P)$, then $!\,\mathfrak{C}(P)$,

d) if $!\,\mathfrak{C}(P)$, then $!\,\mathfrak{B}(P)$.

We shall say that $\mathfrak{B}$ and $\mathfrak{C}$ are *completely equivalent with respect to* Б if they are strongly equivalent with respect to Б and, moreover, the following conditions are satisfied for any word P in Б:

e) given any word Q in the alphabet of the algorithm $\mathfrak{B}$, we have the implication

$$\text{if } \mathfrak{B} : P \Rightarrow Q, \qquad \text{then } \mathfrak{C} : P \Rightarrow Q,$$

f) given any word Q in the alphabet of the algorithm $\mathfrak{C}$, we have the implication

$$\text{if } \mathfrak{C} : P \Rightarrow Q, \qquad \text{then } \mathfrak{B} : P \Rightarrow Q.$$

It is apparent from the definitions that the equivalence relations formulated by us were listed in the order of their successive strengthening.

§27 Normal Algorithms. Normalization Principle

1. The concept of an algorithm in a given alphabet just introduced is not precise enough for yielding the possibility of regarding algorithms as objects of a mathematical theory. For this it must be made precise, which is exactly what we shall turn to now.

First of all, a prescription's being ≪generally understood≫ can not be regarded as a completely precise characterization. A gradual transition from an understandable prescription to an entirely incomprehensible one by means of greater and greater complications is obviously quite possible. It is therefore natural to think about somehow regulating a prescription by splitting it up into rules of a standard predetermined type, whose general comprehensibility would not arouse any doubts. The process of applying an algorithm would then consist of separate elementary steps, each of which would be performed according to one of these rules.

It is natural to strive for as great simplicity as possible in these steps and rules.

Processes whose behavior is determined by a prescription are diverse. It became clear, however, that if one is interested in algorithms only up to an equivalence, then their hardly surveyable diversity can apparently be reduced to a simpler idea.

2. We introduce for consideration verbal algorithms of a special type—the so-called *normal algorithms*. Every normal algorithm $\mathfrak{A}$ will be given by singling out the following three objects: a certain alphabet A, a certain three-letter alphabet, $\alpha\beta\gamma$, having no letters in common with the alphabet A, and a certain γ-scheme Z in the alphabet A$\alpha\beta$ [§25]. We shall call A the *alphabet of the normal algorithm* $\mathfrak{A}$. We shall call Z the scheme of *this algorithm.* Of course, introducing any new letter, δ, playing the role of a separating symbol, we could write down this triple in the form of a δ-system, $\delta A\delta\alpha\beta\gamma\delta Z\delta$, so that in the final analysis every normal algorithm is defined by singling out one unique word.

Substitution formulas [§25.2] which are terms of Z will be called *substitution formulas of the algorithm* $\mathfrak{A}$ (they, as we recall, can be simple or final). Words in the alphabet A will serve as initial data for the performance of $\mathfrak{A}$. Applying the algorithm $\mathfrak{A}$ to a word P in the alphabet A will mean applying $\mathfrak{A}$'s scheme Z to P. This is our plan. We shall leave the details of the definitions until Part 4.

3. The alphabet $\alpha\beta\gamma$, as we have seen, plays an auxiliary role here. Its letters are sort of "punctuation marks" in the scheme: γ separates substitution formulas from each other, while α and β separate the left sides of these formulas from their right sides, simultaneously indicating the substitution formula's *type* (α for simple formulas, β for final ones). These letters can be fixed once and for all in some way, requiring that in the future only alphabets A not containing these letters be considered. Then every normal algorithm will essentially be defined by singling out two objects: the alphabet in which it acts and a certain scheme— the scheme of this algorithm.

A normal algorithm whose alphabet is A will be called a *normal algorithm in the alphabet* A. Therefore, if we hold an alphabet A fixed, every normal algorithm in A will be defined by singling out its scheme, and nothing else. This circumstance permits us to identify, in a certain sense, normal algorithms with the schemes defining them and carry over to normal algorithms all the terminology and notations developed for schemes in §25. Thus, for example, if $\mathfrak{A}$ is the normal algorithm defined by a scheme Z, then the notation

$$\mathfrak{A} : P \models Q, \qquad \mathfrak{A} : P \models\cdot\, Q, \qquad \mathfrak{A} : P\rceil, \qquad \mathfrak{A} : P \Rightarrow Q,$$
$$!\,\mathfrak{A}(P), \qquad (\mathfrak{A} : P),$$

will mean the same thing as the notation

$$Z : P \models Q, \qquad Z : P \models\cdot\, Q, \qquad Z : P\rceil, \qquad Z : P \Rightarrow Q,$$
$$!\,Z(P), \qquad (Z : P)$$

respectively. Other analogous notations will be understood similarly. In various kinds of verbal formulations, we shall often speak of ≪normal algorithms≫ instead of ≪schemes≫. What we are saying also applies to the methods of induction justified in parts 9 and 10 of §25. In the sequel, references to §25 should be understood with due regard for the current remark.

4. Let us now give a detailed definition of a *normal algorithm.*

Every normal algorithm is a prescription, uniquely determined by the objects indicated in Part 1 and uniquely determining, in turn, the course of certain constructive processes of a special type. For a fixed normal algorithm $\mathfrak{A}$, each of these processes is determined by the choice of an initial word, for which one can take any word P in the alphabet A.

In the first step of the process determined by the normal algorithm $\mathfrak{A}$ in the alphabet A and the word P in this alphabet, the word P, itself, is taken. If the word Q were taken in a certain step and if $\mathfrak{A} : Q \vdash R$ for some R, then the word R would be taken in the next step and the process would continue. But if $\mathfrak{A} : Q]$ or $\mathfrak{A} : Q \vdash\cdot\ S$ for some S, then the process would break off; in the first case, its result $\mathfrak{A}(P)$ is regarded as the word Q, while in the second, as the word S.

We shall briefly call the process just described the *process of applying* $\mathfrak{A}$ to P.[1] The reader will convince her or himself without difficulty that the following two statements are true:

4.1. *The process of applying $\mathfrak{A}$ to P breaks off if and only if* $!\mathfrak{A}(P)$ [§25.7].

4.2. *If* $!\mathfrak{A}(P)$, *then* $\mathfrak{A} : P \Rightarrow \mathfrak{A}(P)$[§25.7.4].

5. In what follows, we shall, as a rule, make use of another, non-linear method of writing down schemes of normal algorithms. This method has won a firm place in the literature and has definite technical advantages. It consists in the following. The substitution formulas constituting a normal algorithm's scheme are written down "in a column": the first substitution formula (the kernel of the scheme's first element [§24.7, §24.5]) is written first, under it, the second, etc., up to the last. Then the α in the simple substitution formulas and the β in the final ones are replaced by the letters $\rightarrow$ and $\rightarrow\cdot$, respectively (in this connection, we should, of course, stipulate that these letters do not occur in the algorithm's alphabet). For greater clarity, the formulas making up the scheme

[1]Several natural generalizations of the definition of a normal algorithm were considered in a paper of Nagorny [3]. One of them consists in the fact that an algorithm is defined by *several* schemes and in each step of the process of applying it, instructions as to which of the schemes should be applied in the next step are elaborated. In another generalization, there are "infinitely many" such schemes—they are generated by a certain *normal* algorithm. In a third generalization, the schemes themselves are also "infinite" (in the indicated sense of the word). It was shown in this same paper that these "generalizations" of the *definitions* do not lead to an extension of the corresponding *concepts*: each of the "generalized" algorithms is simulated by some normal algorithm.

are united by a curly bracket to the left. With a view to uniformity, we shall even use a curly bracket when there is only one substitution formula.

Thus, the scheme

$$\gamma ab\alpha ba\gamma ac\alpha ca\gamma a\beta\gamma\alpha a\gamma$$

in the alphabet *abc*, consisting of the four substitution formulas

$$ab\alpha ba, \qquad ac\alpha ca, \qquad a\beta, \qquad \alpha a$$

going in the order indicated, are written down in this fashion as follows:

$$\begin{cases} ab \rightarrow ba \\ ac \rightarrow ca \\ a \rightarrow\cdot \\ \quad \rightarrow a. \end{cases}$$

We should like to emphasize that such a method of writing down schemes of normal algorithms is not an independent definition of them, but is merely a means of making our use of them intuitive and convenient. The reader should remember that γ-schemes in the alphabet $\mathrm{A}\alpha\beta$ are *words* in the alphabet $\mathrm{A}\alpha\beta\gamma$ [§25] and all statements about them should be understood as statements about words of a definite type.

6. As we shall see later on, the concept of a normal algorithm possesses a great degree of generality. In view of this, the following question arises naturally: to what extent does the precise concept of a *normal algorithm* correspond to the more general, but less precise, concept of a verbal algorithm in some alphabet [§25]?

This somewhat vaguely posed question should be made precise. What kind of correspondence can be expected here? Of course, *algorithms in general* are much more varied than *normal algorithms*. It is natural, however, to ask whether it is possible to replace each verbal algorithm in an alphabet A by a normal algorithm which is completely equivalent to it with respect to A [§25.5]. We express our conviction that this question should be answered affirmatively by stating the following *normalization principle for algorithms*:

Every verbal algorithm in an alphabet A *is completely equivalent with respect to* A *to some normal algorithm over* A.

Let us agree to say that a verbal algorithm in an alphabet A is *normalizable* if it is completely equivalent with respect to A to some normal algorithm over A. The normalization principle can then be formulated in the following manner:

Every verbal algorithm is normalizable.

The normalization principle just formulated requires clarification.

First of all, this is not a mathematical theorem, subject to proof. The normalization principle fails to be of such a nature at least because one of the concepts which figures in this principle—the general concept of a verbal algorithm—is not an exact mathematical concept. What the principle expresses is just the making of this concept exact through the concept of a normal algorithm.

The contents of the principle is not, however, limited to merely making the concept of a verbal algorithm precise. The principle also asserts the *suitability* of this. Although the general concept of an algorithm is not sufficiently precise, it is nevertheless a concept which has become extablished and is being used. Settling the question of whether to regard some concrete prescription as an algorithm does not usually give rise to difficulties. The principle asserts that whenever this sort of question is settled positively, i.e., whenever a prescription is recognized as being a verbal algorithm in some alphabet, it is possible to normalize this prescription, i.e., replace it by a normal algorithm equivalent to it with respect to this alphabet. Formulating the normalization principle, we are by the same token making a great number of predictions about the normalizability of algorithms whose construction will be carried out in the future.

The normalization principle is similar in this respect to a physical law. For every physical law is a justification for an enormous number of predictions about future events. For example, the law of the conservation of energy predicts the conservation of the total amount of energy in every closed physical system. In its negative form, it asserts the impossibility of inventing a perpetuum mobile—a physical system doing work indefinitely without an influx of energy. Analogously to this, the normalization principle predicts the normalizability of any verbal algorithm which will ever be invented, and hence the impossibility of inventing one which is not normalizable.

But upon what can be based our confidence in the validity of the normalization principle for algorithms, i.e., in the correctness of those predictions which are made on the basis of it? To a large degree, on the very same thing which our confidence in known physical laws is based—on experience.

And the experience confirming the normalization principle is vast. For people have been engaged in mathematics for quite some time—not less than 4,000 years. Many different algorithms have been devised during this time. And among them, not a single non-normalizable one is known. After all, this is a weighty argument in favor of the normalization principle. No less weighty, let us say, than the experimental confirmation of the law of the conservation of energy.

Another important argument in favor of the normalization principle is the following.

Many ways of combining algorithms—constructing new algorithms from several given ones—are known. All these ways yield normalizable algorithms if the initial algorithms were normalizable. This fact will be established in future sections with respect to the most important methods of combination. It shows that for the construction of a non-normalizable verbal algorithm—if such a construction, contrary to the normalization principle, is somehow possible—it would be necessary to employ some entirely new constructive methods, which can not even be imagined by any contemporary mathematician.

At the present time, however, there are no indications which might encourage us to expect the invention of such methods. But, as is well known, leaps unprepared for by a preceding development do not occur in history.

Arguments of an historical nature should also be borne in mind. In the 1930's there appeared the pioneering papers of Gödel [1], Church [1], Kleene [1], [2], Turing [1], [2] and Post [1], in which certain special kinds of algorithms were defined precisely. For each of these kinds, there arose a certainty in its exhausting *all* algorithms up to an equivalence. In other words, mathematicians were imbued with the conviction that *every* algorithm is equivalent to some algorithm of the given kind. This idea of standardizing algorithms underlies its modern theory. Soon after the first standard concepts of an algorithm were developed, it proved possible to establish that all the proposed standardizations are equivalent to each other, and when certain new standardizations[2] were later proposed, they too turned out equivalent to those defined previously. Therefore, when considering algorithms "up to an equivalence", it makes no difference just which of their standardizations one uses. We shall make use of what we know best, ≪normal algorithms≫. Their equivalence to recursive functions and, by the same token, to the other standardizations was established by Detlovs (see [1]; a detailed presentation is in [2]).

In the literature, the principle asserting the suitability of certain concrete precise concepts of an algorithm is called *Church's Thesis*—in honor of the outstanding American mathematician Alonzo Church, who first enunciated this principle (for λ-definable functions; see his paper [1]). The normalization principle is actually the variant of Church's Thesis pertaining to normal algorithms.

The variant of Church's Thesis pertaining to Turing machines deserves particular attention. By virtue of the specific character of the definition of a Turing machine, it is extremely convincing. As a matter of fact, a reader acquainted with this concept will agree that *in the essentials*, a Turing machine's performance adequately simulates the behavior of a computing mathematician engaged in fulfilling the prescription given her or him: being in a definite "state of mind" and concentrating her or his attention on some definite written information, the mathematician changes this information in the prescribed manner, "shifts" the center of her or his attention and passes into a new "state of mind". A Turing machine is capable of doing the same things. The circumstance that its steps are of a much more local nature than the actions of a live computer is really not too essential: each "global" step in the computer's performance can be simulated by a series of the Turing machine's "local" steps. Taking into account the equivalence of Turing machines to the other known standardizations of the concept of an algorithm, the considerations just presented are perhaps the strongest argument in favor of Church's Thesis, and therefore, also of the normalization principle.

Bearing in mind, however, the possibility that the reader may not share the authors' confidence in the validity of the normalization principle, the authors do not by any means intend to impose this principle on the reader. Consequently, every application of the normalization principle in what follows will be done explicitly.

But what possible applications of the normalization principle are there?

First of all, it can be used for conjecturing correct formulations of one or another mathematical statement. Let us suppose, for example, that there is

---

[2]Kolmogorov [1] proposed the most general approach to making the concept of an algorithm precise. See Kolmogorov and Uspenskii [1], and also Schönhage [1], [2], for concretizations of this approach.

a problem whose solution requires the construction of an algorithm. Let this problem be "stable" in the sense that every algorithm, equivalent to one solving it, also solves it. Assume that we have found *any algorithm* whatsoever solving this problem. Then it is possible to assert with confidence that there also exists a *normal algorithm* solving it. The problem has been reduced to one of merely constructing it. Having accepted the normalization principle, we shall not begin trying to prove that there exists no normal algorithm solving this problem.

Second, in an analogous situation, we might perhaps be able to prove a theorem to the effect that there exists no normal algorithm solving some such "stable" problem. We could *explain* to a sceptically inclined opponent, regarding this result as "too narrow and technical", that it would be pointless to seek some other (non-normalizable) algorithm solving this problem (although we will not, of course, be able to *prove* this mathematically).

Third, the normalization principle can be used for developing exact definitions of those mathematical concepts which in one or another form take into account various considerations of effectiveness (computability, or, as we prefer to say, constructiveness). Restrictions, requiring the presence of definite *algorithms*, usually arise naturally in such definitions. Taking the normalization principle into account, we then speak of *normal algorithms*, as a result of which the definitions acquire characteristics of precise mathematical formulations.

The concept of complete equivalence of algorithms figured in the formulation given above of the normalization principle. It is natural to ask what would happen if we replaced this concept in the principle's formulation by the concept of equivalence. At first sight, it might seem that we would get some new principle, asserting something less than the normalization principle, namely:

Every verbal algorithm in an alphabet A *is equivalent with respect to* A *to some normal algorithm over* A.

We shall soon see, however [§37.6], that this "weakened" normalization principle is in reality only a different formulation of the very same principle. We have a single normalization principle.[3]

§28 Joining Algorithms

1. Let us taken an alphabet A such that the letters $\rightarrow$ and $\rightarrow\cdot$ do not occur in A. Let Q be any fixed word in A. Denote by the symbol $\mathfrak{A}_{\mathrm{A},Q}$ the normal algorithm A with the scheme

$$\{ \rightarrow\cdot\ Q.$$

This scheme has the form $\gamma\beta Q\gamma$ in linear notation.

Earlier [§25.11.1] we proved that

1.1. *For every word P in the alphabet* A,

$$\mathfrak{A}_{\mathrm{A},Q}(P) \coloneqq QP.$$

---

[3]Exactly the same situation would arise if we replaced *complete* equivalence by *strong* equivalence.

2. Let A be the alphabet consisting of the letters a, b and c; let ε be a letter not occurring in A. Denote by the symbol $\mathfrak{B}_{A,Q}$ the normal algorithm in the alphabet Aε with the scheme

$$(1) \qquad \begin{cases} \varepsilon a \rightarrow a\varepsilon \\ \varepsilon b \rightarrow b\varepsilon \\ \varepsilon c \rightarrow c\varepsilon \\ \varepsilon \rightarrow\cdot\ Q \\ \quad \rightarrow\cdot\ \varepsilon. \end{cases}$$

In linear notation, this scheme has the form

$$\gamma\varepsilon a\alpha a\varepsilon\gamma\varepsilon b\alpha b\varepsilon\gamma\varepsilon c\alpha c\varepsilon\gamma\varepsilon\beta Q\gamma\alpha\varepsilon\gamma.$$

Until the end of the section, P, R and S will denote arbitrary words in the alphabet A, and ξ, any letter of this alphabet.

Let us prove

2.1. $\mathfrak{B}_{A,Q} : P \vdash \varepsilon P.$

In fact, the left side of the last substitution formula (the empty word) occurs in P, while the left sides of all the preceding formulas do not occur in P because they contain the letter ε. Consequently, the algorithm $\mathfrak{B}_{A,Q}$ acts on P [§25.4]; moreover, the last formula of the scheme is the substitution formula active on the word P. The result of the action of $\mathfrak{B}_{A,Q}$ on P is the result of the action of the active formula on P, i.e., the result of substituting the right side of this formula (the word ε) in P in place of the first occurrence $**P$ of its left side [§23.6.2], i.e., the word εP [§23.8.5], and, since the substitution formula active on P is simple,

$$\mathfrak{B}_{A,Q} : P \vdash \varepsilon P.$$

Q. E. D.

2.2. $\mathfrak{B}_{A,Q} : P\varepsilon\xi R \vdash P\xi\varepsilon R.$

Indeed, the algorithm $\mathfrak{B}_{A,Q}$ acts on the word $P\varepsilon\xi R$, since two of its scheme's substitution formulas act on this word: one of the first three, having the form $\varepsilon\xi \rightarrow \xi\varepsilon$, and the last. It is clear that the formula $\varepsilon\xi \rightarrow \xi\varepsilon$ will be active on this word, since it precedes the scheme's last formula. There is one and only one occurrence, $P * \varepsilon\xi * R$, of the left side of the formula $\varepsilon\xi \rightarrow \xi\varepsilon$ in the word $P\varepsilon\xi R$. Consequently, it will be the first. The result of substituting the right side of the active formula, i.e., the word $\xi\varepsilon$, in place of this occurrence is the word $P\xi\varepsilon R$ and, since the active substitution formula is simple, we have

$$\mathfrak{B}_{A,Q} : P\varepsilon\xi R \vdash P\xi\varepsilon R.$$

Q. E. D.

2.3. $\mathfrak{B}_{A,Q} : \varepsilon PS \models P\varepsilon S$.

We shall prove this by the method of right induction on P.

The assertion 2.3 is trivially true when $P \rightleftharpoons \Lambda$. Now let the assertion 2.3 be true for some P and an arbitrary S. Then, in particular, for any letter ξ we have

$$\mathfrak{B}_{A,Q} : \varepsilon P\xi S \models P\varepsilon\xi S.$$

But by virtue of 2.2,

$$\mathfrak{B}_{A,Q} : P\varepsilon\xi S \vdash P\xi\varepsilon S,$$

and hence

$$\mathfrak{B}_{A,Q} : \varepsilon P\xi S \models P\xi\varepsilon S,$$

which is just what remained to be proven.

Putting $S \rightleftharpoons \Lambda$ in 2.3, we obtain

2.4. $\mathfrak{B}_{A,Q} : \varepsilon P \models P\varepsilon$.

According to the scheme of the algorithm $\mathfrak{B}_{A,Q}$, we have

2.5. $\mathfrak{B}_{A,Q} : P\varepsilon \vdash\cdot\ PQ$.

Therefore,

2.6. $\mathfrak{B}_{A,Q} : P \models\cdot\ PQ$ [2.1, 2.4, 2.5]

and, consequently,

2.7. $\mathfrak{B}_{A,Q}(P) \rightleftharpoons PQ$.

3. It is convenient to write down the scheme 2(1) of the normal algorithm $\mathfrak{B}_{A,Q}$, whose first three substitution formulas have the "general form" $\varepsilon\xi \to \xi\varepsilon$, where ξ is an arbitrary letter of A, in the following abbreviated form:

$$(2) \qquad \begin{cases} \varepsilon\xi \to \xi\varepsilon \\ \varepsilon \to\cdot\ Q \\ \to \varepsilon, \end{cases}$$

indicating in addition that ξ runs through the alphabet A. In deciphering this notation, we give ξ the values a, b, c in turn, keeping—for the sake of definiteness—to the order of the letters in A (which, incidentally, plays no particular role in the given case). We shall use analogous abbreviated notations in other cases, too, making the necessary clarifications to the extent to which they might be required.

Looking over the proofs of the assertions 2.1–2.7, the reader will easily convince her or himself that they make no use whatsoever of the fact that A is a three-letter alphabet. Here A may be regarded as an arbitrary alphabet. This is just what we shall do, and shall assume that the algorithm $\mathfrak{B}_{A,Q}$ is defined once and for all by the scheme (2), where A is an arbitrary alphabet. We shall then consider the assertions 2.1–2.7 to be proven for precisely this algorithm.

4. It is natural to call the algorithms $\mathfrak{A}_{A,Q}$ and $\mathfrak{B}_{A,Q}$ *joining algorithms*; $\mathfrak{A}_{A,Q}$—the *left joining (the word Q) algorithm over the alphabet* A, $\mathfrak{B}_{A,Q}$—the

right joining (the word Q) algorithm over the alphabet A. A special case of the left joining algorithm over the alphabet A is the *identity normal algorithm* $\mathfrak{I}_{\mathrm{A}}$ *in the alphabet* A, defined as $\mathfrak{A}_{\mathrm{A},\Lambda}$. For it we have

$$\mathfrak{I}_{\mathrm{A}}(P) \asymp P$$

for any word P in the alphabet A [1.1].

5. We shall call the normal algorithm $\mathfrak{L}_{\mathrm{A}}$ in the alphabet A, defined by the scheme

$$\{ \rightarrow,$$

the *empty normal algorithm* in A. We actually have already shown [§25.11.4] that $\mathfrak{L}_{\mathrm{A}}$ is not applicable to any word in the alphabet A.

§29 Abbreviating Algorithms

1. We shall say of a simple substitution formula that it is *abbreviating* if the length of its right side is less than the length of its left side.

We shall say of a normal algorithm that it is *abbreviating* if all its simple substitution formulas are abbreviating.

The following lemma is obvious:

1.1. *If* $\mathfrak{A}$ *is an abbreviating algorithm and* $\mathfrak{A} : P \vdash Q$, *then*

$$[Q^{\delta} < [P^{\delta}.$$

Further, we have

1.2. *Any abbreviating algorithm* $\mathfrak{A}$ *in an alphabet* A *is applicable to every word* P *in the alphabet* A*; moreover, the inequality*

$$(\mathfrak{A} : P) \leqslant [P^{\delta}\mathsf{I} \tag{1}$$

holds.

We shall prove this theorem by the method of induction on the length of the word P, holding the abbreviating algorithm $\mathfrak{A}$ fixed.

The algorithm $\mathfrak{A}$ is applicable to every word of length Λ, and for every such word P, the inequality (1) holds. Indeed, if $[P^{\delta} \asymp \Lambda$, then $P \asymp \Lambda$. No simple substitution formula acts on P on this case, since its left side is non-empty. If the algorithm $\mathfrak{A}$ acts on P, then the formula active on P in $\mathfrak{A}$ is a final substitution formula. Then there is a word Q such that $\mathfrak{A} : P \vdash\cdot Q$, and we have $(\mathfrak{A} : P) \asymp \mathsf{I}$ [§25.8], from which it follows that $\mathfrak{A}$ is applicable to P. But if $\mathfrak{A}$ does not act on P, then $(\mathfrak{A} : P) \asymp \Lambda$ [§25.8.8]. In both cases, the algorithm $\mathfrak{A}$ is applicable to P and (1) holds.

Now assume that for the natural number N, it has been proven that the algorithm $\mathfrak{A}$ is applicable to every word in A of length at most N, and that the inequality (1) holds for every such word. Let P be a word in A of length $N\mathsf{I}$.

One of the following three cases holds: 1) $\mathfrak{A} : P]$; 2) there is a word Q such that $\mathfrak{A} : P \vdash\cdot\ Q$; 3) there is a word Q such that $\mathfrak{A} : P \vdash Q$. In the first two cases, the algorithm $\mathfrak{A}$ is applicable to P and $(\mathfrak{A} : P) \leqslant \mathsf{I}$ [§25.8]. In the third case, $[Q^\delta < [P^\delta$ [1.2] and hence $[Q^\delta \leqslant N$. It follows from this by the inductive assumption that $\mathfrak{A}$ is applicable to Q and that $(\mathfrak{A} : Q) \leqslant [Q^\delta \mathsf{I}$. Consequently, the algorithm $\mathfrak{A}$ is applicable to P and $(\mathfrak{A} : P) \mathrel{=\!\!=} (\mathfrak{A} : Q)\mathsf{I}$ [§25.8.9]. But $[Q^\delta \mathsf{I} \leqslant [P^\delta$. Therefore,

$$(\mathfrak{A} : P) \mathrel{=\!\!=} (\mathfrak{A} : Q)\mathsf{I} \leqslant [Q^\delta \mathsf{II} \leqslant [P^\delta \mathsf{I}.$$

Q. E. D.

2. Let A be an alphabet and let $\mathfrak{C}_\mathrm{A}$ be a normal algorithm in A whose scheme, written down in abbreviated form, is

$$\{\ \xi \to,$$

where ξ runs through A.

The normal algorithm $\mathfrak{C}_\mathrm{A}$ is abbreviating. We therefore have

2.1. *The algorithm $\mathfrak{C}_\mathrm{A}$ is applicable to every word in the alphabet* A [1.2].

Further, it is obvious that

2.2. *The algorithm $\mathfrak{C}_\mathrm{A}$ acts on every non-empty word in the alphabet A.*

We shall now prove that

2.3. *The algorithm $\mathfrak{C}_\mathrm{A}$ changes every word in the alphabet A into the empty word.*

Let P be a word in A. There is a word Q in A such that

(1) $$\mathfrak{C}_\mathrm{A}(P) \mathrel{=\!\!=} Q \qquad [2.1].$$

We therefore have $\mathfrak{C}_\mathrm{A} : P \models\cdot\ Q$ or $\mathfrak{C}_\mathrm{A} : P \models Q]$. The first possibility can be eliminated, because there are no final substitution formulas in the scheme $\mathfrak{C}_\mathrm{A}$. Consequently, $\mathfrak{C}_\mathrm{A} : P \models Q]$. It follows from this that

(2) $$Q \mathrel{=\!\!=} \Lambda \quad [2.2].$$

Therefore, $\mathfrak{C}_\mathrm{A}(P) \mathrel{=\!\!=} \Lambda$ [(1), (2)]. Q. E. D.

In view of 2.3, we shall call the algorithm $\mathfrak{C}_\mathrm{A}$ the *annihilating algorithm in* A.

3. Let us fix a word C in an alphabet A and denote by $\mathfrak{C}_{\mathrm{A},C}$ the normal algorithm in A whose scheme, written down in abbreviated form, is the following:

$$\begin{cases} \xi \to \\ \ \to\cdot\ C, \end{cases}$$

where ξ runs through A. The algorithm $\mathfrak{C}_{\mathrm{A},C}$ is abbreviating. We therefore have

3.1. *The algorithm $\mathfrak{C}_{A,C}$ is applicable to every word in the alphabet* A. [1.2].

3.2. *The algorithm $\mathfrak{C}_{A,C}$ acts on every word in the alphabet* A*; if, in addition, P is a non-empty word in* A*, then the substitution formula active on P in $\mathfrak{C}_{A,C}$ is simple.*

From 3.2 follows

3.3. *A natural end to a process of applying the algorithm $\mathfrak{C}_{A,C}$ is impossible.*

3.4. *If $\mathfrak{C}_{A,C} : R \vdash\cdot\ Q$, then $R \equiv \Lambda$ and $Q \equiv C$.*

In fact, if R is non-empty, then by virtue of 3.2 the algorithm $\mathfrak{C}_{A,C}$ simply converts R into some word S, which is impossible, since $\mathfrak{C}_{A,C} : R \vdash\cdot Q$ by hypothesis. Hence $R \equiv \Lambda$. The only substitution formula acting on Λ is the formula $\rightarrow\cdot\ C$, in view of which $\mathfrak{C}_{A,C} : \Lambda \vdash\cdot\ C$. Therefore, $Q \equiv C$. Q. E. D.

3.5. *The algorithm $\mathfrak{C}_{A,C}$ changes every word in the alphabet* A *into the word C.*

Let P be a word in A. There is a word Q in A such that

(1) $$\mathfrak{C}_{A,C}(P) \equiv Q \quad [3.1].$$

We have $\mathfrak{C}_{A,C} : P \models\cdot Q$ or $\mathfrak{C}_{A,C} : P \models Q]$. The second possibility can be eliminated [3.3]. Consequently, $\mathfrak{C}_{A,C} : P \models\cdot\ Q$, and so there is a word R such that $\mathfrak{C}_{A,C} : R \vdash\cdot\ Q$. It follows from this that

(2) $$Q \equiv C \quad [3.4].$$

Therefore, $\mathfrak{C}_{A,C}(P) \equiv C$ [(1), (2)]. Q. E. D.

4. Let us hold the letter ε of the alphabet A fixed and denote by $\mathfrak{J}_{A\varepsilon}$ the normal algorithm in A with the scheme

$$\begin{cases} \varepsilon\xi \rightarrow \varepsilon \\ \varepsilon \rightarrow, \end{cases}$$

where ξ runs through A. The algorithm $\mathfrak{J}_{A\varepsilon}$ is abbreviating. We therefore have

4.1. *The algorithm $\mathfrak{J}_{A\varepsilon}$ is applicable to every word in the alphabet* A [1.2].

In the next lemmas, the letters X, Y, P will denote words in the alphabet A.

We have

4.2. *$\mathfrak{J}_{A,\varepsilon} : Y]$ if and only if ε does not occur in Y.*

Let P be a word not containing ε. We shall say of a word X that it is a left P-word in the following two cases: 1) when ε occurs in X and P is the left wing of the first occurrence of ε in X; 2) when $X \equiv P$.

The following lemma is easily proven:

4.3. *If X is a left P-word and $\mathfrak{J}_{A,\varepsilon} : X \vdash Y$, then Y is a left P-word.*

By induction on the steps of the performance of $\mathfrak{J}_{A,\varepsilon}$ [§25.10 §27.3], from this we obtain

4.4. *If X is a left P-word and $\mathfrak{J}_{A,\varepsilon} : X \models Y$, then Y is a left P-word.*

Let us now prove

4.5. *The algorithm $\mathfrak{J}_{A,\varepsilon}$ changes every left P-word into P.*

Let X be a left P-word. $\mathfrak{J}_{A,\varepsilon}$ is applicable to X [4.1]. There is a word Y in A such that $\mathfrak{J}_{A,\varepsilon}(X) \risingdotseq Y$. Since the scheme of $\mathfrak{J}_{A,\varepsilon}$ consists of simple substitution formulas, we have $\mathfrak{J}_{A,\varepsilon} : X \models Y\rceil$, and so $\mathfrak{J}_{A,\varepsilon} : X \models Y$ and $\mathfrak{J}_{A,\varepsilon} : Y\rceil$. Consequently, Y is a left P-word and ε does not occur in Y [4.4, 4.2]. Hence $Y \risingdotseq P$. Q.E.D.

Bearing in mind the definition of a left P-word, we can reformulate the assertion 4.5 in the following manner:

4.6. *The algorithm $\mathfrak{J}_{A,\varepsilon}$ changes every word in the alphabet A, not containing ε, into the same word; it changes every word X in the alphabet A, containing ε, into the left wing of ε's first occurrence in X.*

Denote by $\mathfrak{G}_{A,\varepsilon}$ the normal algorithm in the alphabet A with the scheme

$$\begin{cases} \xi\varepsilon \to \varepsilon \\ \varepsilon \to, \end{cases}$$

where ξ runs through A.

We shall call an occurrence K of a word P in a word Q the *last occurrence of the word P in the word Q* if every other occurrence of P in Q precedes K.

By analogy with our proof of 4.6, we can prove

4.7. *The algorithm $\mathfrak{G}_{A,\varepsilon}$ changes every word in the alphabet A, not containing ε, into the same word; it changes every word X in the alphabet A, containing ε, into the right wing of ε's last occurrence in X.*

Now assume that ε is not a letter of the alphabet A. Take the alphabet Aε and construct the normal algorithms $\mathfrak{J}_{A\varepsilon,\varepsilon}$ and $\mathfrak{G}_{A\varepsilon,\varepsilon}$ for it. Recalling the definition of an ε-pair of words in an alphabet A [§22.3], we see that

4.8. *Given any ε-pair of words in an alphabet A, the algorithm $\mathfrak{J}_{A\varepsilon,\varepsilon}$ changes it into the first element of this pair, while the algorithm $\mathfrak{G}_{A\varepsilon,\varepsilon}$, into the second.*

In view of 4.8, we call the algorithms $\mathfrak{J}_{A\varepsilon,\varepsilon}$ and $\mathfrak{G}_{A\varepsilon,\varepsilon}$ in the alphabet Aε the *severing algorithms for the alphabet* A.

Let ε and κ be two distinct letters not occurring in an alphabet A. Construct in the alphabet A$\varepsilon\kappa$ the normal algorithm $\mathfrak{K}_{A,\varepsilon,\kappa}$ with the scheme

$$\begin{cases} \xi\varepsilon \to \varepsilon \\ \kappa\xi \to \kappa \\ \varepsilon \to \\ \kappa \to, \end{cases}$$

where ξ runs through A. The algorithm $\mathfrak{K}_{A,\varepsilon,\kappa}$ is abbreviating. It is therefore applicable to every word in $A\varepsilon\kappa$. The reader will convince her or himself without difficulty that

4.9. *Given any words P, Q, R in an alphabet* A, *the algorithm* $\mathfrak{K}_{A,\varepsilon,\kappa}$ *changes the word* $P\varepsilon Q\kappa R$ *into the word* Q.

(First ε "eats up" P, then κ "eats up" R, after this ε and κ "exit" and the performance of $\mathfrak{K}_{A,\varepsilon,\kappa}$ ends by breaking off naturally with the result Q.)

In view of 4.9, we call the algorithm $\mathfrak{K}_{A,\varepsilon,\kappa}$ the *carving algorithm for the alphabet* A.

§30 Branching Algorithm

1. Suppose that α is not a letter of an alphabet A. Let C, D and E be any fixed words in A. Construct the normal algorithm $\mathfrak{A}_{A,C,D,E}$ in the alphabet $A\alpha$,[4] whose scheme can be written down in the following abbreviated form:

$$(1)\qquad \left\{\begin{array}{r} \xi\alpha \to \alpha\xi \\ \alpha\xi \to \alpha \\ \alpha \to\cdot E \\ C\xi \to \alpha \\ \xi C \to \alpha \\ C \to\cdot D \\ \to \alpha, \end{array}\right.$$

where ξ runs through the alphabet A.

We shall show that it is applicable to every word in A and, moreover,

$$(2)\qquad \mathfrak{A}_{A,C,D,E}(C) = D$$

and

$$(3)\qquad \mathfrak{A}_{A,C,D,E}(P) = E$$

for every word P in A, graphically different from C.

In order to establish (2), let us note that the left sides of $\mathfrak{A}_{A,C,D,E}$'s substitution formulas, corresponding to the first five lines of (1), do not occur in C, since these left sides either contain α or else have length exceeding that of C by one. The left side of the next substitution formula, $C \to\cdot D$, occurs in C. Therefore, this formula is active on the word C in the scheme (1) and the result of its action—and by the same token, the result of the action of the scheme (1)—on

[4] In view of our transition to a non-linear method of writing down schemes, the letters α, β and γ are freed, and we shall use them in the sequel as letters of one or another normal algorithm's alphabet.

the word C, as is easily seen, is the word D, and since the active formula is final in the given case, we have

$$\mathfrak{A}_{A,C,D,E} : C \vdash\cdot D,$$

from which (2) immediately follows.

For the establishment of (3)'s truth, we shall prove the next eight lemmas, 1.1–1.8. In these lemmas, P, Q and R will denote arbitrary words in the alphabet A, and ξ, any letter of this alphabet.

1.1. $\mathfrak{A}_{A,C,D,E} : Q\xi\alpha R \vdash Q\alpha\xi R$.

In fact, exactly one of the substitution formulas, the one represented by the first line of the scheme (1): the formula $\xi\alpha \to \alpha\xi$, acts on the word $Q\xi\alpha R$. It is precisely the one which is active in the given case. The occurrence $Q * \xi\alpha * R$ is the only—and consequently, the first occurrence—of its left side in the word $Q\xi\alpha R$. Hence, the result of the action of the scheme (1) on the word $Q\xi\alpha R$ is $Q\alpha\xi R$. But since the active formula is simple, we have

$$\mathfrak{A}_{A,C,D,E} : Q\xi\alpha R \vdash Q\alpha\xi R.$$

Q.E.D.

Therefore, Lemma 1.1 says that as a result of a single simple step, the letter α in a word of the form $Q\xi\alpha R$ "moves" one space to the left.

1.2. $\mathfrak{A}_{A,C,D,E} : Q\alpha R \models \alpha Q R$.

We shall prove this assertion by right induction on Q. For this we consider the property F of Q, expressed by the condition

$$\ll \text{given any } R, \mathfrak{A}_{A,C,D,E} : Q\alpha R \models \alpha Q R \gg,$$

and prove that any Q possesses this property.

Indeed, the empty word possesses the property F [§25.6.1]. Now let Q possess F, and let ξ be an arbitrary letter. Then

$$\mathfrak{A}_{A,C,D,E} : Q\xi\alpha R \vdash Q\alpha\xi R \quad [1.1]. \tag{4}$$

But ξR is a word in A, and so by the inductive assumption,

$$\mathfrak{A}_{A,C,D,E} : Q\alpha\xi R \vdash \alpha Q\xi R. \tag{5}$$

Consequently,

$$\mathfrak{A}_{A,C,D,E} : Q\xi\alpha R \models \alpha Q\xi R \quad [(4), (5), §25.6.2, §25.6.3]$$

and, therefore, $Q\xi$ possesses F. Thus, any Q really possesses the property F, and this proves our assertion.

1.3. $\mathfrak{A}_{\mathrm{A},C,D,E} : \alpha\xi P \vdash \alpha P$.

In fact, none of the substitution formulas represented by the first line of (1) acts on $\alpha\xi P$. The occurrence $*\alpha\xi * P$ is the only occurrence of a word of the form $\alpha\xi$ in $\alpha\xi P$. Therefore, it follows from an examination of (1) that the formula $\alpha\xi \rightarrow \alpha$ is active on $\alpha\xi P$ in the scheme (1), and hence

$$\mathfrak{A}_{\mathrm{A},C,D,E} : \alpha\xi P \vdash \alpha P.$$

Q.E.D.

1.4. $\mathfrak{A}_{\mathrm{A},C,D,E} : \alpha P \vdash \alpha$.

We shall prove this assertion by left induction on P [§17.3.2]. For this let us consider the property G of P expressed by the condition

$$\ll \mathfrak{A}_{\mathrm{A},C,D,E} : \alpha P \models \alpha \gg$$

and show that any word P possesses this property.

Indeed, the empty word possesses it [§25.6.1]. Now let P possess G, and let ξ be an arbitrary letter. Then

(6) $$\mathfrak{A}_{\mathrm{A},C,D,E} : \alpha\xi P \vdash \alpha P \quad [1.3].$$

But by the inductive assumption,

(7) $$\mathfrak{A}_{\mathrm{A},C,D,E} : \alpha P \models \alpha$$

and hence

$$\mathfrak{A}_{\mathrm{A},C,D,E} : \alpha\xi P \models \alpha \quad [(6),\ (7)],$$

i.e., ξP possesses the property G. Consequently, any P possesses this property.

Q.E.D.

1.5. $\mathfrak{A}_{\mathrm{A},C,D,E} : \alpha \vdash\!\cdot\, E$.

In fact, the substitution formulas represented by the first two lines of the scheme (1) do not act on α, while the formula $\alpha \vdash\!\cdot\, E$ acts on this word and, therefore, is active in the given case. The occurrence $*\alpha*$ is the first occurrence of this formula's left side in the word α. The result of substituting its right side in place of this occurrence is E, and since this formula is final,

$$\mathfrak{A}_{\mathrm{A},C,D,E} : \alpha \rightarrow\!\cdot\, E.$$

Q.E.D.

1.6. *If C occurs in P and $P \neq C$, then there exist words Q and R in* A *such that* $\mathfrak{A}_{\mathrm{A},C,D,E} : P \vdash Q\alpha R$.

Indeed, let P be a word in A, distinct from C, and let C occur in P. Then either one of the words of the form $C\xi$ or one of the words of the form ξC occurs in P. In other words, the left side of at least one of the formulas represented by the fourth and fifth lines of the scheme (1) occurs in P. But the left sides of the formulas represented above these do not occur in P, since they contain α. Hence one of the formulas represented by the fourth or fifth line of (1) is active on P. This formula has the form $S \rightarrow \alpha$, and it is required to apply it to the first occurrence of the word S (equal to $C\xi$ or ξC) in P. Let $Q * S * R$ be this occurrence. Then Q and R, just like P, are words in A, while the result of substituting the right side of the formula is the word $Q\alpha R$. Since the active formula is simple, we have $\mathfrak{A}_{\mathrm{A},C,D,E} : P \vdash Q\alpha R$. Q.E.D.

1.7. *If C does not occur in a word P in the alphabet* A, *then*

$$\mathfrak{A}_{\mathrm{A},C,D,E} : P \vdash \alpha P.$$

As a matter of fact, in this case, among the left sides of the substitution formulas of the algorithm $\mathfrak{A}_{\mathrm{A},C,D,E}$, that of the last formula—the empty word—occurs in P, but all the others fail to occur in P, since either α or C occurs in them. Consequently, it is required to substitute the right side of the last formula, i.e., α, in place of the empty word's first occurrence in P, which yields αP. Since the active formula is simple, we have $\mathfrak{A}_{\mathrm{A},C,D,E} : P \vdash \alpha P$. Q.E.D.

1.8. *If P is a word in the alphabet* A, *distinct from C, then one can find words Q and R in* A, *such that*

$$\mathfrak{A}_{\mathrm{A},C,D,E} : P \vdash Q\alpha R.$$

Indeed, if C occurs in P, then such a Q and an R can be found by 1.6. But if C does not occur in P, then we can take Λ for Q and P for R.

We can now prove the equality (3). Indeed, let P be a word in A, distinct from C. Take the words Q and R in A according to 1.8 in such a way that $\mathfrak{A}_{\mathrm{A},C,D,E} : P \vdash Q\alpha R$. We then have

$$\begin{aligned} \mathfrak{A}_{\mathrm{A},C,D,E} : P &\vdash Q\alpha R \\ &\models \alpha Q R \quad [1.2] \\ &\models \alpha \quad [1.4] \\ &\vdash\cdot E \quad [1.5], \end{aligned}$$

from which we obtain $\mathfrak{A}_{\mathrm{A},C,D,E} : P \models\cdot E$, and hence $\mathfrak{A}_{\mathrm{A},C,D,E}(P) \simeq E$ [§25.7]. Q.E.D.

Therefore, the action of the algorithm $\mathfrak{A}_{\mathrm{A},C,D,E}$ on the word P in A proceeds in the following manner. If $P \simeq C$, then $\mathfrak{A}_{\mathrm{A},C,D,E}$ converts P into D in a single step, and with this its performance ends. But if $P \not\simeq C$, then as a signal, $\mathfrak{A}_{\mathrm{A},C,D,E}$ "introduces" into the word being worked on the letter α [1.8], which then "makes its way" to the beginning of the word [1.2], "erases" it [1.4] and "turns into" E [1.5], after which the performance of $\mathfrak{A}_{\mathrm{A},C,D,E}$ comes to a halt. Hence we call $\mathfrak{A}_{\mathrm{A},C,D,E}$ a *branching algorithm*.

§31 Doubling Algorithm

1. Let α, β and γ denote distinct letters not belonging to an alphabet A. We construct the normal algorithm $\mathfrak{F}_A$ in the alphabet $A\alpha\beta\gamma$, whose scheme can be written down in the following abbreviated form:

$$(1) \qquad \begin{cases} \eta\xi\beta \to \xi\beta\eta \\ \alpha\xi \to \xi\beta\xi\alpha \\ \beta \to \gamma \\ \gamma \to \\ \alpha \to\cdot \\ \quad \to \alpha, \end{cases}$$

where ξ and η run through the alphabet A (for an n-letter alphabet A, this scheme consists of $n^2 + n + 4$ substitution formulas).

Let us show that

$$(2) \qquad \mathfrak{F}_A(P) \simeq PP$$

for every word P in A.

In order to establish the truth of the assertion (2), we introduce two operations, Φ and Ψ, on words in A, and then prove Lemmas 1.1–1.4, from which this assertion will follow. In the sequel, P, Q, R, S will denote arbitrary words in A, and ξ, η, any letters of this alphabet.

Set (see the remark in §17.3.2)

$$(3) \qquad \Phi(\Lambda) \rightleftharpoons \Lambda,$$
$$(4) \qquad \Phi(P\xi) \rightleftharpoons \Phi(P)\xi\beta,$$
$$(5) \qquad \Psi(\Lambda) \rightleftharpoons \Lambda,$$
$$(6) \qquad \Psi(P\xi) \rightleftharpoons \Psi(P)\xi\gamma.$$

It is easy to see that

$$(7) \qquad \Phi(\xi) \simeq \xi\beta \quad [(4), (3)],$$
$$(8) \qquad \Psi(\xi) \simeq \xi\gamma \quad [(6), (5)].$$

Furthermore, it is not hard to show by right induction on Q that

$$(9) \qquad \Phi(PQ) \simeq \Phi(P)\Phi(Q) \quad [(3), (4)]$$

and

$$(10) \qquad \Psi(PQ) \simeq \Psi(P)\Psi(Q) \quad [(5), (6)].$$

We have

1.1. $\mathfrak{F}_A : P \vdash \alpha O$.

As a matter of fact, the left sides of all the substitution formulas of the scheme (1), except the last, fail to occur in P, since α, β and γ are not letters of A. The left side of the last formula, $\to \alpha$, (the empty word) does occur in P, and so this formula is active on P. $**P$ is the first occurrence of this formula's left side in the word P. It is required to substitute the right side of the active formula, i.e., α, in place of this occurrence, which gives us the word αP. Since the active formula is simple, we get

$$\mathfrak{F}_A : P \vdash \alpha P.$$

Q.E.D.

1.2. $\mathfrak{F}_A : Q\alpha \vdash\cdot Q$.

This is proven by an analysis of the scheme (1), analogously to the way in which 1.1 was proven. The formula $\alpha \to\cdot$ is the substitution formula active in (1) on $Q\alpha$.

1.3. $\mathfrak{F}_A : \Phi(P)P\alpha\xi Q \vdash \Phi(P)P\xi\beta\xi\alpha Q$.

This is proven by an analysis of the scheme (1). It is easy to see that none of the formulas represented by the first line of (1) acts on the word $\Phi(P)P\alpha\xi Q$. Nevertheless, among the formulas represented by the second line, there is a formula, $\alpha\xi \to \xi\beta\xi\alpha$, whose left side occurs in $\Phi(P)P\alpha\xi Q$. This formula is active on the given word. Its left side's unique occurrence in the word under consideration has the form $\Phi(P)P * \alpha\xi * Q$. The formula's right side, i.e., the word $\xi\beta\xi\alpha$, is substituted for it, and since the active formula is simple, we have

$$\mathfrak{F}_A : \Phi(P)P\alpha\xi Q \vdash \Phi(P)Q\xi\beta\xi\alpha Q.$$

Q.E.D.

1.4. $\mathfrak{F}_A : \Phi(S)R\eta\xi\beta P\xi\alpha Q \vdash \Phi(S)R\xi\beta\eta P\xi\alpha Q$.

This is proven by analyzing the scheme (1). The formula $\eta\xi\beta \to \xi\beta\eta$, represented by the first line of (1), is active on the word $\Phi(S)R\eta\xi\beta P\xi\alpha Q$.

1.5. $\mathfrak{F}_A : \Phi(S)RP\xi\beta\xi\alpha Q \models \Phi(S)R\xi\beta P\xi\alpha Q$.

We shall prove this assertion by left induction on P. To this end, we introduce the property F of a word P, expressed by the condition

$$\ll \text{given any } Q, R, S \text{ and } \xi,\ \mathfrak{F}_A : \Phi(S)RP\xi\beta\xi\alpha Q \models \Phi(S)R\xi\beta P\xi\alpha Q \gg$$

and show by an ordinary inductive argument that any P possesses this property.

Indeed, the empty word Λ possesses the property F [§25.6.1]. Let us now suppose that P possesses it and show that then the word ηP also possesses it, whatever be the letter η. In fact, $R\eta$ is a word in A and hence by the inductive assumption

$$\mathfrak{F}_A : \Phi(S)R\eta P\xi\beta\xi\alpha Q \models \Phi(S)R\eta\xi\beta P\xi\alpha Q. \tag{11}$$

But according to 1.4,

$$\mathfrak{F}_{\mathrm{A}} : \Phi(S)R\eta\xi\beta P\xi\alpha Q \vdash \Phi(S)R\xi\beta\eta P\xi\alpha Q \tag{12}$$

and, consequently,

$$\mathfrak{F}_{\mathrm{A}} : \Phi(S)R\eta P\xi\beta\xi\alpha Q \models \Phi(S)R\xi\beta\eta P\xi\alpha Q \quad [(11), (12)],$$

which is what remained to be proven.

1.6. $\mathfrak{F}_{\mathrm{A}} : \Phi(P)P\xi\beta\xi\alpha Q \models \Phi(P)\xi\beta P\xi\alpha Q$.

This assertion is the special case of 1.5 for $R \rightleftharpoons \Lambda$ and $S \rightleftharpoons P$.

1.7. $\mathfrak{F}_{\mathrm{A}} : \alpha PQ \models \Phi(P)P\alpha Q$.

We shall prove this assertion by right induction on P. To this end, let us consider the property G of words P, expressed by the condition

$$\ll \text{given any } Q,\ \mathfrak{F}_{\mathrm{A}} : \alpha PQ \models \Phi(P)P\alpha Q \gg$$

and show by inductive reasoning that any word in A possesses this property.

Indeed, the empty word Λ possesses it [(3), §25.6.1]. Now assume that P possesses it, and let ξ be an arbitrary letter. Then by the inductive assumption,

$$\mathfrak{F}_{\mathrm{A}} : \alpha P\xi Q \models \Phi(P)P\alpha\xi Q. \tag{13}$$

But

$$\mathfrak{F}_{\mathrm{A}} : \Phi(P)P\alpha\xi Q \vdash \Phi(P)P\xi\beta\xi\alpha Q \quad [1.3], \tag{14}$$

and so

$$\begin{aligned} \mathfrak{F}_{\mathrm{A}} : \alpha P\xi Q &\models \Phi(P)P\xi\beta\xi\alpha Q && [(13), (14)] \\ &\models \Phi(P)\xi\beta P\xi\alpha Q && [1.6] \\ &\rightleftharpoons \Phi(P\xi)P\xi\alpha Q && [(4)]. \end{aligned}$$

Consequently, $P\xi$ possesses the property G. By the same token, Lemma 1.7 is proven.

1.8. $\mathfrak{F}_{\mathrm{A}} : \alpha P \models \Phi(P)P\alpha$.

This assertion is the special case of 1.7 for $Q \rightleftharpoons \Lambda$.

1.9. $\mathfrak{F}_{\mathrm{A}} : \Psi(P)\xi\beta\Phi(Q)R\alpha \vdash \Psi(P)\xi\gamma\Phi(Q)R\alpha$.

This is proven by means of an analysis of the scheme (1). The formula represented by the scheme's third line is active.

1.10. $\mathfrak{F}_{\mathrm{A}} : \Phi(PQ)PQ\alpha \models \Psi(P)\Phi(Q)PQ\alpha$.

We shall carry out the proof by right induction on P. For this we consider the property H of words P, expressed by the condition

$$\ll \text{given any } Q, \ \mathfrak{F}_{\mathrm{A}} : \Phi(PQ)PQ\alpha \models \Psi(P)\Phi(Q)PQ\alpha \gg,$$

and, arguing as is customary in such cases, show that any word in A possesses it.

Indeed, the empty word Λ possesses this property [(5), §25.6.1]. Now let the word P possess it, and let ξ be an arbitrary letter. Then

$$\begin{aligned} \mathfrak{F}_{\mathrm{A}} : \Phi(P\xi Q)P\xi Q\alpha &\models \Psi(P)\Phi(\xi Q)P\xi Q\alpha \\ &= \Psi(P)\xi\beta\Phi(Q)P\xi Q\alpha \quad [(9), (7)] \\ &\vdash \Psi(P)\xi\gamma\Phi(Q)P\xi Q\alpha \quad [1.9] \\ &= \Psi(P\xi)\Phi(Q)P\xi Q\alpha \quad [(6)] \end{aligned}$$

and, therefore, $P\xi$ also possesses H. Consequently, any word in A actually possesses the property H. Q.E.D.

1.11. $\mathfrak{F}_{\mathrm{A}} : \Phi(P)P\alpha \models \Psi(P)P\alpha$.

(The special case of 1.10 for $Q = \Lambda$.)

1.12. $\mathfrak{F}_{\mathrm{A}} : P\xi\gamma\Psi(Q)P\xi Q\alpha \vdash P\xi\Psi(Q)P\xi Q\alpha$.

This is proven by analyzing the scheme (1). The formula represented by the scheme's fourth line is active.

1.13. $\mathfrak{F}_{\mathrm{A}} : \Psi(PQ)PQ\alpha \models P\Psi(Q)$.

We shall carry out the proof by right induction on P. For this we consider the property K of words P, expressed by the condition

$$\ll \text{given any } Q, \ \mathfrak{F}_{\mathrm{A}} : \Psi(PQ)PQ\alpha \models P\Psi(Q)PQ\alpha \gg,$$

and show by means of the usual reasoning for such cases that any word in A possesses it.

In fact, the empty word possesses it. Now let the word P possess it, and let ξ be any letter. Then

$$\begin{aligned} \mathfrak{F}_{\mathrm{A}} : \Psi(P\xi Q)P\xi Q\alpha &\models P\Psi(\xi Q)P\xi Q\alpha \quad [\text{ind. ass.}] \\ &= P\xi\gamma\Psi(Q)P\xi Q\alpha \quad [(10), (8)] \\ &\vdash P\xi\Psi(Q)P\xi Q\alpha \quad [1.12], \end{aligned}$$

i.e., $P\xi$ also possesses K. Hence, any word in A does in fact possess this property. Q.E.D.

1.14. $\mathfrak{F}_{\mathrm{A}} : \Psi(P)P\alpha \models PP\alpha.$

(Special case of 1.13 for $Q \rightleftharpoons \Lambda$.)

Let us now prove the assertion (2). Indeed,

$$\begin{aligned}\mathfrak{F}_{\mathrm{A}} : P &\vdash \alpha P && [1.1]\\ &\models \Phi(P)P\alpha && [1.8]\\ &\models \Psi(P)P\alpha && [1.11]\\ &\models PP\alpha && [1.14]\\ &\vdash\cdot\, PP && [1.2],\end{aligned}$$

whence $\mathfrak{F}_{\mathrm{A}} : P \models\cdot\, PP$ and, consequently,

$$\mathfrak{F}_{\mathrm{A}}(P) \rightleftharpoons PP.$$

Q.E.D.

Therefore, the normal algorithm $\mathfrak{F}_{\mathrm{A}}$ may be characterized as a doubling algorithm over the alphabet A.

2. The proof just carried out convincingly demonstrates the effectiveness of the method of induction. However, it may seem insufficiently transparent to an unsophisticated reader. In view of this, it is advisable to accompany it by several remarks.

First of all, the performance of the algorithm $\mathfrak{F}_{\mathrm{A}}$ should be roughly described. This is the way it looks. At the beginning, the letter α "appears" in front of the initial word P [1.1]. Then it "runs" through the word P from left to right, making "copies" of its letters [1.3]. These copies, "diluted" by the letters β attached to them, run to the left through the part of P already passed by α [1.6] and, therefore, in front of P there accumulates its "copy" in the form of the word $\Phi(P)$. When α reaches the end of P, the result turns out to be essentially obtained already [1.8]. Only a slight clean-up, the removal of the "superfluous" letters β and α, remains to be carried out. Each β turns into γ [1.11] before its "annihilation", then all the γ's "are erased" [1.14] and, finally, α "is erased" in the final step [1.2]. Here the intermediate step of β's getting "renamed" γ looks somewhat paradoxical. It is easy, however, to convince oneself that the simpler algorithm $\mathfrak{G}_{\mathrm{A}}$ in the alphabet $\mathrm{A}\alpha\beta$, whose scheme can be written down in the abbreviated form

$$\left\{\begin{aligned}\eta\xi\beta &\rightarrow \xi\beta\eta\\ \alpha\xi &\rightarrow \xi\beta\xi\alpha\\ \beta &\rightarrow\\ \alpha &\rightarrow\cdot\\ &\rightarrow \alpha\end{aligned}\right.$$

(ξ and η run through A as before) is not equivalent to $\mathfrak{F}_{\mathrm{A}}$ with respect to A and so can not replace $\mathfrak{F}_{\mathrm{A}}$. Indeed, let ς and χ be distinct letters of A. Then, as is not difficult to see, when applied to the word $\varsigma\chi$, the algorithm $\mathfrak{G}_A$ performs in the following manner:

$$\begin{aligned}\mathfrak{G}_A : \varsigma\chi &\vdash \alpha\varsigma\chi \\ &\vdash \varsigma\beta\varsigma\alpha\chi \\ &\vdash \varsigma\beta\varsigma\chi\beta\chi\alpha \\ &\vdash \varsigma\beta\chi\beta\varsigma\chi\alpha \\ &\vdash \varsigma\chi\beta\varsigma\chi\alpha \\ &\vdash \chi\beta\varsigma\varsigma\chi\alpha \\ &\vdash \chi\varsigma\varsigma\chi\alpha \\ &\vdash\cdot \chi\varsigma\varsigma\chi.\end{aligned}$$

Therefore, $\mathfrak{G}_A(\varsigma\chi) \doteq \chi\varsigma\varsigma\chi \neq \varsigma\chi\varsigma\chi \doteq \mathfrak{F}_A(\varsigma\chi)$. By comparison with $\mathfrak{F}_A$'s performance, an extra application of the formula $\varsigma\chi\beta \to \chi\beta\varsigma$ had to be made here, which is why a transposition of the first two letters took place in the final result. What prevents such unnecessary permutations in the algorithm $\mathfrak{F}_A$ is the introduction of a special intermediate stage in the elimination of β's occurrences: γ's are first substituted for these occurrences.

Lemmas 1.1, 1.2, 1.8, 1.11 and 1.14 are the basic ones in our proof. The first two of them are trivial. The last three are first proved in a strengthened form (Lemmas 1.7, 1.10 and 1.13) and then weakened—we recommend that the reader pay attention to this device. Lemmas 1.3, 1.9 and 1.12 prepare for the inductive steps in 1.7, 1.10 and 1.13, respectively. Lemmas 1.4—1.6 describe the mechanism for accumulating copies of the initial word.

3. The doubling normal algorithm gives us a typical example of how a transition of a global nature—the doubling of a word—is reduced to several transitions of a local nature—substitutions in accordance with the algorithm's scheme. The number of the latter depends, moreover, on the word being doubled. One can calculate that it is equal to

$$\frac{(k+1)(k+4)}{2}$$

for the algorithm $\mathfrak{F}_A$, where k is the length of the word being doubled. The question of a lower bound for the number of steps in the performance of an arbitrary doubling normal algorithm was investigated by Tseitin in [5].

A natural measure of a *normal algorithm's complexity* is the length of its scheme (in linear notation [§25]), which characterizes the "storage space" required for "remembering" this algorithm. Such an approach to evaluating the complexity of normal algorithms was first outlined by Markov[5] in [16], [17] and then developed by him and his disciples.

It is not difficult to calculate that the complexity of $\mathfrak{F}_A$, i.e., the length of its scheme, is equal to $8n^2 + 8n + 14$, where n is the number of letters of the alphabet A. Therefore, it is a quadratic function of n.

---

[5]An analogous approach was employed by Kolmogorov [2] for defining the *complexity of a finite constructive object.*

An original idea for constructing a doubling algorithm with linear complexity was proposed by Ostroukhov in [1] (see also [2]). The complexity of the algorithm constructed by his method is equal to $10n + C$. Starting from this idea, Ostroukhov arrived at a proof of a very important general theorem on the linearization of general methods for constructing algorithms.

Later, by means of an extremely clever perfection of Ostroukhov's construction, Zharov was able to construct in [1] (see also [2]) a doubling algorithm with complexity $5n + C_1$ and show that no further improvement in the constant multiplying n is possible. In addition, he obtained a corresponding strengthening of Ostroukhov's general theorem.

4. We constructed the doubling algorithm $\mathfrak{F}_{\mathrm{A}}$ as a normal algorithm in a three-letter extension of the alphabet A. Later, in §41, we shall show that by means of a comparatively simple device, the number of auxiliary letters can be decreased here to two (and—somewhat more complicated— even to one: see §41.8.1). However, in the case of a non-empty A, it is impossible to do without them entirely. Let us show this.

Let $\mathfrak{A}$ be a normal algorithm in an alphabet A. If $\mathfrak{A}$'s scheme contains final substitution formulas, then for every such formula $U \rightarrow \cdot V$, we calculate the difference $[V^{\delta} - [U^{\delta}$, and denote the largest number so obtained by the symbol $s(\mathfrak{A})$. But if the scheme of the algorithm $\mathfrak{A}$ consists entirely of simple substitution formulas, then define $s(\mathfrak{A})$ "somehow", for example regard $s(\mathfrak{A})$ as equal to zero. $s(\mathfrak{A})$ is a *complexity measure* of the algorithm $\mathfrak{A}$.

Let k be a rational number. We shall call a normal algorithm $\mathfrak{A}$ in an alphabet A *k-dilating* if for every word P in A, such that $\mathfrak{A}(P)$ is defined, we have

$$[\mathfrak{A}(P)^{\delta} \geqslant k[P^{\delta}. \tag{1}$$

(Here and until the end of the section, $\geqslant$, $>$ and $=$ will denote the ordinary relations of order and equality for rational numbers.)

The following assertion holds (Nagorny [2], Theorem 3.1):

4.1. *Let k be a rational number greater than one, and let $\mathfrak{A}$ be a k-dilating normal algorithm in an alphabet* A. *If $\mathfrak{A}$ is applicable to a non-empty word P, then*

$$s(\mathfrak{A}) \geqslant (k-1) \cdot [P^{\delta}.$$

Indeed, let $k > 1$ be a rational number, and let the k-dilating normal algorithm $\mathfrak{A}$ in the alphabet A be applicable to the non-empty word P. Then for some word W we have

$$\mathfrak{A}(P) \simeq W. \tag{2}$$

One of the following two possibilities holds: either

$$\mathfrak{A} : P \models W], \tag{3}$$

or else

(4) $$\mathfrak{A} : P \models\cdot W.$$

If (3) holds, then

(5) $$\mathfrak{A}(W) \coloneqq W$$

and, since W is a word in A,

(6) $$[W^\delta \geqslant k \cdot [W^\delta \quad [(5), (1)].$$

But

$$[W^\delta \geqslant k \cdot [P^\delta > 0 \quad [(2), (1)],$$

and hence $1 \geqslant k$ [(6)], which, however, can not be, since by hypothesis $k > 1$.

Thus, (4) holds and, therefore, there exists a word Q in A such that

$$\mathfrak{A} : P \models Q$$

and

(7) $$\mathfrak{A} : Q \vdash\cdot W.$$

Consider the final formula

$$U \rightarrow V,$$

active on Q in the scheme of the algorithm $\mathfrak{A}$. There exist words R and S in the alphabet A, such that

(8) $$Q \coloneqq RUS,$$
(9) $$W \coloneqq RVS.$$

Since $[V^\delta - [U^\delta \leqslant s(\mathfrak{A})$, we have

$$[W^\delta - [Q^\delta \leqslant s(\mathfrak{A}) \quad [(9), (8)],$$

and hence

(10) $$[W^\delta \leqslant [Q^\delta + s(\mathfrak{A}).$$

On the other hand,

$$\mathfrak{A}(Q) \coloneqq W \quad [(7)],$$

and so

(11) $$[W^\delta \geqslant k \cdot [Q^\delta \quad [(1)].$$

Consequently,

$$k[Q^\delta \leqslant [Q^\delta + s(\mathfrak{A}) \quad [(11), (10)]$$

and

$$[Q^\delta \leqslant \frac{s(\mathfrak{A})}{k-1},$$

so that

$$(13) \qquad \begin{aligned} [W^\delta &\leqslant \frac{s(\mathfrak{A})}{k-1} + s(\mathfrak{A}) \quad [(10), (12)] \\ &= \frac{k \cdot s(\mathfrak{A})}{k-1}. \end{aligned}$$

But

$$\mathfrak{A}(P) \simeq W \quad [(2)],$$

and hence

$$(14) \qquad [W^\delta \geqslant k[P^\delta \quad [(1)],$$

which yields

$$k \cdot [P^\delta \leqslant \frac{k \cdot s(\mathfrak{A})}{k-1} \quad [(14), (13)]$$

and

$$(k-1) \cdot [P^\delta \leqslant s(\mathfrak{A}).$$

Q.E.D.

In particular, it follows from 4.1 that it is impossible to construct a normal algorithm $\mathfrak{A}$ in a non-empty alphabet A, doubling each word in A. In fact, such an algorithm would be 2-dilating and, therefore, for any non-empty word P in A the inequality $[P^\delta \leqslant s(\mathfrak{A})$ would be satisfied [4.1], which is obviously impossible.

§32 Reversing Algorithm

1. We shall construct a normal algorithm over an alphabet A, changing every word in A into this word's reversal [§17.3.2]. Let α and β denote two distinct letters not belonging to A. Let us present the normal algorithm $\mathfrak{H}_\mathrm{A}$ in the alphabet $\mathrm{A}\alpha\beta$ by means of the following scheme, written in abbreviated form:

$$(1) \qquad \left\{ \begin{aligned} \alpha\alpha &\rightarrow \beta\alpha \\ \beta\alpha &\rightarrow \beta \\ \beta\xi &\rightarrow \xi\beta \\ \beta &\rightarrow\cdot \\ \alpha\xi\eta &\rightarrow \eta\alpha\xi \\ &\rightarrow \alpha, \end{aligned} \right.$$

where ξ and η run through the alphabet A.

This is just the desired reversing algorithm over the alphabet A, i.e.,

$$(2) \qquad \mathfrak{H}_\mathrm{A}(P) \simeq [P^\smile$$

for every word P in A.

This fact can be proven easily along the lines of the proof carried out in the preceding section. We shall therefore merely give a brief outline of the proof here, leaving the details to the reader. In general, to the degree that the reader accumulates experience, we shall make our proofs less detailed, without, incidentally, going to extremes in this respect.

As in the preceding section, P, Q, R will denote arbitrary words in A, and ξ, η, any letters of this alphabet.

We shall need the operation θ on words in A, which we define by the equalities

(3) $$\theta(\Lambda) \rightleftharpoons \Lambda,$$

(4) $$\theta(P\xi) \rightleftharpoons \theta(P)\alpha\xi.$$

It is easy to see that

(5) $$\theta(\xi) \boldsymbol{=} \alpha\xi \quad [(4), (3)].$$

Furthermore, it is not hard to show by right induction on Q that

(6) $$\theta(PQ) \boldsymbol{=} \theta(P)\theta(Q) \quad [(3), (4)].$$

We have

1.1. $\mathfrak{H}_{\mathrm{A}} : P\theta(Q) \vdash \alpha P\theta(Q)$.

In fact, only the last line of the scheme (1) acts on the word $P\theta(Q)$.

1.2. $\mathfrak{H}_{\mathrm{A}} : P\beta \vdash\cdot P$.

Indeed, the formula $\beta \to\cdot$ is the first one acting on $P\beta$ in the scheme (1).

1.3. $\mathfrak{H}_{\mathrm{A}} : Q\alpha\xi\eta P\theta(R) \vdash Q\eta\alpha\xi P\theta(R)$.

In fact, the formulas represented by the first four lines of (1) do not act on the word $Q\alpha\xi\eta P\theta(R)$. The formula $\alpha\xi\eta \to \eta\alpha\xi$ from the scheme's fifth line is active.

1.4. $\mathfrak{H}_{\mathrm{A}} : \alpha\xi QP\theta(R) \models Q\alpha\xi P\theta(R)$.

This is proven by right induction on Q with the aid of 1.3.

1.5. $\mathfrak{H}_{\mathrm{A}} : \alpha\xi Q\theta(R) \models Q\alpha\xi\theta(R)$.

The special case of 1.4. for $P \boldsymbol{=} \Lambda$.

1.6. $\mathfrak{H}_{\mathrm{A}} : PQ \models Q\theta([P^{\smile})$.

This is proven by right induction on P with the aid of 1.1 and 1.5.

Indeed, when $P \boldsymbol{=} \Lambda$, the graphic equality of the words PQ and $Q\theta([P^{\smile})$ holds, and the required assertion, all the more so.

Let us further assume that for some P,

$$\mathfrak{H}_{\mathrm{A}} : PQ \models Q\theta([P^{\smile})$$

holds for any Q, and let ξ be an arbitrary letter. Then

$$\begin{aligned} \mathfrak{H}_{\mathrm{A}} : P\xi Q &\models \xi Q\theta([P^{\smile}) && \text{[ind. ass.]} \\ &\vdash \alpha\xi Q\theta([P^{\smile}) && [1.1] \\ &\models Q\alpha\xi\theta([P^{\smile}) && [1.5] \\ &\equiv Q\theta(\xi[P^{\smile}) && [(5), (6)] \\ &\equiv Q\theta([P\xi^{\smile}) && [\S 17.3.2], \end{aligned}$$

i.e., in this case

$$\mathfrak{H}_{\mathrm{A}} : P\xi Q \models Q\theta([P\xi^{\smile})$$

for any Q. Therefore, both premises of an application of the method of right induction are fulfilled.

1.7. $\mathfrak{H}_{\mathrm{A}} : P \models \theta([P^{\smile})$.

The special case of 1.6 for $Q \equiv \Lambda$.

1.8. *If* $P \neq \Lambda$, *then* $\mathfrak{H}_{\mathrm{A}} : \alpha\theta(P) \vdash \beta\theta(P)$.

In fact, let $P \neq \Lambda$. Then $P \equiv \xi Q$, $\theta(P) \equiv \theta(\xi)\theta(Q) \equiv \alpha\xi\theta(Q)$ and $\alpha\theta(P) \equiv \alpha\alpha\xi\theta(P)$. It is easy to see that

$$\mathfrak{H}_{\mathrm{A}} : \alpha\alpha\xi\theta(Q) \vdash \beta\alpha\xi\theta(Q).$$

But $\beta\alpha\xi\theta(Q) \equiv \beta\theta(\xi Q) \equiv \beta\theta(P)$.

1.9. $\mathfrak{H}_{\mathrm{A}} : \beta\theta(PQ) \models P\beta\theta(Q)$.

This is proven by right induction on P.

Indeed, when $P \equiv \Lambda$, the graphic equality of the words $\beta\theta(PQ)$ and $P\beta\theta(Q)$ holds, and the required assertion, all the more so. Further, let

$$\mathfrak{H}_{\mathrm{A}} : \beta\theta(PQ) \models P\beta\theta(Q)$$

hold for some P and any Q, and let ξ be an arbitrary letter. Then

$$\begin{aligned} \mathfrak{H}_{\mathrm{A}} : \beta\theta(P\xi Q) &\models P\beta\theta(\xi Q) && \text{[ind. ass.]} \\ &\equiv P\beta\alpha\xi\theta(Q) && [(6), (5)] \\ &\vdash P\beta\xi\theta(Q) && [(1), \text{line } 2] \\ &\vdash P\xi\beta\theta(Q) && [(1), \text{line } 3] \end{aligned}$$

and, consequently, in this case

$$\mathfrak{H}_{\mathrm{A}} : \beta\theta(P\xi Q) \models P\xi\beta\theta(Q)$$

for any Q. Therefore, both premises of an application of the method of right induction are fulfilled.

1.10. $\mathfrak{H}_{\mathrm{A}} : \beta\theta(P) \models P\beta \quad [1.9,\ Q\theta \eqcirc \Lambda]$.

1.11. $\mathfrak{H}_{\mathrm{A}} : \Lambda \models\cdot \Lambda$.

In fact,

$$\begin{aligned} \mathfrak{H}_{\mathrm{A}} : \Lambda &\vdash \alpha \\ &\vdash \alpha\alpha \\ &\vdash \beta\alpha \\ &\vdash \beta \\ &\vdash\cdot \Lambda. \end{aligned}$$

We shall now prove the assertion (2). Let P be an arbitrary word in A. If P is empty, then $[P^{\smile}$ is also empty, and so the validity of (2) follows from 1.11. But if $P \neq \Lambda$, then

$$\begin{aligned} \mathfrak{H}_{\mathrm{A}} : P &\models \theta([P^{\smile}) && [1.7] \\ &\vdash \alpha\theta([P^{\smile}) && [1.1] \\ &\vdash \beta\theta([P^{\smile}) && [1.8] \\ &\models [P^{\smile}\beta && [1.10] \\ &\vdash\cdot [P^{\smile} && [1.2], \end{aligned}$$

from which

$$\mathfrak{H}_{\mathrm{A}} : P \models\cdot [P^{\smile}$$

and, consequently,

$$\mathfrak{H}_{\mathrm{A}}(P) \eqcirc [P^{\smile},$$

which is what finishes our proof.

It is easy to perceive that when applied to an initial word P in A, the performance of the algorithm $\mathfrak{H}_{\mathrm{A}}$ can be described in the following manner.

At first the letter α appears in front of P. It "attaches itself" to the word's first letter and "drags" it to the right through all letters of the alphabet A. Then in front of the word obtained in such a manner, there again appears α. If the word P has not yet been used up, then α "attaches itself" to its second letter and carries it through all letters of A up to the nearest letter α. This cycle is repeated until P is entirely exhausted. It turns out that by this time, the reversal of the word P has actually been obtained already [1.7], true, in a slightly encoded form.

The decoding (throwing away the "superfluous" occurrences of α) begins with the appearance of α once more in front of the word being transformed. It turns out that there are now two of them alongside of each other, and the first of them changes into β [1.8]. After this, β "runs" through the word from left to right, jumping past letters of A [(1), line 3] and erasing each letter α [(1), line 2]. Having reached the end of the word and, by the same token, having finished cleaning up, β disappears, and with this, $\mathfrak{H}_{\mathrm{A}}$'s performance ends.

(We actually considered the general case of a non-empty word P in our commentary. The case of an empty P is considered in Lemma 1.11.)

The complexity, i.e., the length of the scheme of $\mathfrak{H}_A$, is not hard to calculate, and is equal to $8n^2 + 6n + 18$, where n is the number of letters of the alphabet A. The reversing algorithm constructed by means of Zharov's methods[6] has a scheme of length $5n + C$ (by Ostrouknov's, $10n + C_1$).

3. Two auxiliary letters, α and β, are used in the scheme of the algorithm $\mathfrak{H}_A$. It is easy to understand how the second letter's role can be delegated to the word $\alpha\alpha$, taking the scheme

$$\begin{cases} \alpha\alpha\alpha \rightarrow \alpha\alpha \\ \alpha\alpha\xi \rightarrow \xi\alpha\alpha \\ \alpha\alpha \rightarrow\cdot \\ \alpha\xi\eta \rightarrow \eta\alpha\xi \\ \quad\ \rightarrow \alpha, \end{cases}$$

where ξ and η run through the alphabet A, instead of the scheme 1(1). This scheme contains only one auxiliary letter and also defines a reversing algorithm. The following question naturally arises: can we not do entirely without auxiliary letters here, i.e., can we not construct a normal algorithm $\mathfrak{A}$ in an alphabet A, such that for any word P in A we will have

$$\mathfrak{A}(P) \eqcirc [P^{\smile}?$$

The answer to this question turns out, generally speaking, to be negative, as the following assertion shows:

3.1. *If an alphabet* A *contains more than one letter, then there exists no normal algorithm* $\mathfrak{A}$ *in* A*, which would satisfy the condition*

$$\mathfrak{A}(P) \eqcirc [P^{\smile}$$

for each word P in A.

Indeed, let $\mathfrak{A}$ be such a normal algorithm. We shall show that

3.2. *If a word P is such that $P \not\eqcirc [P^{\smile}$, then $\mathfrak{A} : P \vdash\cdot [P^{\smile}$.*

In fact, let $P \not\eqcirc [P^{\smile}$. Then $\mathfrak{A} : P \models [P^{\smile}]$ or $\mathfrak{A} : P \models\cdot [P^{\smile}$. In the first case, on one hand, $\mathfrak{A}([P^{\smile}) \eqcirc [P^{\smile}$, but on the other hand, $\mathfrak{A}(P^{\smile}) \eqcirc [[P^{\smile\smile}$. But $[[P^{\smile\smile} \eqcirc P$ and, therefore, $P \eqcirc [P^{\smile}$, which contradicts the hypothesis. Consequently, $\mathfrak{A} : P \models\cdot [P^{\smile}$, and so there exists a word Q such that $\mathfrak{A} : P \models Q$ and $\mathfrak{A} : Q \vdash\cdot [P^{\smile}$ [§25.7]. Take such a Q. $\mathfrak{A} : Q \models\cdot [P^{\smile}$ and hence $\mathfrak{A}(Q) \eqcirc [P^{\smile}$. On the other hand, $\mathfrak{A}$ is a reversing algorithm, and therefore $\mathfrak{A}(Q) \eqcirc [Q^{\smile}$. Thus, $[P^{\smile} \eqcirc [Q^{\smile}$, and so $P \eqcirc Q$, whence $\mathfrak{A} : P \vdash\cdot [P^{\smile}$. Q.E.D.

Let us now continue the proof of assertion 3.1.

---

[6]See §31.1

Since the alphabet A consists of more than one letter, there are words P in A such that $P \not\equiv [P^{\smile}$. It follows from this that there are final substitution formulas in the scheme of $\mathfrak{A}$. Take the first of them. Let this be the formula

$$U \rightarrow\cdot V.$$

Choose two letters in the alphabet A, ξ and η, such that U does not begin with ξ and $\xi \not\equiv \eta$. Consider the word $\xi U \eta$.

It is easy to see that $[\xi U \eta^{\smile} \equiv \eta [U^{\smile} \xi$ and hence $[\xi U \eta^{\smile} \not\equiv \xi U \eta$, so that $\mathfrak{A} : \xi U \eta \vdash\cdot \eta [U^{\smile} \xi$, where the formula active on the word $\xi U \eta$ is precisely $U \rightarrow\cdot V$, since its left side occurs in $\xi U \eta$, while all the preceding substitution formulas are simple.

Suppose that $U \equiv \Lambda$. Then $\xi U \eta \equiv \xi \eta$ and $\mathfrak{A} : \xi \eta \vdash\cdot V \xi \eta$. From this we get $\eta \xi \equiv V \xi \eta$, and hence $\xi \equiv \eta$, which contradicts the choice of ξ and η. Consequently, $U \not\equiv V$.

But if $U \not\equiv \Lambda$, then $\xi * U * \eta$ is the first occurrence of the word U in the word $\xi U \eta$, and so $\mathfrak{A} : \xi U \eta \vdash\cdot \xi V \eta$. Consequently, $\eta [U^{\smile} \xi \equiv \xi V \eta$ and, therefore, $\xi \equiv \eta$, which again contradicts the choice of ξ and η. Consequently, it is not the case that $U \not\equiv \Lambda$.

Therefore, we proved, on the one hand, that $U \not\equiv \Lambda$, but on the other hand, we refuted the assumption that $U \not\equiv \Lambda$. Hence, it is not the case that $\mathfrak{A}(P) \equiv [P^{\smile}$ for any P in A, which proves 3.1.

Note that if A is a one-letter or empty alphabet, then the identity normal algorithm in A, defined by the scheme

$$\{\rightarrow\cdot ,$$

is reversing.

§33 Letter-by-Letter Coding and Double Projecting Algorithms

1. To each letter ξ of an alphabet A, assign a definite word K_ξ in this alphabet. Replacing, in an arbitrary word P in A, each letter ξ by the word K_ξ assigned to it, we obtain a word in the alphabet A—the *result of replacing the letters ξ by the words K_ξ in the word P*.

We shall denote the result of replacing the letters ξ by the words K_ξ in the word P by the symbol $\Phi(P)$. We shall give a more precise—"working"—definition of this operation by means of the equalities

$$\Phi(\Lambda) \rightleftharpoons \Lambda,$$
$$\Phi(P\xi) \rightleftharpoons \Phi(P) K_\xi,$$

where P denotes an arbitrary word in A, and ξ, any letter of the alphabet.

It is easy to see (this is proven by the method of right induction) that for any two words P and Q in the alphabet A,

$$\Phi(PQ) \equiv \Phi(P)\Phi(Q).$$

Furthermore,

$$\Phi(\xi) \rightleftharpoons K_\xi.$$

It is not hard to construct a normal algorithm over the alphabet A, carrying out the replacement of the letters ξ by the words K_ξ, i.e., changing every word P in A into the word $\Phi(P)$.

In fact, assume that α does not belong to A. Let $\mathfrak{K}_A$ be the normal algorithm in Aα, whose scheme can be written down in the following abbreviated form:

(1)
$$\begin{cases} \alpha\xi \to K_\xi\alpha \\ \alpha \to\cdot \\ \quad \to \alpha, \end{cases}$$

where ξ runs through A.

We shall show that for any P in A,

(2)
$$\mathfrak{K}_A(P) \rightleftharpoons \Phi(P).$$

From an analysis of the scheme (1), we have

1.1. $\mathfrak{K}_A : P \vdash \alpha P.$

1.2. $\mathfrak{K}_A : \Phi(P)\alpha\xi Q \vdash \Phi(P\xi)\alpha Q.$

1.3. $\mathfrak{K}_A : P\alpha \vdash\cdot P.$

Further, we have

1.4. $\mathfrak{K}_A : \alpha PQ \models \Phi(P)\alpha Q.$

This is proven by right induction on P, using 1.2.

1.5. $\mathfrak{K}_A : \alpha P \models \Phi(P)\alpha.$

The special case of 1.4 for $Q \rightleftharpoons \Lambda$.

(2) follows from 1.1, 1.5, and 1.3. Indeed,

$$\begin{aligned} \mathfrak{K}_A : P &\vdash \alpha P & [1.1] \\ &\models \Phi(P)\alpha & [1.5] \\ &\vdash\cdot \Phi(P) & [1.3], \end{aligned}$$

i.e.,

$$\mathfrak{K}_A : P \models \Phi(P),$$

whence

$$\mathfrak{K}_A(P) \rightleftharpoons \Phi(P).$$

Q.E.D.

Consider the case when | is a letter of the alphabet A and the word | is assigned to each letter ξ as the K_ξ. It is obvious that in this case

$$\mathfrak{K}_A(P) \rightleftharpoons |P^\delta$$

for any word P in A.

This same result is given by the simpler normal algorithm in A, defined by the scheme which can be written down in the following abbreviated form:

$$\{\xi \to \mathsf{I},$$

where ξ runs through the alphabet $\text{A}\backslash\mathsf{I}$.

2. An important special case of the replacement of letters by words is the deletion of certain letters. This is the case when $K_\xi \rightleftharpoons \Lambda$ for certain letters ξ and $K_\xi \rightleftharpoons \xi$ for the remaining ξ.

We shall employ the following terminology in the sequel in connection with this operation.

Let A be an extension of the alphabet Б. We shall call the result of the deletion of the letters of the alphabet A\Б from a word P in A the *projection of the word P on the alphabet* Б, and shall denote it by

$$[P^{\text{Б}}$$

(see §19.3).

We shall call the construction the projections of the words in A on Б the *projection operation from* A *onto* Б.

The corresponding special case of the algorithm $\mathfrak{K}_{\text{A}}$ is the normal algorithm in Aα, whose scheme can be written down in the following abbreviated form:

$$\begin{cases} \alpha\xi \to \xi\alpha \\ \alpha\eta \to \alpha \\ \alpha \to\cdot \\ \quad \to \alpha, \end{cases}$$

where ξ runs through the alphabet Б, and η, the alphabet A\Б.

However, the simpler normal algorithm in A, whose scheme can be written down in the following abbreviated form:

$$\{\xi \to$$

where ξ runs through A\Б, obviously achieves the same end—it changes every word in the alphabet A into the projection of this word on the alphabet Б and, consequently, is completely equivalent to $\mathfrak{K}_{\text{A}}$ with respect to A.

3. Let A and Б be alphabets with no letters in common. For every word P in the union of these alphabets, it is possible to construct the projection $[P^{\text{A}}$ on the alphabet A, as well as the projection $[P^{\text{Б}}$ on the alphabet Б. We shall show that the normal algorithm $\mathfrak{L}_{\text{A},\text{Б}}$ in the alphabet AБ, whose scheme can be written down in the following abbreviated form:

$$\{\eta\xi \to \xi\eta,$$

where ξ runs through A and η runs through Б, changes each word P in the alphabet AБ into the word $[P^{\text{A}}[P^{\text{Б}}$.

For this, let us first of all agree to call the number of occurrences in P of words of the form $\eta Q\xi$, where ξ is a letter of A, η is a letter of Б and Q is any word in AБ, the *height of the word P* in the alphabet AБ. We shall denote the height of P by the symbol $[P^{\text{B}}$.

We shall prove several lemmas.

3.1. *If R is a word in* АБ *and* ς, *a letter of* Б, *then*[7]

$$(1) \qquad [R\varsigma^{\text{B}} = [R^{\text{B}}$$

Indeed, to each occurrence

$$(2) \qquad S * \eta Q\xi * T$$

of a word of the form $\eta Q\xi$ (ξ is a letter of A, η is a letter of Б) in R, assign the occurrence

$$(3) \qquad S * \eta Q\xi * T\varsigma$$

of the same word in $R\varsigma$. We shall call the occurrence (3) the *image* of the occurrence (2). It is obvious that different occurrences of words of the form $\eta Q\xi$ (ξ is a letter of A, η is a letter of Б) in R have different images.

Consider, on the other hand, any occurrences

$$(4) \qquad U * \eta Q\xi * V$$

of a word of the form $\eta Q\xi$ (ξ is a letter of A, η is a letter of Б) in $R\xi$. We have

$$(5) \qquad U\eta Q\xi V \rightleftharpoons R\varsigma.$$

Here $\xi \neq \varsigma$, since ξ is a letter of A, ς is a letter of Б. Hence $V \not\rightleftharpoons \Lambda$ and V ends with the letter ς, i.e.,

$$(6) \qquad V \rightleftharpoons W\varsigma.$$

We have

$$U\eta Q\xi W \rightleftharpoons R \quad [(5), (6)],$$

from which it follows that

$$(7) \qquad U * \eta Q\xi * W$$

is an occurrence of $\eta Q\xi$ in R. By virtue of (6), the occurrence (4) is the image of the occurrence (7). Therefore, every occurrence of a word of the form $\eta Q\xi$ (ξ is a letter of A, η is a letter of Б) in $R\varsigma$ is the image of some occurrence of this word in R.

Consequently, if to each occurrence of a word of the form $\eta Q\xi$ (ξ is a letter of A, η is a letter of Б) in R, we assign its image, then by the same token we establish a one-to-one correspondence between the occurrences of such words in R, on the one hand, and their occurrences in $R\varsigma$, on the other. Hence we have the equality (1). Q.E.D.

[7]See §18.10 regarding the arithmetical symbolism employed below.

3.2. *If R is a word in* АБ *and ς is a letter of* А, *then*

$$[R\varsigma^{\text{Б}} = [R^{\text{Б}} + [[R^{\text{Б}\delta}. \tag{8}$$

Indeed, define the image of an occurrence of a word of the form $\eta Q\xi$ (ξ is a letter of A, η is a letter of Б) in R as in the proof of the preceding lemma. Occurrences of words of this form in $R\xi$ which are images of occurrences of such words in R will be called occurrences of the first kind. They are in one-to-one correspondence with the occurrences of words of the form $\eta Q\xi$ (ξ is a letter of A, η is a letter of Б) in R, and the number of them is therefore equal to $[R^{\text{Б}}$.

Further, to each occurrence

$$S * \chi * T \tag{9}$$

of any letter χ of the alphabet Б in the word R, assign the occurrence

$$S * \chi T\xi * \tag{10}$$

of the word $\chi T\varsigma$ in $R\varsigma$. The occurrence (10) will also be called the *image* of the occurrence (9). Since ς is a letter of A, the image of every occurrence of a letter of the alphabet Б in R is an occurrence of a word of the form $\eta Q\xi$ (ξ is a letter of A, η is a letter of Б) in $R\varsigma$. Different occurrences of letters of the alphabet Б in R obviously have different images, i.e., the correspondence between these occurrences and their images is one-to-one. We shall call images of occurrences of letters of the alphabet Б in the word R occurrences of the second kind. The number of occurrences of the second kind is equal to the number of occurrences of letters of the alphabet Б in R, i.e., it equals $[[R^{\text{Б}\delta}$.

Not a single occurrence of the first kind can be an occurrence of the second kind, since occurrences of the first kind have a non-empty right wing, while occurrences of the second kind have an empty right wing.

Every occurrence of a word of the form $\eta Q\xi$ (ξ is a letter of A, η is a letter of Б) in $R\varsigma$ is an occurrence of the first or second kind. Indeed, consider any occurrence (4) of a word of this kind in $R\varsigma$. If $V \neq \Lambda$, then arguing as in the proof of the preceding lemma, we convince ourselves that (4) is the image of some occurrence of the same word in R, i.e., an occurrence of the first kind. If $V = \Lambda$, then (4) is obviously the image of the occurrence $U * \eta * Q$ of the letter η of the alphabet Б in R, i.e., an occurrence of the second kind.

Therefore, the number of occurrences of words of the form $\eta Q\xi$ (ξ—a letter of A, η—a letter of Б) in $R\xi$ is equal to the sum of the numbers of occurrences of the first kind and the second kind, i.e., the sum of the numbers $[R^{\text{B}}$ and $[[R^{\text{Б}\delta}$, which is just what the equality (8) asserts.

3.3. *If R is a word in* АБ *and ς is a letter of this alphabet, then*

$$[R\varsigma^{\text{B}} = [R^{\text{B}} + ([[R^{\text{Б}\delta} \times [[\varsigma^{\text{A}\delta}).$$

In fact, $[[\varsigma^{\text{A}\delta}$ is 1 if ς is a letter of A, and 0 if ς is a letter of Б. By virtue of this, 3.3 follows from 3.1 and 3.2.

3.4. *If P and R are words in АБ, then*

$$[PR^{\mathrm{B}} = [P^{\mathrm{B}} + [R^{\mathrm{B}} + ([[P^{\mathrm{Б}\delta} \times [[R^{\mathrm{A}\delta}). \tag{11}$$

Let us hold P fixed and prove by right induction on R that every word R in АБ satisfies the equality (11).

The empty word satisfies it since we obviously have

$$[\Lambda^{\mathrm{B}} = 0, \ [[\Lambda^{\mathrm{A}\delta} = 0.$$

In order to prove the lemma, we need therefore only prove that (11)'s validity for some word R in АБ implies the validity of

$$[PR\varsigma^{\mathrm{B}} = [P^{\mathrm{B}} + [R\varsigma^{\mathrm{B}} + ([[P^{\mathrm{Б}\delta} \times [[R\varsigma^{\mathrm{A}\delta})$$

for any letter ς of АБ.

Thus, let us assume that (11) is satisfied, where R is a word in АБ. Let us also assume that ς is a letter of АБ. We have

$$[[R\varsigma^{\mathrm{A}\delta} = [[R^{\mathrm{A}\delta} + [[\varsigma^{\mathrm{A}\delta}, \tag{12}$$

$$[[PR^{\mathrm{Б}\delta} = [[P^{\mathrm{Б}\delta} + [[R^{\mathrm{Б}\delta} \tag{13}$$

$$\begin{aligned}[PR\varsigma^{\mathrm{B}} &= [PR^{\mathrm{B}} + ([[PR^{\mathrm{Б}\delta} \times [[\varsigma^{\mathrm{A}\delta}) && [3.3]\\ &= [P^{\mathrm{B}} + [R^{\mathrm{B}} + ([[P^{\mathrm{Б}\delta} \times [[R^{\mathrm{A}\delta} + (([[P^{\mathrm{Б}\delta}+ \\ &\qquad [[R^{\mathrm{Б}\delta}) \times [[\varsigma^{\mathrm{A}\delta}) && [(11),\ (13)]\\ &= [P^{\mathrm{B}} + [R^{\mathrm{B}} + ([[R^{\mathrm{Б}\delta} \times [[\varsigma^{\mathrm{A}\delta}) \\ &\qquad + ([[P^{\mathrm{Б}\delta} \times ([[R^{\mathrm{A}\delta} + [[\varsigma^{\mathrm{A}\delta})) \\ &= [P^{\mathrm{B}} + [R\varsigma^{\mathrm{B}}([[P^{\mathrm{Б}\delta} \times [[R\varsigma^{\mathrm{A}\delta}) && [(3.3),\ (12)].\end{aligned}$$

Q.E.D.

3.5. *For any words P, Q and R in АБ,*

$$\begin{aligned}[PQR^{\mathrm{B}} = [P^{\mathrm{B}} + [Q^{\mathrm{B}} + [R^{\mathrm{B}} + ([[P^{\mathrm{Б}\delta} \times [[Q^{\mathrm{A}\delta}) \\ + ([[P^{\mathrm{Б}\delta} \times [[R^{\mathrm{A}\delta}) + ([[Q^{\mathrm{Б}\delta} \times [[R^{\mathrm{A}\delta}).\end{aligned}$$

In fact, we have

$$\begin{aligned}[PQR^{\mathrm{B}} &= [PQ^{\mathrm{B}} + [R^{\mathrm{B}} + ([[PQ^{\mathrm{Б}\delta} \times [[R^{\mathrm{A}\delta}) && [3.4]\\ &= [P^{\mathrm{B}} + [Q^{\mathrm{B}} + ([[P^{\mathrm{Б}\delta} \times [[Q^{\mathrm{A}\delta})+ \\ &\qquad [R^{\mathrm{B}} + ([[P^{\mathrm{Б}}[Q^{\mathrm{Б}\delta} \times [[R^{\mathrm{A}\delta}) && [3.4]\\ &= [P^{\mathrm{B}} + [Q^{\mathrm{B}} + [R^{\mathrm{B}} + ([[P^{\mathrm{Б}\delta} \times [[Q^{\mathrm{A}\delta} + \\ &\qquad (([[P^{\mathrm{Б}\delta} + [[Q^{\mathrm{Б}\delta}) \times [[R^{\mathrm{A}\delta}) \\ &= [P^{\mathrm{B}} + [Q^{\mathrm{B}} + [R^{\mathrm{B}} + ([[P^{\mathrm{Б}\delta} \times A^{\mathrm{A}\delta})+ \\ &\qquad ([[P^{\mathrm{Б}\delta} \times [[R^{\mathrm{A}\delta}) + ([[Q^{\mathrm{Б}\delta} \times [[R^{\mathrm{A}\delta}).\end{aligned}$$

3.6. *If* $[P^{\text{Б}} = 0$, *then there exist words* Q *and* R *in the alphabets* A *and* Б, *respectively, such that* $P \rightleftharpoons QR$.

Let, in fact, $[P^{\text{Б}} = 0$ for some word P in the alphabet AБ. If no letters of Б occur in this word, then P is a word in A and we may set $Q \rightleftharpoons P$, $R \rightleftharpoons \Lambda$. Assume that letters of Б occur in P.

Then let $Q * \eta * S$ be the first occurrence of a letter of Б in P. Q is then a word in A, η is a letter of Б. If $T * \xi * U$ were an occurrence of a letter ξ of A in S, then we would have

$$\begin{aligned} P &\rightleftharpoons Q\eta S \\ &\rightleftharpoons Q\eta T\xi U \end{aligned}$$

and the word $\eta T\xi$, where ξ is a letter of A and η is a letter of Б, would occur in P, contrary to $[P^{\text{Б}} = 0$. Consequently, no letter of A occurs in S, and so S is a word in Б. It remains to set $R \rightleftharpoons \eta S$ in order to have $P \rightleftharpoons QR$, where Q is a word in A and R is a word in Б.

3.7. *If* Q *is a word in* A *and* R *is a word in* Б, *then*

$$\begin{aligned} [QR^{\text{A}} &\rightleftharpoons Q, \\ [QR^{\text{Б}} &\rightleftharpoons R. \end{aligned}$$

This is obvious from the definition of projections.

3.8. *If* $[P^{\text{Б}} = 0$, *then* $P \rightleftharpoons [P^{\text{A}}[P^{\text{Б}}$.

This follows from the two preceding lemmas.

3.9. *If* $[P^{\text{Б}} = 0$, *then* $\mathfrak{L}_{\text{A,Б}} : P]$.

Indeed, if $[P^{\text{Б}} = 0$, then no words of the form $\eta\xi$ (ξ is a letter of A, η is a letter of Б), i.e., no left sides of $\mathfrak{L}_{\text{A,Б}}$'s substitution formulas occur in P. This word is then not subject to $\mathfrak{L}_{\text{A,Б}}$.

3.10. *If* $[P^{\text{Б}} \neq 0$, *then there exists a word* Q *such that* $\mathfrak{L}_{\text{A,Б}} : P \vdash Q$.

In fact, let P be a word in AБ for which $[P^{\text{Б}} \neq 0$. Then a word of the form $\eta R\xi$ occurs in P, where ξ is a letter of A, η is a letter of Б. Let $S * \eta R\xi * T$ be an occurrence of such a word in P. A letter of A occurs in $\eta R\xi$. Let $U * \varsigma * V$ be the first occurrence of a letter of the alphabet A in $\eta R\xi$. Then $U \not\rightleftharpoons \Lambda$, since $\varsigma \neq \eta$. U is a word in the alphabet Б. Consequently, U has the form $W\chi$, where χ is a letter of Б. We therefore have

$$\begin{aligned} P &\rightleftharpoons S\eta R\xi T \\ &\rightleftharpoons SU\varsigma VT \\ &\rightleftharpoons SW\chi\varsigma VT. \end{aligned}$$

Thus, the word $\chi\varsigma$, where ς is a letter of A and χ is a letter of Б, occurs in P, i.e., the left side of one of $\mathfrak{L}_{\text{A,Б}}$'s substitution formulas occurs in P. Since all of this algorithm's substitution formulas are simple, the algorithm simply converts P into a word Q. Q.E.D.

3.11. *If* $\mathfrak{L}_{\text{A,Б}} : P \vdash Q$, *then* $[Q^{\text{B}} = [P^{\text{B}} - 1$, $[Q^{\text{A}} \risingdotseq [P^{\text{A}}$ *and* $[Q^{\text{Б}} \risingdotseq [P^{\text{Б}}$.

Indeed, let $\mathfrak{L}_{\text{A,Б}} : P \vdash Q$. Then the first step in the application of the algorithm $\mathfrak{L}_{\text{A,Б}}$ to the word P consists in substituting the right side $\xi\eta$ of one of the algorithm's substitution formulas in place of the first occurrence of its left side $\eta\xi$ (ξ is a letter of A, η is a letter of Б). Let $R * \eta\xi * S$ be the first occurrence of $\eta\xi$ in P. Then

$$P \risingdotseq R\eta\xi S,$$
$$Q \risingdotseq R\xi\eta S.$$

Bearing in mind the fact that

$$[\eta\xi^{\text{B}} = 1,$$
$$[\xi\eta^{\text{B}} = 0$$

by the definition of height and that

$$[[\xi\eta^{\text{A}\delta} = [[\xi\eta^{\text{Б}\delta} = [[\eta\xi^{\text{A}\delta} = [[\eta\xi^{\text{Б}\delta} = 1,$$

we obtain, according to 3.5,

$$[P^{\text{B}} = [R^{\text{B}} + 1 + [S^{\text{B}} + [[R^{\text{Б}\delta} + ([[R^{\text{Б}\delta} \times [[S^{\text{A}\delta}) + [[S^{\text{A}\delta},$$
$$[Q^{\text{B}} = [R^{\text{B}} + [S^{\text{B}} + [[R^{\text{Б}\delta} + ([[R^{\text{Б}\delta} \times [[S^{\text{A}\delta}) + [[S^{\text{A}\delta},$$

whence $[Q^{\text{B}} = [P^{\text{B}} - 1$.

Also bearing in mind the fact that

$$[\xi\eta^{\text{A}} \risingdotseq [\eta\xi^{\text{A}} \risingdotseq \xi,$$

we get

$$[P^{\text{A}} \risingdotseq [R^{\text{A}}\xi[S^{\text{A}} \risingdotseq [Q^{\text{A}}.$$

We convince ourselves that $[P^{\text{Б}} \risingdotseq [Q^{\text{Б}}$ in an analogous manner.

3.12. *The algorithm* $\mathfrak{L}_{\text{A,Б}}$ *is applicable to any word in* AБ.

In fact, according to 3.9, it is applicable to any word of height zero. Now let N be any natural number. Assume that $\mathfrak{L}_{\text{A,Б}}$ is applicable to any word Q in AБ, such that $[Q^{\text{B}} = N$, and let P be a word in AБ, such that $[P^{\text{B}} = N + 1$. We shall show that $\mathfrak{L}_{\text{A,Б}}$ is applicable to P. Indeed, $[P^{\text{B}} \neq 0$ and, therefore, there exists a word Q such that $\mathfrak{L}_{\text{A,Б}} : P \vdash Q$ [3.10]. According to 3.11, $[Q^{\text{B}} = N$ and hence the algorithm $\mathfrak{L}_{\text{A,Б}}$ is applicable to Q. But then it is also applicable to P. [§25.8.9]. Consequently, $\mathfrak{L}_{\text{A,Б}}$ is applicable to any word in AБ. Q.E.D.

Now let us show that

$$\mathfrak{L}_{\text{A,Б}}(P) \risingdotseq [P^{\text{A}}[P^{\text{Б}} \qquad (14)$$

for every word P in AБ.

Indeed, let P be a word in АБ. The algorithm $\mathfrak{L}_{\text{А,Б}}$ is applicable to P [3.12]. Consequently, there exists a Q in АБ, such that $\mathfrak{L}_{\text{А,Б}}(P) \coloneqq Q$. It is clear that

$$(15) \qquad \mathfrak{L}_{\text{А,Б}} : P \models Q]$$

and that Q is a word of height zero [3.10]. It is easy to show by induction on the steps in the performance of the algorithm $\mathfrak{L}_{\text{А,Б}}$ and by using 3.11 that

$$[P^{\text{А}} \coloneqq [Q^{\text{А}}$$
$$[P^{\text{Б}} \coloneqq [Q^{\text{Б}}.$$

Since $[Q^{\text{B}} = 0$, we have

$$(18) \qquad Q \coloneqq [Q^{\text{А}}[Q^{\text{Б}} \quad [3.8].$$

Therefore,

$$\begin{aligned} \mathfrak{L}_{\text{А,Б}}(P) &\coloneqq Q \\ &\coloneqq [Q^{\text{А}}[Q^{\text{Б}} \quad [(18)] \\ &\coloneqq [P^{\text{А}}[P^{\text{Б}} \quad [(16), (17)]. \end{aligned}$$

Consequently, $\mathfrak{L}_{\text{А,Б}}(P) \coloneqq [P^{\text{А}}[P^{\text{Б}}$. Q.E.D.

Let us formulate here a corollary which will be applied in the sequel.

3.13. *If Q is a word in* А *and R is a word in* Б, *then*

$$(19) \qquad \mathfrak{L}_{\text{А,Б}}(RQ) \coloneqq QR.$$

In fact, under these conditions

$$[RQ^{\text{А}} \coloneqq Q,$$
$$[RQ^{\text{Б}} \coloneqq R,$$

and hence (19) follows from (14).

The relations (14) allow us to characterize $\mathfrak{L}_{\text{А,Б}}$ as the *double projecting algorithm.*

§34 Some Arithmetical Algorithms

1. We shall now construct several normal algorithms which perform on natural numbers and systems of natural numbers. From here until the end of the section, M, N, Q, R, S denote variables for natural numbers represented as words in the alphabet **I**.

Let us give the normal algorithm $\mathfrak{A}_1$ in the alphabet **I**$*$ by the scheme

$$(1) \qquad \begin{cases} \mathsf{I} * \mathsf{I} & \rightarrow * \\ * & \rightarrow . \end{cases}$$

This algorithm changes a pair of numbers $N * M$ into the absolute value of the difference between these two numbers, i.e.,

$$(2) \qquad \mathfrak{A}_1(N * M) \coloneqq |N - M|$$

for any two natural numbers N and M.

In fact, from an analysis of the scheme (1) we have

1.1. $\mathfrak{A}_1 : M\mathsf{I} * \mathsf{I}N \vdash M * N$.

1.2. $\mathfrak{A}_1 : M* \vdash M$.

1.3. $\mathfrak{A}_1 : *M \vdash M$.

1.4. $\mathfrak{A}_1 : M\urcorner$.

Furthermore, the following assertions can be proven by right induction on N with the aid of 1.1 and the equality $N\mathsf{I} \coloneqq \mathsf{I}N$:

1.5. $\mathfrak{A}_1 : MN * N \models M*$.

1.6. $\mathfrak{A}_1 : N * NM \models *M$.

And now the assertion (2) is proven in the following manner.
If $N \leqslant M$, then $N * M \coloneqq N * N|N - M|$ and

$$\begin{aligned} \mathfrak{A}_1 : N * N|N - M| &\models *|N - M| \quad [1.6] \\ &\vdash |N - M|\urcorner \quad [1.3,\ 1.4]. \end{aligned}$$

But if $N > M$, then $N * M \coloneqq |N - M|M * M$ and

$$\begin{aligned} \mathfrak{A}_1 : |N - M|M * M &\models |N - M|* \quad [1.5] \\ &\models |N - M|\urcorner \quad [1.2,\ 1.4]. \end{aligned}$$

In both cases, consequently,

$$\mathfrak{A}_1(N * M) \coloneqq |N - M|.$$

Q.E.D.

In particular, $\mathfrak{A}_1(M * N) \coloneqq \Lambda$ if and only if $M \coloneqq N$.

2. Further, let us give the normal algorithm $\mathfrak{A}_2$ in the alphabet I by the scheme

$$\{\mathsf{IIIII} \rightarrow .$$

We shall show that it changes every natural number into the remainder resulting from its division by 5.

Indeed, the validity of the following lemmas can be immediately discovered by examining the scheme (1):

2.1. $\mathfrak{A}_2 : N + 5 \vdash N$.

2.2. *If* $N < 5$, *then* $\mathfrak{A}_2 : N\urcorner$.

From 2.1 follows

2.3. $\mathfrak{A}_2 : R + 5 \times Q \models R.$

Now let N be an arbitrary natural number. Represent it in the form $R + 5 \times Q$, where R is the remainder and Q is the partial quotient resulting from the division of N by 5. Since

$$N = R + 5 \times Q$$

and $R < 5$, the algorithm $\mathfrak{A}_2$ performs in the following way when applied to the word N:

$$\mathfrak{A}_2 : N \models R\rceil \quad [2.3,\ 2.2],$$

whence

$$\mathfrak{A}_2(N) \asymp R.$$

Q.E.D.

In particular, $\mathfrak{A}_2(N) \asymp \Lambda$ if and only if N is divisible by 5.

3. We now give the normal algorithm $\mathfrak{A}_3$ in the alphabet $\mathsf{I}*$ by the scheme

(1) $$\left\{\begin{array}{ll} *\mathsf{IIIII} & \rightarrow \mathsf{I}* \\ *\mathsf{I} & \rightarrow * \\ * & \rightarrow\cdot \\ & \rightarrow *. \end{array}\right.$$

Let us show that it changes every natural number N into the partial quotient resulting from the division of N by 5, i.e., into $[\frac{N}{5}]$.[8]

Indeed, the truth of the following lemmas is easily discerned by examining the scheme (1):

3.1. $\mathfrak{A}_3 : N * (M + 5) \vdash (N + 1) * M.$

3.2. *If* $0 < M < 5$, *then*

$$\mathfrak{A}_3 : N * M \vdash N * (M - 1).$$

3.3. $\mathfrak{A}_3 : N* \vdash\cdot N.$

3.4. $\mathfrak{A}_3 : N \vdash *N.$

From 3.1 follows

3.5. $\mathfrak{A}_3 : N * (R + 5 \times Q) \models (N + Q) * R.$

From 3.2 follows

[8] Square brackets are employed here for denoting the integral part.

3.6. *If $R < 5$, then*

$$\mathfrak{A}_3 : Q * R \models Q * .$$

Now let N be an arbitrary natural number. Represent it in the form $R + 5 \times Q$, where R is the remainder and Q is the partial quotient resulting from the division of N by 5. Since $N = R + 5 \times Q$ and $R < 5$, the algorithm $\mathfrak{A}_3$ performs in the following manner when applied to the word N:

$$\begin{aligned} \mathfrak{A}_3 : N &\vdash *(R + 5 \times Q) && [3.4] \\ &\models Q * R && [3.5] \\ &\models Q* && [3.6] \\ &\vdash\cdot\, Q. && [3.3] \end{aligned}$$

Therefore, $\mathfrak{A}_3 : N \models\cdot\, Q$, whence $\mathfrak{A}_3(N) \simeq Q$. Q.E.D.

4. It is also easy to construct a normal algorithm in $\mathsf{I}*$, changing every natural number N into the pair of natural numbers $Q * R$, where Q is the partial quotient and R is the remainder resulting from the division of N by 5. In fact, let us give the normal algorithm $\mathfrak{A}_4$ in $\mathsf{I}*$ by the scheme

$$\begin{cases} *\mathsf{IIIII} \rightarrow \mathsf{I}* \\ \quad * \rightarrow \cdot * \\ \quad\ \ \rightarrow *. \end{cases}$$

This algorithm performs in the manner just indicated, as the reader can prove without difficulty.

The divisor 5 in the three preceding examples was taken only for definiteness. But such normal algorithms can obviously be constructed for division by any other natural number different from zero. For this, it is only necessary to replace the five strokes in the left side of the first formula of each of the three schemes just considered by a different, appropriate number of strokes. However, if we tried to construct analogous algorithms for division by zero, throwing away all the strokes in the left side of these schemes' first formulas, then, as is not hard to see, we would obtain algorithms inapplicable to any natural number, which, of course, completely corresponds to the essence of the matter.

5. We shall now construct two somewhat more complicated normal algorithms. The first of them—the *algorithm finding the greatest common divisor*—will change every pair of natural numbers into the greatest common divisor of these numbers; the second—the *multiplication algorithm*—will change every pair of natural numbers into the product of these numbers. Both algorithms will be algorithms over the alphabet $\mathsf{I}*$, but not in this alphabet.

We give the normal algorithm $\mathfrak{A}_5$ in the alphabet $| * abc$ by the scheme

$$(1) \qquad \begin{cases} |a \to a| \\ | * | \to a* \\ |* \to *b \\ b \to | \\ a \to c \\ c \to | \\ * \to . \end{cases}$$

We shall show that $\mathfrak{A}_5$ changes every pair of numbers $N*M$ into the greatest common divisor of these numbers.

The following lemmas are immediately obtained by an analysis of the scheme (1):

5.1. $\mathfrak{A}_5 : a^Q N a M * R \vdash a^Q (N-1) a (M+1) * R \ (N > 0)$.

5.2. $\mathfrak{A}_5 : a^Q N * R \vdash a^Q (N-1) a * (R-1) \ (N > 0, R > 0)$.

5.3. $\mathfrak{A}_5 : a^Q N * b^R \vdash a^Q (N-1) * b^{R+1} \ (N > 0)$.

5.4. $\mathfrak{A}_5 : a^Q * N b^R \vdash a^Q * (N+1) b^{R-1} \ (R > 0)$.

5.5. $\mathfrak{A}_5 : c^M a^Q * N \vdash c^{M+1} a^{Q-1} * N \ (Q > 0)$.

5.6. $\mathfrak{A}_5 : Q c^M * N \vdash (Q+1) c^{M-1} * N \ (M > 0)$.

5.7. $\mathfrak{A}_5 : *N \vdash N$.

5.8. $\mathfrak{A}_5 : N\rceil$.

With their aid, the following lemmas can then be proven:

5.9. $\mathfrak{A}_5 : a^Q N a * R \models a^{Q+1} N * R$ [5.1].

5.10. $\mathfrak{A}_5 : a^Q N * R \models a^{Q+1} (N-1) * (R-1) \ (N > 0, R > 0)$ [5.2, 5.9].

5.11. $\mathfrak{A}_5 : N * R \models a^R (N \dot- R) * \ (N \geqslant R)$ [5.10].

5.12. $\mathfrak{A}_5 : N * R \models a^N * (R - N) \ (N < R)$ [5.10].

5.13. $\mathfrak{A}_5 : a^Q N* \models a^Q * b^N$ [5.3].

5.14. $\mathfrak{A}_5 : a^Q * b^N \models a^Q * N$ [5.4].

5.15. $\mathfrak{A}_5 : a^Q * N \models c^Q * N$ [5.5].

5.16. $\mathfrak{A}_5 : c^Q * N \models Q * N$ [5.6].

5.17. $\mathfrak{A}_5 : N * R \models R * (N - R) \ (N \geqslant R)$ [5.11, 5.13, 5.14, 5.15, 5.16].

5.18. $\mathfrak{A}_5 : N * R \models N * (R - N) \ (N < R)$ [5.12, 5.15, 5.16].

5.19. $\mathfrak{A}_5 : *N \models N\rceil$ [5.7, 5.8].

5.20. $\mathfrak{A}_5 : N* \models N\rceil$ [5.17, 5.19].

Let us now agree to say that the pairs of natural numbers $N * R$ and $M * Q$ are *equivalent* if every common divisor of N and R is a common divisor of M

and Q, and conversely. When $N \geqslant R$, the pair $N * R$ is obviously equivalent to the pair $R * (N - R)$, while for $N < R$, it is equivalent to the pair $N * (R - N)$. Lemmas 5.17 and 5.18 show, therefore, that *every pair $N*R$ of numbers different from zero is simply transformed by the algorithm $\mathfrak{A}_5$ into an equivalent pair of numbers $N * Q$, such that $M + Q < N + R$.* It then follows from this that *every pair of numbers is simply transformed by the algorithm $\mathfrak{A}_5$ into an equivalent pair of numbers, containing zero as one of its elements, i.e., into a pair of the form $*Q$ or $Q*$.*

For pairs of this form, the number Q is obviously the greatest common divisor, i.e., the common divisor which is divisible by every common divisor. By virtue of the equivalence between the initial pair of numbers and the pair of this form obtained from it, Q is also the greatest common divisor of the initial pair's numbers. On the other hand, according to 5.19 and 5.20, $\mathfrak{A}_5$ naturally transforms every pair of the form $*Q$ or $Q*$ into Q. Therefore, $\mathfrak{A}_5$ naturally transforms the initial pair of numbers into these numbers' greatest common divisor. Consequently, $\mathfrak{A}_5$ *changes every pair of numbers $N * R$ into the greatest common divisor of these numbers.* Q.E.D.

The algorithm $\mathfrak{A}_6$ is given in the alphabet $\mathsf{I} * ab$ by the scheme

$$(2) \qquad \begin{cases} b\mathsf{I} \to \mathsf{I}b \\ a\mathsf{I} \to \mathsf{I}ba \\ a \to \\ \mathsf{I}* \to *a \\ *\mathsf{I} \to * \\ * \to \\ b \to \mathsf{I}. \end{cases}$$

We shall show that $\mathfrak{A}_6$ changes every pair of $M*R$ numbers into the product of these numbers.

The truth of the following lemmas can be immediately perceived from the scheme (2).

5.21. $\mathfrak{A}_6 : M * Nb^Q\mathsf{I}b^RP \vdash M * Nb^{Q-1}\mathsf{I}b^{R+1}P$ *($Q > 0$, P is a word in $\mathsf{I} * ab$).*

5.22. $\mathfrak{A}_6 : M * Nb^QaRb\mathsf{I}^S \vdash M * Nb^Q\mathsf{I}ba(R-1)b^S$ *($R > 0$).*

5.23. $\mathfrak{A}_6 : M * Nb^Qab^R \vdash M * Nb^{Q+R}$.

5.24. $\mathfrak{A}_6 : M * Nb^Q \vdash (M-1) * aNb^Q$ *($M > 0$).*

5.25. $\mathfrak{A}_6 : *Nb^Q \vdash *(N-1)b^Q$ *($N > 0$).*

5.26. $\mathfrak{A}_6 : *b^Q \vdash b^Q$.

5.27. $\mathfrak{A}_6 : Rb^Q \vdash (R+1)b^{Q-1}$ *($Q > 0$).*

5.28. $\mathfrak{A}_6 : R\rceil$.

The following lemmas are implied by them:

5.29. $\mathfrak{A}_6 : M * Nb^Q|bP \models M * (N+1)b^{Q+1}P$ *(P is a word in $*ab$)* [5.21].

5.30. $\mathfrak{A}_6 : M * Nb^Q aRb^S \models M * (N+1)b^{Q+1}a(R-1)b^S$ *($R > 0$).*

Indeed, for $R > 0$, we have

$$\begin{aligned} \mathfrak{A}_6 : M * Nb^Q aRb^S &\models M * Nb^Q|ba(R-1)b^S & [5.22] \\ &\models M * (N+1)b^{Q+1}a(R-1)b^S & [5.29]. \end{aligned}$$

5.31. $\mathfrak{A}_6 : M * Nb^Q Rb^S \models M * (N+R)b^{Q+R}ab^S$ [5.30].

5.32. $\mathfrak{A}_6 : M * aRb^S \models M * Rb^R ab^S$ [5.31].

5.33. $\mathfrak{A}_6 : M * Rb^S \models (M-1) * Rb^{R+S}$ *($M > 0$).*

In fact, when $M > 0$ we have

$$\begin{aligned} \mathfrak{A}_6 : M * Rb^S &\vdash (M-1) * aRb^S & [5.24] \\ &\models (M-1) * Rb^R ab^S & [5.32] \\ &\vdash (M-1) * Rb^{R+S} & [5.23]. \end{aligned}$$

5.34. $\mathfrak{A}_6 : M * Rb^S \models (M-Q) * Rb^{(Q\times R)+S}$ *($Q \leqslant M$).*

Indeed, the lemma is trivial for $Q = 0$. Assume that it has been established for $Q = N$, where $N < M$, and let us prove it for $Q = N+1$. We have

$$\begin{aligned} \mathfrak{A}_6 : M * Rb^S &\models (M-N) * Rb^{(N\times R)+S} & \\ &\models (M-N-1) * Rb^{R+(N\times R)+S} & [5.33] \\ &\fallingdotseq (M-(N+1)) * Rb^{((N+1)\times R)+S}, & \end{aligned}$$

from which follows what we are proving.

5.35. $\mathfrak{A}_6 : *Rb^S \models b^S$ [5.25].

5.36. $\mathfrak{A}_6 : b^S \models S$ [5.27].

5.37. $\mathfrak{A}_6 : M * R \models (M \times R)\rceil$.

In fact,

$$\begin{aligned} \mathfrak{A}_6 : M * R &\models *Rb^{(M\times R)} & [5.34] \\ &\models *b^{(M\times R)} & [5.35] \\ &\vdash b^{(M\times R)} & [5.26] \\ &\models (M \times R)\rceil & [5.36, 5.28]. \end{aligned}$$

By virtue of 5.37 we obtain

$$\mathfrak{A}_6 : (M * R) \fallingdotseq M \times R.$$

Q.E.D.

CHAPTER IV

COMBINING NORMAL ALGORITHMS

In §§35–41, we determine a number of constructions permitting us to build up new normal algorithms, starting with given ones. The presence of such constructions shows that the natural ways of combining algorithms—such as the successive application of two algorithms, the repeated application of one algorithm up to the time when it yields a result which "satisfies" the other algorithm in a known manner, etc.— leads from normalizable algorithms to normalizable algorithms. Making up the apparatus of the general theory of algorithms, the results presented in these sections also serve as a serious argument in favor of the normalization principle.

§35 Extensions of an Algorithm

In what follows we shall often make use of a symbol for conditional equality, ≪$\simeq$≫. Placing such a symbol between two expressions, we shall be asserting by the same token that these expressions denote one and the same word, as soon as at least one of them makes sense. In this connection, one or another side condition, imposed on component parts of the expressions under consideration, will ordinarily be written in parentheses.

In this notation, the complete equivalence of algorithms $\mathfrak{A}$ and $\mathfrak{B}$ with respect to an alphabet A is expressed as follows:

$$\mathfrak{A}(P) \simeq \mathfrak{B}(P) \quad (P \text{ is a word in A}).$$

Conditional equality is obviously reflexive, symmetric and transitive.

1. Let $\mathfrak{A}$ be an algorithm in an alphabet A, and let Б be an extension of this alphabet. Let us agree to say of an algorithm $\mathfrak{B}$ in the alphabet Б that it is an *extension of the algorithm* $\mathfrak{B}$ *to the alphabet* Б if

$$\mathfrak{A}(P) \simeq \mathfrak{B}(P) \quad (P \text{ is a word in A}), \tag{1}$$

i.e., if $\mathfrak{B}$ is completely equivalent to $\mathfrak{A}$ with respect to A [§26.6].

It is clear that an algorithm in A can have several extensions to Б, since the condition (1) does not impose any restrictions on $\mathfrak{B}$'s performance on words in Б which are not words in A. We shall now consider two special kinds of extensions of *normal* algorithms.

2. Let $\mathfrak{A}$ be a normal algorithm in an alphabet A, and let Б be an extension of this alphabet. Let us present a normal algorithm in Б, taking the scheme of $\mathfrak{A}$ as its scheme, which is, of course, permissible, since every word in A is simultaneously a word in Б.

Then, as we shall soon see, the following conditional equality holds:

$$(1) \qquad \mathfrak{B}(PR) \simeq \mathfrak{A}(P)R \qquad (P \text{ is a word in A}, \; R \text{ is a word in Б} \setminus \text{A}),$$

from which, in particular (for $R \leftrightharpoons \Lambda$), 1(1) follows.

Let us first of all fix a word P in A, and a word R in Б\A and agree to say that the *image* of the occurrence

$$(2) \qquad S * T * U$$

in the word P is the occurrence

$$(3) \qquad S * T * UR.$$

We shall prove several lemmas.

2.1. *The image of every occurrence of a word T in P is an occurrence of T in PR.*

Indeed, if (2) is an occurrence in the word P, then

$$(4) \qquad STU \leftrightharpoons P,$$
$$STUR \leftrightharpoons PR \quad [(4)],$$

from which it follows that (3), i.e., the image of the occurrence (2), is an occurrence in the word PR.

2.2. *Every occurrence of a non-empty word T in the alphabet A in the word PR is the image of some occurrence of T in P.*

Indeed, let T be a non-empty word in the alphabet A, and let $S * T * V$ be its occurrence in PR. Then

$$(5) \qquad STU \leftrightharpoons PR.$$

Since $T \not\leftrightharpoons \Lambda$, T can be represented in the form $W\xi$, where ξ is a letter of A. We have

$$(6) \qquad T \leftrightharpoons W\xi,$$
$$SW\xi V \leftrightharpoons PR \quad [(5), (6)].$$

Therefore, ξV and R are terminal segments of one and the same word. Hence R ends with the word ξV or V ends with the word R. The former is impossible, since the letter ξ of the alphabet A does not occur in a word R in Б\A. Consequently, V ends with the word R, i.e., there exists a word U, such that

(7) $$V \coloneqq UR.$$

We now have

(8)
$$STUR \coloneqq PR \quad [(5), (7)],$$
$$STU \coloneqq P \quad [(8)],$$
$$S * T * V \coloneqq S * T * UR \quad [(7)].$$

Consequently, (2) is an occurrence of T in P, and the occurrence $S * T * V$ under consideration is its image. With this the lemma is proven.

2.3. *If V and W are occurrences of one and the same word in P, then the image of V precedes the image of W if and only if V precedes W.*

According to the definition of precedence [§23.5], this immediately follows from the coincidence of the left wings of the occurrences in P with the left wings of these images' occurrences.

2.4. *If T occurs in P, then the image of the first occurrence of T in P is the first occurrence of T in PR.*

Indeed, let V be the first occurrence of T in P, and let W be the image of V. If $T \coloneqq V$, then $V \coloneqq * * P$, whence $W \coloneqq * * PR$. Consequently, in this case W is the first occurrence of T in PR. Now let $T \not\coloneqq \Lambda$.

Consider any occurrence W_1 of T in PR, which differs from W. According to 2.2, it is the image of some occurrence V_1 of T in P. $V_1 \not\coloneqq V$, since $W_1 \not\coloneqq W$. Since V is the first occurrence of T in P, V precedes V_1. Hence W precedes W_1 [2.3]. Consequently, W precedes every occurrence of T in PR, which differs from it, i.e., is the first occurrence of T in PR. Q.E.D.

2.5. *A word T in the alphabet A occurs in P if and only if it occurs in PR.*

This is true of $T \coloneqq \Lambda$, since Λ occurs in P and in PR; this immediately follows from lemmas 2.1 and 2.2 for $T \not\coloneqq \Lambda$.

2.6. *If V is an occurrence of a word T in P and Q is the result of substituting the word W in place of V, then QR is the result of substituting the word W in place of V's image.*

Indeed, let

(9) $$V \coloneqq S * T * U.$$

Then

(10) $$Q \coloneqq SWU \quad [(9)],$$
(11) $$QR \coloneqq SWUR \quad [(10)].$$

Therefore, QR is the result of substituting W in place of $S * T * UR$ [(11)], i.e., in place of V's image [(9)]. Q.E.D.

2.7. *If* $\mathfrak{A} : P \vdash Q$, *then* $\mathfrak{B} : PR \vdash QR$.

In fact, let $\mathfrak{A} : P \vdash Q$. Let F be the formula active on P in the scheme of $\mathfrak{A}$. The left sides of the formulas preceding F in $\mathfrak{A}$'s scheme do not occur in P, and hence do not occur in PR either, since they are words in A [2.5]. But the left side of F occurs in PR, since it occurs in P. In view of the coincidence of the schemes of the algorithms $\mathfrak{A}$ and $\mathfrak{B}$, F is therefore the formula of the scheme of $\mathfrak{B}$, which is active on the word PR.

Let V be the first occurrence of F's left side in P. Then the image of V is the first occurrence of F's left side in PR [2.4]. When applied to the word PR, the algorithm $\mathfrak{B}$ therefore prescribes the substitution of F's right side in place of V's image. But the result of substituting F's right side in place of V is the word Q, since $\mathfrak{A} : P \vdash Q$. Consequently, the result of substituting F's right side in place of V's image is QR [2.6].

Moreover, the formula F is simple, since $\mathfrak{A} : P \vdash Q$. Consequently, $\mathfrak{B} : PR \vdash QR$. Q.E.D.

2.8. *If* $\mathfrak{A} : P \vdash\cdot QR$, *then* $\mathfrak{B} : PR \vdash\cdot QR$.

The proof of this lemma is completely analogous to that of the preceding lemma.

2.9. *If* $\mathfrak{A} : P]$, *then* $\mathfrak{B} : PR]$.

This is easily proven with the aid of 2.5.

The words P and R have so far been held fixed by us, which has been of considerable significance in the definition of an occurrence's image. However, the concept of an occurrence's image plays no role in Lemmas 2.7–2.9, proven above. These lemmas are therefore applicable to any words P, Q and R, such that P is a word in A and R is a word in Б\A. In what follows, they will be applied in precisely this fashion.

In Lemmas 2.10–2.20, we also assume that P is a word in A, and R, in Б\A.

2.10. *If* $\mathfrak{B} : PR \vdash S$, *then there exists a word* Q *in* A, *such that* $S \coloneqq QR$ *and* $\mathfrak{A} : P \vdash Q$.

Indeed, let $\mathfrak{B} : PR \vdash S$. One of the following three assertions holds: either there exists a word Q such that $\mathfrak{A} : P \vdash Q$, or there exists a word Q, such that $\mathfrak{A} : P \vdash\cdot Q$, or $\mathfrak{A} : P]$ [§25.4.7]. However, in the second case we have $\mathfrak{B} : PR \vdash\cdot QR$ [2.8], while in the third, $\mathfrak{B} : PR]$ [2.9]. Since neither one nor the other is compatible with the assumption that $\mathfrak{B} : PR \vdash S$, these cases can be eliminated. Therefore, there exists a word Q, such that $\mathfrak{A} : P \vdash Q$. Q is a word in A, since $\mathfrak{A}$ is a normal algorithm in A. Since $\mathfrak{A} : P \vdash Q$, we have $\mathfrak{B} : PR \vdash QR$ [2.7]. But since $\mathfrak{B} : PR \vdash S$ by hypothesis, $S \coloneqq QR$, which is just what remained to be proven.

The following two lemmas are proven in an analogous manner:

2.11. *If* $\mathfrak{B} : PR \vdash\cdot S$, *then there exists a word* Q *in* A, *such that* $S \coloneqq QR$ *and* $\mathfrak{A} :\vdash\cdot Q$.

2.12. *If* $\mathfrak{B} : PR]$, *then* $\mathfrak{A} : P]$.

Let us prove

2.13. *If* $\mathfrak{A} : P \models Q$, *then* $\mathfrak{B} : PR \models QR$.

This assertion is easily proven—with the aid of Lemma 2.7—by the method of induction on the steps of the performance of the algorithm $\mathfrak{A}$ [25.10]. Because this sort of reasoning is typical, we shall carry it out with maximal care this, the first, time, in spite of the situation's utter simplicity.

Thus, having fixed P and R, consider the property F of words Q in the alphabet A, expressed by the condition $\ll \mathfrak{B} : PR \models QR \gg$. Our task consists in proving that, firstly, P possesses the property F and that, secondly, for any Q and S, such that $\mathfrak{A} : P \models Q \vdash S$, from Q's possession of the property F follows its possession by S. Let us show this.

The word P possesses the property F trivially [§25.6.1]. Now let Q and S be words for which $\mathfrak{A} : P \models Q$ and $\mathfrak{A} : Q \vdash S$, and let $\mathfrak{B} : PR \models QR$. We shall show that $\mathfrak{B} : PR \models SR$. But this really is the case, because from $\mathfrak{A} : Q \vdash S$ follows $\mathfrak{B} : QR \vdash SR$ [2.7], which in combination with $\mathfrak{B} : PR \models QR$, yields $\mathfrak{B} : PR \models SR$ [§25.6.2, §25.6.3].

Therefore, both requirements for the applicability of the method of induction are fulfilled in our case. The inductive conclusion in the situation under consideration asserts that for fixed P and R, the implication 2.13 is true for any Q. But P and R were fixed arbitrarily, so that 2.13 holds for any P, Q and R. Q.E.D.

2.14. *If* $\mathfrak{A} : P \models\cdot Q$, *then* $\mathfrak{B} : PR \models\cdot QR$.

In fact, let $\mathfrak{A} : P \models\cdot Q$. Then there exists an S such that $\mathfrak{A} : P \models S$ and $\mathfrak{A} : S \vdash\cdot Q$. But then $\mathfrak{B} : PR \models SR$ [2.13] and $\mathfrak{B} : SR \vdash\cdot QR$ [2.8], whence $\mathfrak{B} : PR \models\cdot QR$. Q.E.D.

2.15. *If* $\mathfrak{A} P \models Q\rceil$, *then* $\mathfrak{B} : PR \models QR\rceil$.

Indeed, let $\mathfrak{A} : P \models Q\rceil$. This means that $\mathfrak{A} : P \models Q$ and $\mathfrak{A} : Q\rceil$. But then $\mathfrak{B} : PR \models QR$ [2.13] and $\mathfrak{B} : QR\rceil$ [2.9]; consequently, $\mathfrak{B} : PR \models QR\rceil$. Q.E.D.

2.16. *If* $\mathfrak{B} : PR \models S$, *then there exists a word* T *such that* $S \fallingdotseq TR$ *and* $\mathfrak{A} : P \models T$.

Let us hold P and R fixed and prove this assertion by the method of induction on the steps of the algorithm $\mathfrak{B}$'s performance. To this end, consider the property F of words S in the alphabet A, expressed by the condition

(12) $\ll$there exists a word T in A such that $S \fallingdotseq TR$ and $\mathfrak{A} : P \models T \gg$.

We shall show that every word S, such that $\mathfrak{B} : PR \models S$, possesses this property.

In fact, the word PR possesses the property F, since one may take P as T in (12). Now let S and Q be words for which

(13) $$\mathfrak{B} : PR \models S \vdash Q$$

and S possesses the property F. We shall show that then Q also possesses it. Indeed, there exists a T such that $S \fallingdotseq TR$ and $\mathfrak{A} : P \models T$. But then there exists a U such that $Q \fallingdotseq UR$ and $\mathfrak{A} : T \vdash U$ [(13), 2.10]. Consequently, $Q \fallingdotseq UR$ and $\mathfrak{A} : P \models U$. By the same token, Lemma 2.16 is proven.

2.17. *If* $\mathfrak{B} : PR \models\cdot S$, *then the algorithm* $\mathfrak{A}$ *is applicable to* P.

Indeed, let $\mathfrak{B} : PR \models\cdot S$. Then there exists a word T such that $\mathfrak{B} : PR \models T \vdash\cdot S$. But then there exists a word U such that $T \rightleftharpoons UR$ and $\mathfrak{A} : P \models U$ [2.16]. If there would exist a word V such that $\mathfrak{A} : U \vdash V$, then we would have $\mathfrak{B} : UR \vdash VR$ [2.7], which contradicts $\mathfrak{B} : UR \vdash\cdot S$. Analogously, $\mathfrak{A} : U\urcorner$ is impossible [2.9]. Consequently, we have $\mathfrak{A} : U \vdash\cdot W$ for some W. But then $\mathfrak{A} : P \models\cdot W$ and, therefore, $!\mathfrak{A}(P)$. Q.E.D.

The following lemma is proven analogously:

2.18. *If* $\mathfrak{B} : PR \models S\urcorner$, *then the algorithm* $\mathfrak{A}$ *is applicable to* P.

Let us now prove

2.19. *If the algorithm* $\mathfrak{A}$ *is applicable to* P, *then*

$$\mathfrak{B}(PR) \rightleftharpoons \mathfrak{A}(P)R.$$

Assume, in fact, that the algorithm $\mathfrak{A}$ is applicable to P and that the following equality holds:

$$\mathfrak{A}(P) \rightleftharpoons Q. \tag{14}$$

Then $\mathfrak{A} : P \models\cdot Q$ or $\mathfrak{A} : P \models Q\urcorner$. In the first case, $\mathfrak{B} : PR \models\cdot QR$ [2.14], in the second, $\mathfrak{B} : PR \models QR\urcorner$ [2.15]. In both cases,

$$\begin{aligned} \mathfrak{B}(PR) &\rightleftharpoons QR \\ &\rightleftharpoons \mathfrak{A}(P)R. \end{aligned}$$

Q.E.D.

2.20. *If the algorithm* $\mathfrak{B}$ *is applicable to* PR, *then the algorithm* $\mathfrak{A}$ *is applicable to* P.

Indeed, if the algorithm $\mathfrak{B}$ is applicable to PR, then there exists a word S such that $\mathfrak{B} : PR \models\cdot S$ or $\mathfrak{B} : PR \models S\urcorner$. In both cases, the algorithm $\mathfrak{A}$ is applicable to P [2.17, 2.18]. Q.E.D.

The result of interest to us, namely, the conditional equality (1), immediately follows from Lemmas 2.19 and 2.20. It yields the conditional equality 1(1). With this it is proven that $\mathfrak{B}$ is an extension of $\mathfrak{A}$ to Б.

3. We shall call the extension of an algorithm $\mathfrak{A}$ to an alphabet Б, obtained in the manner just indicated, the *natural extension of this algorithm to the alphabet* Б.

We summarize the preceding in the following fashion:

3.1. *The natural extension of a normal algorithm is a normal algorithm.*

3.2. *The scheme of a normal algorithm's natural extension coincides with the scheme of this algorithm.*

3.3. *Given any normal algorithm $\mathfrak{A}$ in an alphabet* A *and any extension* Б *of this alphabet, there exists a natural extension of the algorithm* $\mathfrak{A}$ *to the alphabet* Б.

3.4. *If* $\mathfrak{B}$ *is the natural extension to an alphabet* Б *of a normal algorithm* $\mathfrak{A}$ *in an alphabet* A, *then the conditional equality* 2(1) *holds.*

4. Let $\mathfrak{A}$ be a normal algorithm in an alphabet A. If Б is a proper extension of this alphabet, then the natural extension of the algorithm $\mathfrak{A}$ to the alphabet Б can turn out to be applicable not only to words in A. It will, for example, be applicable to every word in Б\A if the scheme of the algorithm $\mathfrak{A}$ contains no substitution formulas with an empty left side.

In fact, the left sides of all substitution formulas of the algorithm $\mathfrak{A}$ in this case are non-empty words in the alphabet A and, as such, do not occur in a word in the alphabet Б\A. In view of the fact that $\mathfrak{A}$'s scheme coincides with that of its natural extension $\mathfrak{B}$ to the alphabet Б, this means that

$$\mathfrak{B} : P\rceil$$

for any word P in Б\A. It follows from this, however, that

$$\mathfrak{B}(P) = P$$

for every such word P.

It proves desirable in certain cases to so construct a normal algorithm in the alphabet Б, that it be an extension the algorithm $\mathfrak{A}$ to this alphabet and that, in addition, it be applicable only to words in the alphabet A (and, therefore, to those and only those words to which $\mathfrak{A}$ is applicable). This is achieved in the following manner.

At the top of the scheme of the algorithm $\mathfrak{A}$, we add all possible formulas of the form

$$\xi \to \xi, \tag{1}$$

where ξ is an arbitrary letter of the alphabet Б\A. Let the normal algorithm $\mathfrak{C}$ in Б be given by the scheme obtained in this way. Then, as it is not hard to see, $\mathfrak{C}$ is also an extension of $\mathfrak{A}$ to Б.

In fact, the addition of formulas of the form (1) will obviously not be reflected in the first step of the algorithm's application to a word in the alphabet A, since the left sides of these formulas do not occur in such a word. The following lemmas will therefore hold:

4.1. *If* $\mathfrak{A} : P \vdash Q$, *then* $\mathfrak{C} : P \vdash Q$.

4.2. *If* $\mathfrak{A} : P \vdash\cdot\, Q$, *then* $\mathfrak{C} : P \vdash\cdot\, Q$.

4.3. *If* $\mathfrak{A} : P\rceil$, *where* P *is a word in* A, *then* $\mathfrak{C} : P\rceil$.

On the basis of them, the following lemmas are easily proven with the aid of Lemma §25.4.7:

4.4. *If P is a word in* A *and* $\mathfrak{C} : P \vdash Q$, *then* $\mathfrak{A} : P \vdash Q$ *and* Q *is a word in* A.

4.5. *If P is a word in* A *and* $\mathfrak{C} : P \vdash\cdot Q$, *then* $\mathfrak{A} : P \vdash\cdot Q$ *and* Q *is a word in* A.

4.6. *If P is a word in* A *and* $\mathfrak{C} : P\rceil$, *then* $\mathfrak{A} : P\rceil$.

Further, the following lemmas are proven:

4.7. *If* $\mathfrak{A} : P \models\cdot Q$, *then* $\mathfrak{C} : P \models\cdot Q$ [4.1, 4.2].

4.8. *If* $\mathfrak{A} : P \models Q\rceil$, *then* $\mathfrak{C} : P \models Q\rceil$ [4.1, 4.3].

4.9. *If* $\mathfrak{A}(P) \asymp Q$, *then* $\mathfrak{C}(P) \asymp Q$ [4.7, 4.8].

4.10. *If the algorithm* $\mathfrak{A}$ *is applicable to the word* P, *then the algorithm* $\mathfrak{C}$ *is also applicable to* P *and* $\mathfrak{A}(P) \asymp \mathfrak{C}(P)$ [4.9].

4.11. *If P is a word in* A *and* $\mathfrak{C} : P \models\cdot Q$, *then* $\mathfrak{A} : P \models\cdot Q$ [4.4, 4.5].

4.12. *If P is a word in* A *and* $\mathfrak{C} : P \models Q\rceil$, *then* $\mathfrak{A} : P \models Q\rceil$ [4.4, 4.6].

4.13. *If P is a word in* A *and* $\mathfrak{C}(P) \asymp Q$, *then* $\mathfrak{A}(P) \asymp Q$ [4.11, 4.12].

4.14. *If the algorithm* $\mathfrak{C}$ *is applicable to a word* P *in the alphabet* A, *then the algorithm* $\mathfrak{A}$ *is also applicable to* P *and* $\mathfrak{A}(P) \asymp \mathfrak{C}(P)$ [4.13].

From Lemmas 4.10 and 4.14 follows the conditional equality

(2) $$\mathfrak{A}(P) \simeq \mathfrak{C}(P) \quad (P \text{ is a word in A}),$$

meaning that $\mathfrak{C}$ is an extension of $\mathfrak{A}$ to Б.

Let us attempt, on the other hand, to apply the algorithm $\mathfrak{C}$ to some word P in the alphabet Б, which is not a word in the alphabet A, i.e., which contains occurrences of letters of the alphabet Б\A. At least one of the left sides of the formulas (1), heading the algorithm $\mathfrak{C}$'s scheme, occurs in P. Let

(3) $$\eta \rightarrow \eta$$

be the first of those formulas (1) whose left sides occur in P. Then (3) will obviously also be the first substitution formula of the algorithm $\mathfrak{C}$, acting on P. When applied to the word P, the algorithm $\mathfrak{C}$ prescribes the substitution this formula's right side in place of its left side's first occurrence in P, which, of course, again yields P. Since the applied formula is simple, we have

$$\mathfrak{C} : P \vdash P,$$

from which follows the impossibility of there being an end to the process of applying the algorithm $\mathfrak{C}$ to the word P. Consequently, this process is inapplicable to P.

We see, therefore, that our extension of the algorithm is trivial: it is applicable to those and only those words to which $\mathfrak{A}$ is applicable. The side condition, P is a word in A, may be deleted from the conditional equality (2). Indeed, if P is not a word in A, then neither the symbol $\mathfrak{A}(P)$ nor the symbol $\mathfrak{C}(P)$ makes sense.

5. We shall call the extension of the normal algorithm $\mathfrak{A}$ to the normal algorithm Б, obtained as was just described, a *formal extension of this algorithm to* Б. In the case when the alphabet Б$\setminus$A contains more than one letter, we shall obtain different formal extensions of $\mathfrak{A}$ to Б if we choose different orders for writing down the formulas (1). However, the differences between them will be of a purely external nature: the results of the performance of different formal extensions of $\mathfrak{A}$ to Б will always coincide for one and the same initial word.

The following assertions summarize the preceding:

5.1. *Every formal extension of a normal algorithm is a normal algorithm.*

5.2. *The scheme of every formal extension of a normal algorithm $\mathfrak{A}$ is obtained from this algorithm's scheme by adding substitution formulas of the form* 4(1) *to the top, where ξ runs through the letters of the formal extension's alphabet, not belonging to the alphabet of the algorithm $\mathfrak{A}$.*

5.3. *Given any normal algorithm $\mathfrak{A}$ in an alphabet* A *and any extension* Б *of this alphabet, there exists a formal extension of the algorithm $\mathfrak{A}$ to the alphabet* Б.

4.4. *For every formal extension $\mathfrak{C}$ of a normal algorithm $\mathfrak{A}$, the conditional equality*

$$\mathfrak{A}(P) \simeq \mathfrak{C}(P)$$

holds.

The methods just presented permit us to "reduce", in a certain sense, any two normal algorithms "to a common denominator". Indeed, let $\mathfrak{A}$ and $\mathfrak{C}$ be normal algorithms in the alphabets A and Б, respectively. We construct the alphabet A∪Б, which is an extension of both the alphabets A and Б, and formally extend the algorithms $\mathfrak{A}$ and $\mathfrak{B}$ with alphabets A and Б, respectively, to the alphabet A∪Б. The algorithms so obtained are algorithms in one and the same alphabet A∪Б. But as for the results of their performance, the algorithms so obtained do not differ from the original ones at all in this respect.

§36 Closure of an Algorithm

1. We shall say that a normal algorithm is closed if its scheme contains a formula with an empty left side.

1.1. *If a normal algorithm $\mathfrak{A}$ in an alphabet* A *is closed, then $\mathfrak{A}$ acts on every word in* A.

In fact, a closed algorithm's scheme contains a formula with an empty left side, acting on every word in the algorithm's alphabet.

1.2. *If a normal algorithm $\mathfrak{A}$ is closed, then a natural breaking off of the process of applying it is impossible.*

This follows from the preceding lemma.

1.3. *If a normal algorithm $\mathfrak{A}$ is closed, then $\mathfrak{A}(P) \eqcirc Q$ if and only if $\mathfrak{A} : P \models\cdot Q$.*

Indeed, if the algorithm $\mathfrak{A}$ is closed, then according to 1.2, $\mathfrak{A}$ can not transform P into Q naturally. According to §25.7, 1.3 follows from this.

2. Let $\mathfrak{A}$ be a normal algorithm in an alphabet A. The normal algorithm in A with the scheme

$$\left\{ \begin{array}{l} \mathfrak{A} \\ \rightarrow\cdot \end{array} \right.$$

will be called the *closure of the algorithm* $\mathfrak{A}$.

We employ the following symbolism here and in the sequel: the symbol for a normal algorithm ($\mathfrak{A}$ in the given case), written after a curly bracket, denotes the scheme of this algorithm, written without its curly bracket.

The scheme of the closure of a normal algorithm $\mathfrak{A}$ is therefore obtained from the scheme $\mathfrak{A}$ by adding the formula

$$\rightarrow\cdot$$

to the bottom.

2.1. *The closure of every normal algorithm is closed.*

This is obvious from the definitions.

We shall denote the closure of the normal algorithm $\mathfrak{A}$ by $\mathfrak{A}^{\cdot}$.

The validity of the following lemmas is easily perceived by comparing the schemes of an algorithm $\mathfrak{A}$ and of its closure:

2.2. *If $\mathfrak{A} : P \vdash Q$, then $\mathfrak{A}^{\cdot} : P \vdash Q$.*

2.3. *If $\mathfrak{A} : P \vdash\cdot Q$, then $\mathfrak{A}^{\cdot} : P \vdash\cdot Q$.*

2.4. *If $\mathfrak{A} : P\rceil$, then $\mathfrak{A}^{\cdot} : P \vdash\cdot P$.*

The following lemmas follow from them

2.5. *If $\mathfrak{A}^{\cdot} : P \vdash Q$, then $\mathfrak{A} : P \vdash Q$* [2.2, 2.3, 2.4, §25.4.7].

2.6. *If $\mathfrak{A}^{\cdot} : P \vdash\cdot Q$, then either $\mathfrak{A} : P \vdash\cdot Q$ or else $\mathfrak{A} : P\rceil$ and $P \eqcirc Q$* [2.2, 2.3, 2.4, §25.4.7].

Using Lemma 2.2, we obtain by induction on the steps of $\mathfrak{A}$'s performance

2.7. *If $\mathfrak{A} : P \models Q$, then $\mathfrak{A}^{\cdot} : P \models Q$.*

Using Lemma 2.5, we obtain by induction on the steps of $\mathfrak{A}$'s performance

2.8. *If $\mathfrak{A}^{\cdot} : P \models Q$, then $\mathfrak{A} : P \models Q$.*

We further have

2.9. *If $\mathfrak{A} : P \models\cdot Q$, then $\mathfrak{A}^{\cdot} : P \models\cdot Q$* [2.7, 2.3].

2.10. *If $\mathfrak{A} : P \models Q\rceil$, then $\mathfrak{A}^{\cdot} : P \models\cdot Q$* [2.7, 2.4].

2.11. *If $\mathfrak{A}^{\cdot} : P \models\cdot Q$, then $\mathfrak{A} : P \models\cdot Q$ or $\mathfrak{A} : P \models Q\rceil$* [2.8, 2.6].

2.12. *If $\mathfrak{A}(P) \eqcirc Q$, then $\mathfrak{A}^{\cdot}(P) \eqcirc Q$* [2.9, 2.10, §25.7].

2.13. *If $\mathfrak{A}^{\cdot}(P) \eqcirc Q$, then $\mathfrak{A}(P) \eqcirc Q$* [2.11, 1.3, 2.1, §25.7].

2.14. $\mathfrak{A}(P) \simeq \mathfrak{A}^{\cdot}(P)$ [2.11, 2.13].

We have therefore proven the following theorem:

2.15. *Every normal algorithm is completely equivalent to its closure with respect to its alphabet.*

By virtue of Theorems 2.15 and 2.1, the study of arbitrary algorithms essentially reduces to the study of closed algorithms, which turns out to be convenient in many cases, in view of Theorem 1.3.

§37 Composition of Algorithms

1. One often has to combine two algorithms in the following manner. Starting with arbitrary initial data, one is ordered to first apply the first algorithm, and then to the result of its performance, the second. This prescription constitutes a new algorithm—the ≪composition≫ of the two given algorithms.

It is natural to ask whether the composition of normalizable algorithms also is a normalizable algorithm. The normalization principle suggests an affirmative answer to this question, and we shall in fact prove the following theorem in this section:

1.1. COMPOSITION THEOREM. *Given any normal algorithms $\mathfrak{A}$ and $\mathfrak{B}$, one can construct a normal algorithm $\mathfrak{C}$ over the union* A *of their alphabets, such that for any word P in* A, *we have the conditional equality*

$$\mathfrak{C}(P) \simeq \mathfrak{B}(\mathfrak{A}(P)). \tag{1}$$

The proof of this theorem[1] makes up the main contents of this section.

2. We shall begin with the special case of the Composition Theorem when both the given algorithms $\mathfrak{A}$ and $\mathfrak{B}$ are normal algorithms in one and the same alphabet A. We shall prove the following theorem:

2.1. *Given any normal algorithms $\mathfrak{A}$ and $\mathfrak{B}$ in an alphabet* A, *one can construct a normal algorithm $\mathfrak{C}$ over* A, *satisfying the condition* 1(1).

In fact, let $\mathfrak{A}$ and $\mathfrak{B}$ be normal algorithms in an alphabet A. To each letter ξ of this alphabet, we assign a new letter, where different new letters are assigned to different letters of A. The letter assigned to ξ will be called this *letter's double* and will be denoted by the symbol $\bar{\xi}$. The doubles of A's letters constitute an *alphabet of doubles*, $\bar{\text{A}}$, containing just as many letters as A and having no letters in common with A.

Further, let α and β denote two additional new letter, distinct from each other, from the letters of the alphabet A and from their doubles.

Let us compose the alphabet Б= $\text{A}\bar{\text{A}}\alpha\beta$. Let us construct the system of formulas $\mathfrak{A}^{\alpha}$ by replacing all dots by letters α in the algorithm $\mathfrak{A}^{\cdot}$'s scheme. Let us construct the system of formulas $\bar{\mathfrak{B}}^{\beta}_{\alpha}$ by replacing all letters of A by their doubles and all dots by β's in $\mathfrak{B}^{\cdot}$'s scheme, followed by the replacement of all formulas of the form

$$\rightarrow \text{B} \tag{1}$$

[1]See the commentary to the proof in Part 5.

by the formulas

$$\alpha \to \alpha \mathrm{B} \tag{2}$$

with the same B.

Let us present the normal algorithm $\mathfrak{C}$ is the alphabet Б by the scheme which can be abbreviated as follows:

$$\left\{\begin{array}{l} \xi\alpha \to \alpha\xi \\ \alpha\xi \to \alpha\bar{\xi} \\ \bar{\xi}\eta \to \bar{\xi}\bar{\eta} \\ \bar{\xi}\beta \to \beta\bar{\xi} \\ \beta\bar{\xi} \to \beta\xi \\ \xi\bar{\eta} \to \xi\eta \\ \alpha\beta \to\cdot \\ \mathfrak{B}^{\beta}_{\alpha} \\ \mathfrak{A}^{\alpha}, \end{array}\right. \tag{3}$$

where ξ and η run through the alphabet A.

We shall show that it satisfies the condition 1(1).

For this, note first of all that between the scheme of the algorithm $\mathfrak{A}^{\cdot}$ and the system of formulas $\mathfrak{A}^{\alpha}$ there is a natural one-to-one correspondence, under which a formula of the system $\mathfrak{A}^{\alpha}$, corresponding to a given formula F of $\mathfrak{A}^{\cdot}$'s scheme, is obtained from F in the same way that the entire system $\mathfrak{A}^{\alpha}$ is obtained from $\mathfrak{A}^{\cdot}$'s scheme—by replacing dots by letters α. Bearing in mind that simple formulas contain no dots at all, while final formulas contain exactly one occurrence each of a dot—between the arrow and the right side—we convince ourselves of the truth of the following assertions:

2.2. *All the formulas of the system $\mathfrak{A}^{\alpha}$ are simple.*

2.3. *Every formula of the system $\mathfrak{A}^{\alpha}$, corresponding to a simple formula of $\mathfrak{A}^{\cdot}$'s scheme, coincides with this formula.*

2.4. *The left side of every formula of the system $\mathfrak{A}^{\alpha}$ coincides with the left side of the corresponding formula of $\mathfrak{A}^{\cdot}$'s scheme.*

2.5. *The right side of every formula of the system $\mathfrak{A}^{\alpha}$, corresponding to a final formula F of $\mathfrak{A}^{\cdot}$'s scheme, is obtained from the right side of F by adding the letter α to the left.*

Taking into account the fact that the algorithm $\mathfrak{A}^{\cdot}$'s scheme contains the formula $\ll\to\cdot\gg$, we obtain

2.6. *The system $\mathfrak{A}^{\alpha}$ contains a formula with an empty left side.*

Bearing in mind, then, the scheme (3) of the algorithm $\mathfrak{C}$, we conclude from this that

2.7. *The algorithm $\mathfrak{C}$ is closed.*

Let us agree, then, to call the word obtained from P by replacing each letter by its double, the double of the word P in the alphabet A. We shall denote the double of the word P by the symbol

$$[P^{-}.$$

This is a word in the alphabet $\bar{\text{A}}$. It is obvious that

$$(4) \qquad [\Lambda^{-} \doteq \Lambda,$$

$$[\xi^{-} \doteq \bar{\xi}$$

$$(5) \qquad [PQ^{-} \doteq [P^{-}[Q^{-}$$

for any words P and Q in A and for any letter ξ of this alphabet.

The next lemma is also obvious:

2.8. *Every word in the alphabet $\bar{\text{A}}$ is a double of one and only one word in* A.

Let us now note that between the algorithm $\mathfrak{B}^{\cdot}$'s scheme and the system of formulas $\mathfrak{B}_{\alpha}^{\beta}$, there is also a natural one-to-one correspondence. The formula of the system $\mathfrak{B}_{\alpha}^{\beta}$, corresponding to a given formula F of $\mathfrak{B}^{\cdot}$'s scheme, is obtained from F by replacing a dot by β, the letters of A by their doubles and, finally, if a formula of the form (1) has been obtained in this way, replacing it by the formula (2). Bearing in mind that the right and left sides of the formulas of $\mathfrak{B}^{\cdot}$'s scheme are words in the alphabet A, we convince ourselves of the truth of the following assertions:

2.9. *All the formulas of the system $\mathfrak{B}_{\alpha}^{\beta}$ are simple.*

2.10. *The left side of every formula of the system $\mathfrak{B}_{\alpha}^{\beta}$ is the double of the left side of the corresponding formula of $\mathfrak{B}^{\cdot}$'s scheme or is equal to α. The latter holds if and only if the left side of the corresponding formula of $\mathfrak{B}^{\cdot}$'s scheme is empty.*

2.11. *The right side of every formula of the system $\mathfrak{B}_{\alpha}^{\beta}$, corresponding to a simple formula F of $\mathfrak{B}^{\cdot}$'s scheme, is the double of the right side of the formula F or is obtained from this double by adding α to the left. The latter holds if and only if the left side of the formula F is empty.*

2.12. *The right side of every formula of the system $\mathfrak{B}_{\alpha}^{\beta}$, corresponding to a final formula F of $\mathfrak{B}^{\cdot}$'s scheme, is obtained from the double of F's right side by adding the word β or the word $\alpha\beta$ to the left. The latter holds if and only if the left side of the formula F is empty.*

Further, we have the following lemma:

2.13. *The left side of every formula of the system $\mathfrak{B}_{\alpha}^{\beta}$ contains a letter of the alphabet $\bar{\text{A}}_{\alpha}$.*

This follows immediately from 2.10.

We shall now demonstrate that when applied to any word P in the alphabet A, the algorithm $\mathfrak{C}$ performs in the following manner:

1st stage. $\mathfrak{C}$ "does the work of the algorithm $\mathfrak{A}$". All transformations required by the algorithm $\mathfrak{A}$ are carried out in succession. In order to do this, formulas of the system $\mathfrak{A}^{\alpha}$ are applied. The stage comes to an end if the algorithm $\mathfrak{A}$ is applicable to P, in which case, however, we have as a result, not $\mathfrak{A}(P)$, but the word obtained from $\mathfrak{A}(P)$ by means of an insertion of the letter α, "jumping in" at the last step of the 1st stage.

2nd stage. The letter α "runs" to the left until it reaches the beginning of the word. Formulas represented by the 1st line of the scheme (3) are applied. As a result, $\alpha\mathfrak{A}(P)$ is obtained.

3rd stage. The letter α "converts" its neighbor, the first letter of the word $\mathfrak{A}(P)$, into this letter's double. One of the formulas represented by the 2nd line of the scheme (3) is applied. (This stage drops out if $\mathfrak{A}(P) \rightleftharpoons \Lambda$.)

4th stage. The process of replacing letters of the alphabet A by their doubles spreads from left to right. Formulas represented by the 3rd line of the scheme (3) are applied. As a result, the word $\alpha[\mathfrak{A}(P)^{-}$ is obtained. (This stage drops out if $[\mathfrak{A}(P)^{\delta} \leqslant 1$.)

5th stage. $\mathfrak{C}$ "does the work of the algorithm $\mathfrak{B}$", but in the alphabet A of doubles and with the letter α, attached on the left to the word being transformed. Formulas of the system $\mathfrak{B}_{\alpha}^{\beta}$ are applied. The stage comes to an end if the algorithm $\mathfrak{B}$ is applicable to the word $\mathfrak{A}(P)$, in which case we have as a result the word obtained from $[\mathfrak{B}(\mathfrak{A}(P))^{-}$ by means of an insertion of the letter β and the addition of the letter α on the left.

6th stage. The letter β "runs" to the left until it reaches the letter α. Formulas represented by the 4th line of the scheme (3) are applied. As a result, $\alpha\beta[\mathfrak{B}(\mathfrak{A}(P))^{-}$ is obtained.

7th stage. The letter β "converts" its neighbor, the first letter of the word $[\mathfrak{B}(\mathfrak{A}(P))^{-}$, into the first letter of the word $\mathfrak{B}(\mathfrak{A}(P))$. One of the formulas represented by the 5th line of the scheme (3) is applied. (This stage drops out if $\mathfrak{B}(\mathfrak{A}(P)) \rightleftharpoons \Lambda$.)

8th stage. The process of replacing the doubles of A's letters by these letters, themselves, spreads from left to right. Formulas represented by the 6th line of the scheme (3) are applied. As a result, the word $\alpha\beta\mathfrak{B}(\mathfrak{A}(P))$ is obtained. (This stage drops out if $[\mathfrak{B}(\mathfrak{A}(P))^{\delta} \leqslant 1$.)

9th and last stage. The word $\alpha\beta$ disappears. The 7th line of the scheme (3), which is a final formula, is applied. The entire process ends on this, and its result is the word $\mathfrak{B}(\mathfrak{A}(P))$.

The reason for introducing the doubles of the alphabet A's letters is clear from this description of the algorithm $\mathfrak{C}$'s performance. They were needed in order that, having translated $\mathfrak{B}$''s scheme into the alphabet of doubles, we might be able to unite the scheme of both given algorithms into one scheme in such a way that they did not interfere with each other. Our next lemmas constitute a precise mathematical recounting of the intuitive description just given for the process of applying the algorithm $\mathfrak{C}$.

2.14. *If* $\mathfrak{A}^{\cdot} : P \vdash Q$, *then* $\mathfrak{C} : P \vdash Q$.

Indeed, let $\mathfrak{A}^{\cdot} : P \vdash Q$. The left sides of $\mathfrak{C}$'s substitution formulas represented by the first seven lines of (3) do not occur in P, since each of them contains a letter of the alphabet $\bar{\text{A}}\alpha$. Neither do the left sides of the system $\mathfrak{B}_{\alpha}^{\beta}$'s formulas, also containing this alphabet's letters [2.13], occur in P.

Lower down in the scheme (3) are the formulas of the system $\mathfrak{A}^{\alpha}$, listed in the same order as the corresponding formulas of the algorithm $\mathfrak{A}^{\cdot}$'s scheme. Consider the formula F of $\mathfrak{A}^{\cdot}$'s scheme, active on P. This formula is simple, since $\mathfrak{A}^{\cdot} : P \vdash Q$. According to 2.3, F also occurs in the system $\mathfrak{A}^{\alpha}$ as a formula corresponding to itself. Above it in the scheme of the algorithm $\mathfrak{A}^{\cdot}$ are formulas whose left sides do not occur in P. According to 2.4, it follows from this that neither do the left sides of the formulas above F in the system $\mathfrak{A}^{\alpha}$ occur in P. Bearing in mind that the left side of F occurs in P, while the left sides of $\mathfrak{C}$'s substitution formulas represented by the first eight lines of (3) do not, we conclude that F is the first substitution formula of $\mathfrak{C}$ with a left side occurring in P. When applied to the word P, the algorithm $\mathfrak{C}$ prescribes, therefore, that F's right side be substituted for its left side's first occurrence in P, i.e., it gives the very same order as the algorithm $\mathfrak{A}^{\cdot}$. Consequently, $\mathfrak{C} : P \vdash Q$. Q.E.D.

2.15. *If* $\mathfrak{A}^{\cdot} : P \vdash\cdot Q$, *then there exist words* R *and* S, *such that*

(6) $$Q \coloneqq RS,$$

(7) $$\mathfrak{C} : P \vdash R\alpha S.$$

In fact, let $\mathfrak{A}^{\cdot} : P \vdash\cdot Q$. As in the proof of the preceding lemma, we then convince ourselves that the left sides of the algorithm $\mathfrak{C}$'s substitution formulas represented by the first eight lines of the scheme (3) do not occur in P.

Consider the formula F of the algorithm $\mathfrak{A}^{\cdot}$'s scheme, which is active on the word P. This formula is final. Consequently, it has the form

$$U \rightarrow\cdot V,$$

where U and V are words in A.

Let G denote the corresponding formula of the system $\mathfrak{A}^{\alpha}$. Then G is the formula

$$U \rightarrow \alpha V.$$

As in the proof of the preceding lemma, we convince ourselves that G is the first substitution formula of $\mathfrak{C}$ with a left side occurring in P.

Let

(8) $$R * U * T$$

be the first occurrence of U in P. When applied to P, the algorithm $\mathfrak{C}$ prescribes the substitution of G's right side, i.e., αV, in place of this occurrence. Since G is a simple formula, we have

$$\mathfrak{C} : P \vdash R\alpha VT.$$

On the other hand, when applied to P, the algorithm $\mathfrak{A}^{\cdot}$ prescribes the substitution of F's right side, i.e., V, in place of the occurrence (8). Since $\mathfrak{A}^{\cdot} : P \vdash\cdot Q$, we have

$$Q \coloneqq RVT.$$

Putting $S \coloneqq VT$, we shall, consequently, have (6) and (7). Q.E.D.

2.16. $\mathfrak{C} : R\alpha S \models \alpha RS$.

Here it is convenient to make use of a device already applied in §§31 and 32. We first prove, by left induction on R, the more general assertion

$$\mathfrak{C} : QR\alpha S \models Q\alpha RS$$

and then set $Q \eqcirc \Lambda$ in it.

2.17. *If* $\mathfrak{A}^{\cdot} : P \vdash\cdot Q$, *then* $\mathfrak{C} : P \models \alpha Q$.

Indeed, let $\mathfrak{A}^{\cdot} : P \vdash\cdot Q$. Then there exist words R and S satisfying the conditions (6) and (7) [2.15]. Q is a word in A, since $\mathfrak{A}^{\cdot}$ is an algorithm in A and $\mathfrak{A}^{\cdot} : P \vdash\cdot Q$. Hence, according to (6), R and S are also words in A, and Lemma 2.16 is applicable. We have, therefore,

$$\begin{aligned} \mathfrak{C} : P &\vdash R\alpha S \quad [(7)] \\ &\models \alpha RS \quad [2.16] \\ &\eqcirc \alpha Q \quad [(6)]. \end{aligned}$$

Consequently, $\mathfrak{C} : P \models \alpha Q$. Q.E.D.

2.18. *If* $\mathfrak{A}^{\cdot} : P \models Q$, *then* $\mathfrak{C} : P \models Q$.

This is proven by induction on the steps of the performance of the algorithm $\mathfrak{A}^{\cdot}$, using Lemma 2.14.

2.19. *If* $\mathfrak{A}^{\cdot} : P \models\cdot Q$, *then* $\mathfrak{C} : P \models \alpha Q$.

Indeed, let $\mathfrak{A}^{\cdot} : P \models\cdot Q$. Then there exists a word R such that

(9) $$\mathfrak{A}^{\cdot} : P \models R,$$

(10) $$\mathfrak{A}^{\cdot} : R \vdash\cdot Q.$$

We therefore have

$$\begin{aligned} \mathfrak{A} : P &\models R \quad [(9), 2.18] \\ &\models \alpha Q \quad [(10), 2.17]. \end{aligned}$$

Consequently, $\mathfrak{C} : P \models \alpha Q$. Q.E.D.

2.20. *If P and Q are words in the alphabet* A, *such that* $\mathfrak{C} : P \vdash Q$, *then* $\mathfrak{A}^{\cdot} : P \vdash Q$.

In fact, let P and Q be words in A, and let

(11) $$\mathfrak{C} : P \vdash Q.$$

By virtue of the fact that the algorithm $\mathfrak{A}^{\cdot}$ is closed [§36.2.1], there exists a word T such that $\mathfrak{A}^{\cdot} : P \vdash T$ or $\mathfrak{A}^{\cdot} : P \vdash\cdot T$.

In the second case, there would be, however, words R and S such that

(12) $$\mathfrak{C} : P \vdash R\alpha S \quad [2.15].$$

We would then have $Q \leftrightharpoons R\alpha S$ [(11), (12)], which is impossible, since Q is a word in A. Therefore,

(13) $$\mathfrak{A}^{\cdot} : P \vdash T,$$
(14) $$\mathfrak{C} : P \vdash T \quad [(13), 2.14],$$
(15) $$T \leftrightharpoons Q \quad [(11), (14)],$$
$$\mathfrak{A}^{\cdot} : P \vdash Q \quad [(13), (15)].$$

Q.E.D.

2.21. *If P is a word in* A, *then* $\mathfrak{C}$ *simply converts P into some word.*

Indeed, let P be a word in A. Arguing as in the proof of the preceding lemma, we convince ourselves of the existence of a word T such that $\mathfrak{A}^{\cdot} : P \vdash T$ or $\mathfrak{A}^{\cdot} : P \vdash\cdot T$. In both cases, $\mathfrak{C}$ simply converts P into some word [2.14, 2.15].

2.22. *If the algorithm* $\mathfrak{C}$ *is applicable to a word P in* A, *then the algorithm* $\mathfrak{A}^{\cdot}$ *is also applicable to P.*

In fact, let the algorithm $\mathfrak{C}$ be applicable to a word P, and let $\mathfrak{C}(P) \leftrightharpoons U$. Then by virtue of the fact that the algorithm $\mathfrak{C}$ is closed [2.7], we have $\mathfrak{C} : P \models\cdot U$ [§36.1.3]. Hence there exists a word V such that $\mathfrak{C} : P \models V \vdash\cdot U$ and, therefore, there exists a converting system X of $\mathfrak{C}$'s scheme, connecting P to V. It is easy to see that V is not a word in A, since $\mathfrak{C} : V \vdash\cdot U$ [2.21]. Consequently, the first term of X is a word in A, but its last term is not. The arithmetical predicate ≪the term of the system X whose number is N is not a word in A≫ is obviously decidable. Therefore, on the basis of The Least Number Theorem [§21.2.1], there exists a term W of the system X, such that W is not a word in A, but all terms of X with smaller number are. Then the system X has the form YWZ, where Y is a converting system of $\mathfrak{C}$'s scheme, all of whose terms are words in A. The first term of Y is the word P. Denote the last term of Y by Q. Then $\mathfrak{C} : Q \vdash W$. It is clear that Y is a converting system for $\mathfrak{A}^{\cdot}$'s scheme too [2.20]. Hence $\mathfrak{A}^{\cdot} : P \models Q$. There exists an R such that $\mathfrak{A}^{\cdot} : Q \vdash R$ or $\mathfrak{A}^{\cdot} : Q \vdash\cdot R$. The first case, however, is impossible, since we would then have $\mathfrak{C} : Q \vdash R$ [2.14] and $R \leftrightharpoons W$. Nevertheless, R is a word in A, but W is not. Consequently, $\mathfrak{A}^{\cdot} : Q \vdash\cdot R$. But then $\mathfrak{A}^{\cdot} : P \models Q \vdash\cdot R$ and, therefore, $\mathfrak{A}^{\cdot}$ is applicable to P. Q.E.D.

Let us now agree to the following terminology. If R, T and S are words in A and $T \not\leftrightharpoons \Lambda$, then the occurrence

$$\alpha[R^{-} * [T^{-} * [S^{-}$$

in the alphabet $\bar{\mathrm{A}}\alpha$ will be called the *image of the occurrence*

$$R * T * S$$

in the alphabet A. Moreover, the occurrence

$$*\alpha * [S^-$$

in the alphabet $\bar{A}\alpha$ will be called the *image of the occurrence*

$$* * S$$

in the alphabet A. By the same token, the image is defined for the occurrences in A with a non-empty base and also for the first occurrences of the empty word in the words in this alphabet.

2.23. *The image of an occurrence in a word P is an occurrence in the word $\alpha[P^-$.*

2.24. *The image of an occurrence of a non-empty word T is an occurrence of the word $[T^-$.*

This follows immediately from the definition of an image.

2.25. *If P is a word in the alphabet A, then every occurrence of the double of a non-empty word T is the alphabet A in the word $\alpha[P^-$ is the image of some occurrence of the word T in the word P.*

Indeed, let P be a word in A and let

$$U * [T^- * W \tag{16}$$

be an occurrence of the double of a non-empty word T in the alphabet A in the word $\alpha[P^-$. Then

$$U[T^- W \equiv \alpha[P^-, \tag{17}$$

from which it follows that the non-empty word $U[T^-$ begins with the letter α. Since $[T^-$ is a word in the alphabet A^-, not containing the letter α, U begins with this letter. Therefore, there exists a word X such that

$$U \equiv \alpha X. \tag{18}$$

We now have

$$\alpha X[T^- W \equiv \alpha[P^- \quad [(17), (18)], \tag{19}$$

$$X[T^- W \equiv [P^- \quad [(19)]. \tag{20}$$

Since $[P^-$ is a word in $\bar{A}$, X and W are also words in $\bar{A}$ [(20)]. Hence there exist words R and S in A, such that

$$X \equiv [R^-, \tag{21}$$

$$W \equiv [S^-. \tag{22}$$

Consequently,

$$U * [T^- * W \eqcirc \alpha[R^- * [T^- * [S^- \quad [(18), (21), (22)],$$

but this means that the occurrence (16) under consideration is the image of the occurrence $R * T * S$ of the non-empty word T.

Finally,

$$(23) \qquad [R^-[T^-[S^- \eqcirc [P^- \quad [(20), (21), (22)],$$

$$(24) \qquad [RTS^- \eqcirc [P^- \quad [(23), (5)],$$

$$RTS \eqcirc P \quad [(24), 2.8],$$

i.e., $R*T*S$ is an occurrence in P. Therefore, the occurrence under consideration is the image of some occurrence of T in P, which is just what remained to be proven.

2.26. *A non-empty word T in the alphabet* A *occurs in a word P in this alphabet if and only if its double occurs in the word $\alpha[P^-$.*

In fact, if T occurs in P, then there exists an occurrence of T in P. Since $T \neq \Lambda$, the image of this occurrence is an occurrence of $[T^-$ in $\alpha[P^-$ [2.23, 2.24]. Consequently, T's double occurs in $\alpha[P^-$.

Conversely, if $[T^-$ occurs in $\alpha[P^-$, then there exists an occurrence of $[T^-$ in $\alpha[P^-$. According to 2.25, this occurrence is the image of some occurrence of T in P. Consequently, T occurs in P. Q.E.D.

2.27. *If P is a word in the alphabet* A *and an occurrence K of a non-empty word T in P precedes an occurrence L of the word T in P, then the image of K precedes the image of L.*

Indeed, let R and U denote the left wings of the occurrences K and L, respectively. If K precedes L, then R is a proper initial segment of U. Hence there exists a non-empty X such that

$$(25) \qquad U \eqcirc RX.$$

We have

$$\alpha[U^- \eqcirc \alpha[R^-[X^- \quad [(25), (5)],$$

where $[X^- \neq \Lambda$. Consequently, $\alpha[R^-$ is a proper initial segment of $\alpha[U^-$. But by the definition of an image, $\alpha[R^-$ and $\alpha[U^-$ are the left wings of the occurrences K and L, respectively. Consequently, K's image precedes L's image. Q.E.D.

2.28. *If a non-empty word T occurs in a word P in the alphabet* A, *then the image of the first occurrence of T in P is the first occurrence of $[T^-$ in $\alpha[P^-$.*

In fact, let K be the first occurrence of T in P. Then the image of K is an occurrence of $[T^-$ in $\alpha[P^-$ [2.23, 2.24]. Now consider an arbitrary occurrence M, distinct from K's image, of $[T^-$ in $\alpha[P^-$. M is the image of some occurrence L of T in P [2.25]. $L \neq K$, since the image of L differs from that of K. Since K is the first occurrence of T in P, K precedes L. Therefore, K's image precedes L's image [2.27], i.e., M. Consequently, K's image precedes every occurrence, different from itself, of $[T^-$ in $\alpha[P^-$, i.e., is the first occurrence of $[T^-$ in $\alpha[P^-$. Q.E.D.

2.29. *Let F be a substitution formula of the algorithm $\mathfrak{B}^{\cdot}$, G, the corresponding formula of the system $\mathfrak{B}^{\beta}_{\alpha}$, and let P be a word in* A. *Then the left side of F occurs in P if and only if the left side of G occurs in $\alpha[P^{-}$.*

In fact, if the left side of F is a non-empty word, then the left side of G is this word's double [2.10], and the lemma's conclusion follows from 2.26. But if the left side of F is the empty word, then the left side of G is the word α [2.10]. The lemma's conclusion is also true in this case, since Λ occurs in P and α occurs in $\alpha[P^{-}$.

2.30. *If, under the hypothesis of the preceding lemma, the left side of F occurs in P, then the first occurrence of G's left side in $\alpha[P^{-}$ is the image of the first occurrence of F's left side in P.*

Indeed, let the hypothesis of the preceding lemma be satisfied, and let F's left side occur in P. If the left side of F is a non-empty word, then the left side of G is this word's double, and the lemma's conclusion follows from 2.28. But if F's left side is the empty word, then G's left side is the word α [2.10]. The first occurrence of F's left side in P is $* * P$, while the first (and only) occurrence of G's left side in $\alpha[P^{-}$ is $*\alpha * [P^{-}$. Since according to the definition, $*\alpha * [P^{-}$ is the image of $* * P$, the lemma is proven.

We shall now prove several lemmas which relate the performance of the algorithm $\mathfrak{C}$ to that of the algorithm $\mathfrak{B}$.

2.31. *If $\mathfrak{B}^{\cdot} : P \vdash Q$, then $\mathfrak{C} : \alpha[P^{-} \vdash \alpha[Q^{-}$.*

In fact, let $\mathfrak{B}^{\cdot} : P \vdash Q$. Among $\mathfrak{B}^{\cdot}$'s substitution formulas, there are formulas whose left side occurs in P. Let F be the first of them. It is simple, since $\mathfrak{B}^{\cdot} : P \vdash Q$. Let U and V denote the left and right sides, respectively, of F. Let $R * U * S$ be the first occurrence of U in P.

Since the algorithm $\mathfrak{B}^{\cdot}$ prescribes the substitution of V in place of $R * U * S$ when applied to P, and $\mathfrak{B}^{\cdot} : P \vdash Q$, we have

$$Q \leftrightharpoons RVS. \tag{26}$$

Further, let G be the formula of the system $\mathfrak{B}^{\beta}_{\alpha}$, corresponding to F. Its left side occurs in $\alpha[P^{-}$, since F's left side occurs in P [2.29]. On the other hand, the left sides of the formulas of the system $\mathfrak{B}^{\beta}_{\alpha}$, which are above G, do not occur in $\alpha[P^{-}$, since these formulas correspond to formulas of the scheme $\mathfrak{B}^{\cdot}$, which are above F, but the left sides of the latter do not occur in P [2.29]. Therefore G is the first formula of the system $\mathfrak{B}^{\beta}_{\alpha}$, whose left side occurs in $\alpha[P^{-}$.

Neither do the left sides of the formulas represented by the first seven lines of the scheme (3) occur in the word $\alpha[P^{-}$, since each of these left sides contains a letter of the alphabet $\text{A}\beta$, not occurring in $\alpha[P^{-}$. Consequently, G is the first substitution formula of the algorithm $\mathfrak{C}$, whose left side occurs in $\alpha[P^{-}$. When applied to this word, the algorithm $\mathfrak{C}$ prescribes, therefore, the substitution of G's right side in place of the first occurrence of its left side in $\alpha[P^{-}$. This occurrence is the image of the first occurrence of the left side of F in P [2.30], i.e., the occurrence $R * U * S$. Therefore, when applied to the word $\alpha[P^{-}$, the algorithm $\mathfrak{C}$ prescribes the substitution of G's right side in place of the image of the occurrence $P * U * S$. In what follows, we shall distinguish two cases: $U \neq \Lambda$ and $U \eqcirc \Lambda$.

a) $U \neq \Lambda$. In this case, the image of $R * U * S$ is the occurrence $\alpha[R^- * [U^- * [S^-$, and the right side of the formula G is the double of F's right side [2.11], i.e., the word $[V^-$. Bearing in mind that G is simple [2.9], we have

$$\begin{aligned} \mathfrak{C} : \alpha[P^- &\vdash \alpha[R^-[V^-[S^- \\ &\eqcirc \alpha[RVS^- \qquad [(5)] \\ &\eqcirc \alpha[Q^- \qquad [(26)]. \end{aligned}$$

b) $U \eqcirc \Lambda$. In this case,

$$R \eqcirc \Lambda, \tag{27}$$

since $R * U * S*$ is the first occurrence of U in P. The image of $R * U * S$ is therefore the occurrence $*\alpha*[S^-$ and the right side of the formula G is the word $\alpha[V^-$ [2.11]. Bearing in mind that G is simple [2.9], we have

$$\begin{aligned} \mathfrak{C} : \alpha[P^- &\vdash \alpha[V^-[S^- \\ &\eqcirc \alpha[VS^- \qquad [(5)] \\ &\eqcirc \alpha[RVS^- \qquad [(27)] \\ &\eqcirc \alpha[Q^- \qquad [(26)]. \end{aligned}$$

Therefore, $\mathfrak{C} : \alpha[P^- \vdash \alpha[Q^-$ in both cases. Q.E.D.

2.32. *If $\mathfrak{B}^\cdot : P \vdash\cdot Q$, then there exist words R and T such that*

$$Q \eqcirc RT,$$
$$\mathfrak{C} : \alpha[P^- \vdash \alpha[R^-\beta[T^-.$$

Indeed, let $\mathfrak{B}^\cdot : P \vdash\cdot Q$. As in the proof of the preceding lemma, consider the first among the substitution formulas of the algorithm $\mathfrak{B}^\cdot$, having left sides occurring in P. Let F denote this formula. This time F is a final formula, since $\mathfrak{B}^\cdot : P \vdash\cdot Q$. Let U and V denote its left and right sides, respectively. Let $R * U * S$ be the first occurrence of U in P. As in the proof of the preceding lemma, $Q \eqcirc RVS$.

Let G be the formula of the system $\bar{\mathfrak{B}}^\beta_\alpha$, corresponding to F. As in the proof of the preceding lemma, we convince ourselves that when the algorithm $\mathfrak{C}$ is applied to the word $\alpha[P^-$, it prescribes the substitution of the right side of the formula G in place of the image of the occurrence $R * U * S$. We again distinguish two cases: $U \neq \Lambda$ and $U \eqcirc \Lambda$.

a) $U \neq \Lambda$. In this case, the image of $R * U * S$ is the occurrence $\alpha[R^- * [U^- * [S^-$ and the right side of the formula G is the word $\beta[V^-$ [2.12]. Bearing in mind that the formula G is simple [2.9], we have

$$\begin{aligned} \mathfrak{C} : \alpha[P^- &\vdash \alpha[R^-\beta[V^-[S^- \\ &\eqcirc \alpha[R^-\beta[VS^- \qquad [(5)]. \end{aligned}$$

b) $U \leftrightharpoons \Lambda$. In this case, we have the equality (27), since $R * U * S$ is the first occurrence of U in P. The image of the occurrence $R * U * S$ is therefore the occurrence $*\alpha * [S^-$ and the right side of the formula G is the word $\alpha\beta[V^-$ [2.12]. Bearing in mind that the formula G is simple [2.9], we have

$$\mathfrak{C} : \alpha[P^- \vdash \alpha\beta[V^-[S^-$$
$$\leftrightharpoons \alpha\beta[VS^- \qquad [(5)]$$
$$\leftrightharpoons \alpha[R^-\beta[VS^- \qquad [(27), (4)].$$

Therefore

$$\mathfrak{C} : \alpha[P^- \vdash \alpha[R^-\beta[VS^- \tag{28}$$

in both cases. According to (26) and (28), we have $Q \leftrightharpoons RT$ and $\mathfrak{C} : \alpha[P^- \vdash \alpha[R^-\beta[T^-$, where $T \leftrightharpoons VS$, which proves the lemma.

2.33. *If R and S are words in* A, *then*

$$\mathfrak{C} : \alpha[R^-\beta[S^- \models \alpha\beta[RS^-.$$

The proof is the usual one for this sort of assertion. We first note that

$$\mathfrak{C} : \alpha[R^-\bar{\xi}\beta[S^- \vdash [R^-\beta\bar{\xi}[S^-,$$

and then carry out right induction on R.

2.34. *If $\mathfrak{B}^{\cdot} : P \vdash^{\cdot} Q$, then $\mathfrak{C} : \alpha[P^- \models \alpha\beta[Q^-$.*

The proof is based on Lemmas 2.32 and 2.33. It is carried out by analogy with the proof of Lemma 2.17.

2.35. *If $\mathfrak{B}^{\cdot} : P \models Q$, then $\mathfrak{C} : \alpha[P^- \models \alpha[Q^-$.*

This is proven with the aid of Lemma 2.31 by induction on the steps of the performance of the algorithm $\mathfrak{B}$.

2.36. *If $\mathfrak{B}^{\cdot} : P \models^{\cdot} Q$, then $\mathfrak{C} : \alpha[P^- \models \alpha\beta[Q^-$.*

The proof is based on Lemmas 2.35 and 2.34. It is carried out analogously to the way in which Lemma 2.19 was proven.

The next three lemmas deal only with the algorithm $\mathfrak{C}$.

2.37. *If Q is a word in* A, *then $\mathfrak{C} : \alpha Q \models \alpha[Q^-$.*

For the proof, we first note that

$$\mathfrak{C} : \alpha\xi R \vdash \alpha\bar{\xi}R \tag{29}$$

and

$$\mathfrak{C} : \alpha[Q^-\bar{\eta}\xi R \vdash \alpha[Q^-\bar{\eta}\bar{\xi}R \tag{30}$$

(these assertions are obtained directly from an examination of the scheme (3)). Then, with the aid of (29) and (30), we prove by right induction on Q that

$$\mathfrak{C} : \alpha QR \models \alpha[Q^-R, \tag{31}$$

after which, setting $R \leftrightharpoons \Lambda$ in (31), we get what is required.

2.38. *If Q is a word in* A, *then* $\mathfrak{C}: \alpha\beta[Q^- \models \alpha\beta Q$.

The proof is analogous to that of Lemma 2.37.

By examining the scheme (3), we first convince ourselves that

$$(32) \qquad \mathfrak{C}: \alpha\beta\bar{\xi}[R^- \vdash \alpha\beta\xi[R^-$$

and

$$(33) \qquad \mathfrak{C}: \alpha\beta Q\eta\bar{\xi}[R^- \vdash \alpha\beta Q\eta\xi[R^-.$$

Then, with the aid of (32) and (33), we show by right induction on Q that

$$(34) \qquad \mathfrak{C}: \alpha\beta[QR^- \models \alpha\beta Q[R^-,$$

after which, putting $R \rightleftharpoons \Lambda$ in (34), we obtain what is desired.

2.39. *If Q is a word in* A, *then* $\mathfrak{C}: \alpha\beta Q \vdash\cdot Q$.

This is obtained directly from an examination of the scheme (3). The active formula in the given situation is

$$\alpha\beta \rightarrow\cdot,$$

the only final formula of the algorithm $\mathfrak{C}$.

2.40. *If* $\mathfrak{B}^\cdot : P \models\cdot Q$, *then* $\mathfrak{C}: \alpha P \models\cdot Q$.

Indeed, let $\mathfrak{B}^\cdot : P \models\cdot Q$. Then

$$\begin{aligned} \mathfrak{C}: \alpha P &\models \alpha[P^- && [2.37] \\ &\models \alpha\beta[Q^- && [2.36] \\ &\models \alpha\beta Q && [2.38] \\ &\vdash\cdot Q && [2.39] \end{aligned}$$

and, consequently, $\mathfrak{C}: \alpha P \models\cdot Q$. Q.E.D.

2.41. *If P and Q are words in the alphabet* A, *such that* $\mathfrak{C}: \alpha[P^- \vdash \alpha[Q^-$, *then* $\mathfrak{B}^\cdot : P \vdash Q$.

In fact, let

$$(35) \qquad \mathfrak{C}: \alpha[P^- \vdash \alpha[Q^-,$$

where P and Q are words in A. Since the algorithm $\mathfrak{B}^\cdot$ is closed [§36.2.1], the word P is subject to it [§36.1.1]. Hence there exists a word R such that one of the following to assertions holds:

$$(36) \qquad \mathfrak{B}^\cdot : P \vdash R$$

or

$$\mathfrak{B}^{\cdot} P \vdash\cdot R.$$

If the second of them held, then by 2.32 there would exist words S and T such that

$$\mathfrak{C} : \alpha[P^{-} \vdash \alpha[S^{-}\beta[T^{-} \tag{37}$$

and we would have

$$\begin{aligned} \alpha[Q^{-} &\eqsim \alpha[S^{-}\beta[T^{-} && [(35), (37)], \\ [Q^{-} &\eqsim [S^{-}\beta[T^{-} && [(38)], \end{aligned} \tag{38}$$

which is impossible, since β does not occur in $[Q^{-}$.

Consequently, (36) holds and

$$\begin{aligned} &(39) & \mathfrak{C} : \alpha[P^{-} &\vdash \alpha[R^{-} && [(36), 2.31], \\ &(40) & \alpha[Q^{-} &\eqsim \alpha[R^{-} && [(35), (39)], \\ &(41) & [Q^{-} &\eqsim [R^{-} && [(40)], \\ &(42) & Q &\eqsim R && [(41), 2.8], \\ & & \mathfrak{B}^{\cdot} : P &\vdash Q && [(36), (42)]. \end{aligned}$$

Q.E.D.

Let us now agree to call words of the form $\alpha[P^{-}$, where P is a word in A, *special*. We shall also call systems *special* if all their terms are special. For special systems X, we introduce the following operation $[X^{\nabla}$:

$$[\gamma^{\nabla} \rightleftharpoons \gamma$$
$$[X\alpha[P^{-}\gamma^{\nabla} \rightleftharpoons [X^{\nabla}P\gamma.$$

(Here X is a special system and P is a word in A.)

It is clear that if X is a special system, then $[X^{\nabla}$ is a system of words in A; moreover, X connects $\alpha[P^{-}$ to $\alpha[Q^{-}$ if and only if $[X^{\nabla}$ connects P to Q.

2.42. *The algorithm $\mathfrak{C}$ simply converts every special word into a word.*

Indeed, let $\alpha[P^{-}$ be a special word. Consider the word P. By virtue of the fact that the normal algorithm $\mathfrak{B}^{\cdot}$ in the alphabet A is closed [§36.2.1], this algorithm simply or finally converts P into some word [§36.1.1]. According to Lemmas 2.31 and 2.32, in the first, as well as in the second case, $\mathfrak{C}$ simply converts $\alpha[P^{-}$ into a word. Q.E.D.

2.43. *If a word P in A is such that the algorithm $\mathfrak{C}$ is applicable to $\alpha[P^{-}$, then the algorithm $\mathfrak{B}^{\cdot}$ is applicable to P.*

Indeed, let P be a word in A, such that $\mathfrak{C}$ is applicable to P and let $\mathfrak{C}(\alpha[P^{-}) \eqsim U$. Then by virtue of the fact that $\mathfrak{C}$ is closed [2.7], we have $\mathfrak{C} : \alpha[P^{-} \vDash\cdot U$. Hence there exists a word V such that $\mathfrak{C} : \alpha[P^{-} \vDash\cdot V \vdash U$

and, therefore, there exists a converting system X of $\mathfrak{C}$'s scheme, connecting the special word $\alpha[P^-$ to the word V. It is easy to see that V is not a special word, since $\mathfrak{C} : V \vdash\cdot U$ [2.42]. Therefore, the first term of the system X is a special word, but its last term is not. The arithmetical predicate ≪the term of the system X whose number is N is not a special word≫ is obviously decidable. Hence, on the basis of the Least Number Theorem [§21.2.1], there exists a term W of the system X, such that W is not a special word, but all the terms of X with smaller numbers are. The system X has the form YWZ, where Y is a special converting system of the algorithm $\mathfrak{C}$'s scheme. The first term of Y is the word $\alpha[P^-$. Let the word R be such that $\alpha[P^-$ is the last term of Y. Then $\mathfrak{C} : \alpha[P^- \vdash W$.

The system $[Y^\nabla$ is a converting system of the algorithm $\mathfrak{B}^\cdot$ [2.41]. It connects the word P to the word R. Hence $\mathfrak{B}^\cdot : P \models R$. There exists a word Q such that $\mathfrak{B}^\cdot : R \vdash Q$ or $\mathfrak{B}^\cdot : R \vdash\cdot Q$. Let $\mathfrak{B}^\cdot : R \vdash Q$. Then $\mathfrak{C} : \alpha[R^- \vdash \alpha[Q^-$ [2.31] and $W \eqcirc \alpha[Q^-$, which is impossible by virtue of the choice of W. Therefore, $\mathfrak{B}^\cdot : R \vdash\cdot Q$ and $\mathfrak{B}^\cdot : P \models\cdot Q$, i.e., $\mathfrak{B}^\cdot$ is applicable to P. Q.E.D.

2.44. *If a word P in the alphabet* A *is such that the algorithm $\mathfrak{C}$ is applicable to αP, then the algorithm $\mathfrak{B}^\cdot$ is applicable to P.*

Indeed, let the algorithm $\mathfrak{C}$ be applicable to αP. Then the algorithm $\mathfrak{C}$ is applicable to $\alpha[P^-$ [2.37], from which it follows that the algorithm $\mathfrak{B}^\cdot$ is applicable to P [2.43].

2.45. *If the expression $\mathfrak{B}(\mathfrak{A}(P))$ makes sense, then the algorithm $\mathfrak{C}$ is applicable to the word P and*

$$\mathfrak{C}(P) \eqcirc \mathfrak{B}(\mathfrak{A}(P)).$$

In fact, let the expression $\mathfrak{B}(\mathfrak{A}(P))$ make sense. Then the algorithm $\mathfrak{A}$ is applicable to the word P, and the algorithm $\mathfrak{B}$, to the word $\mathfrak{A}(P)$. Set

(43) $$Q \rightleftharpoons \mathfrak{A}(P),$$

(44) $$R \rightleftharpoons \mathfrak{B}(\mathfrak{A}(P)).$$

We have

(45) $\mathfrak{B}(Q) \eqcirc R$ [(43), (44)],

(46) $\mathfrak{A}^\cdot(P) \eqcirc Q$ [(43), §36.2.14],

(47) $\mathfrak{B}^\cdot(Q) \eqcirc R$ [(45), §36.2.14],

(48) $\mathfrak{A}^\cdot : P \models\cdot Q$ [(46), §36.2.1, §36.1.3],

(49) $\mathfrak{B}^\cdot : Q \models\cdot R$ [(47),§36.2.1, §36.1.3],

$\mathfrak{C} : P \models \alpha Q$ [(48), 2.19]

$\models\cdot R$ [(49), 2.40].

Consequently,

(50) $$\mathfrak{C} : P \models\cdot R,$$

$\mathfrak{C}(P) \eqcirc R$ [(50)]

$\eqcirc \mathfrak{B}(\mathfrak{A}(P))$ [(44)].

Q.E.D.

In order to complete the proof of Theorem 2.1, it now remains for us to prove that the expression $\mathfrak{B}(\mathfrak{A}(P))$ makes sense for a word P in the alphabet A as soon as $\mathfrak{C}(P)$ makes sense, i.e., as soon as the algorithm $\mathfrak{C}$ is applicable to the word P.

2.46. *If the algorithm $\mathfrak{C}$ is applicable to a word P in the alphabet* A, *then the expression $\mathfrak{B}(\mathfrak{A}(P))$ makes sense.*

Indeed, let the algorithm $\mathfrak{C}$ be applicable to the word P in A. Then the algorithm $\mathfrak{A}^{\cdot}$ is applicable to P [2.22]. Put

(51) $$Q \rightleftharpoons \mathfrak{A}^{\cdot}(P).$$

Then Q is a word in A, since $\mathfrak{A}^{\cdot}$ is an algorithm in A. Further, we have

(52) $$\mathfrak{A}^{\cdot} : P \models\cdot Q \quad [(51), §36.2.1, §36.1.3],$$

(53) $$\mathfrak{C} : P \models \alpha Q \quad [(52), 2.19].$$

Since, according to our hypothesis, the algorithm $\mathfrak{C}$ is applicable to P, it is also applicable to αQ [(53)]. It follows from this that the algorithm $\mathfrak{B}^{\cdot}$ is applicable to the word Q [2.44]. Set

(54) $$R \rightleftharpoons \mathfrak{B}^{\cdot}(Q).$$

Then

(55) $$\mathfrak{A}(P) \eqcirc Q \quad [(51). §36.2.11],$$

(56) $$\mathfrak{B}(Q) \eqcirc R \quad [(54). §36.2.11],$$

$$\mathfrak{B}(\mathfrak{A}(P)) \eqcirc R \quad [(55), (56)].$$

Consequently, $\mathfrak{B}(\mathfrak{A}(P))$ makes sense. Q.E.D.

The proof of the last lemma completes the proof of Theorem 2.1.

3. Let us now prove the Composition Theorem in its general form. Let $\mathfrak{A}$ and $\mathfrak{B}$ be arbitrary normal algorithms. Let Б and B, respectively, denote their alphabets, and let

(1) $$\text{A} \rightleftharpoons \text{Б} \cup \text{B}.$$

A is an extension of each of the alphabets Б and B. Hence there exist formal extensions of the algorithms $\mathfrak{A}$ and $\mathfrak{B}$ to the alphabet A [§35.5.3]. Let $\mathfrak{A}_1$ denote the formal extension of $\mathfrak{A}$ to A, and $\mathfrak{B}_1$, the formal extension of $\mathfrak{B}$ to A. $\mathfrak{A}_1$ and $\mathfrak{B}_1$ are normal algorithms in A [§35.5.1]. The theorem just proven, Theorem 2.1, is applicable to them, and according to it, one can construct a normal algorithm $\mathfrak{C}$ over A, such that

(2) $$\mathfrak{C}(P) \simeq \mathfrak{B}_1(\mathfrak{A}_1(P)) \qquad (P \text{ is a word in A}).$$

We shall show that the normal algorithm $\mathfrak{C}$ over A so constructed satisfies the condition 1(1).

In fact, since $\mathfrak{A}_1$ and $\mathfrak{B}_1$ are formal extensions of the normal algorithms $\mathfrak{A}$ and $\mathfrak{B}$, respectively, by §35.5.4 we have

$$\mathfrak{A}_1(P) \simeq \mathfrak{A}(P), \tag{3}$$

$$\mathfrak{B}_1(Q) \simeq \mathfrak{B}(Q). \tag{4}$$

Consequently, for words P in A,

$$\begin{aligned} \mathfrak{C}(P) &\simeq \mathfrak{B}(\mathfrak{A}_1(P)) \quad [(2), (4)] \\ &\simeq \mathfrak{B}(\mathfrak{A}(P)) \quad [(3)], \end{aligned}$$

and so the condition 1(1) is fulfilled. Q.E.D.

4. The construction just carried out [2, 3] yields, for any two given normal algorithms $\mathfrak{A}$ and $\mathfrak{B}$, a normal algorithm $\mathfrak{C}$ over the union A of their alphabets, satisfying the condition 1(1). The only and, obviously, inessential element of arbitrariness in this construction is the introduction of the "new" letters, i.e., the letters belonging to $\mathfrak{C}$'s alphabet, but not belonging to A. These letters, i.e., the letters playing the roles of α, β and the doubles, can be chosen arbitrarily, as long as they differ from each other and do not belong to the alphabet A. But their roles in the construction of the algorithm $\mathfrak{C}$ is completely determined by the scheme 2(3), where only the roles of $\mathfrak{A}$ and $\mathfrak{B}$ are played by the formal extensions $\mathfrak{A}_1$ and $\mathfrak{B}_1$ of these algorithms.

We shall call the normal algorithm $\mathfrak{C}$, constructed as just described and in essence uniquely determined by the normal algorithms $\mathfrak{A}$ and $\mathfrak{B}$, the *normal composition* of the algorithms $\mathfrak{A}$ and $\mathfrak{B}$. We shall denote the normal composition of the algorithms $\mathfrak{A}$ and $\mathfrak{B}$ by the symbol

$$(\mathfrak{B} \circ \mathfrak{A}).$$

In line with this, the equality

$$\mathfrak{C} \rightleftharpoons (\mathfrak{B} \circ \mathfrak{A})$$

will denote that $\mathfrak{C}$ is the normal composition of the algorithms $\mathfrak{A}$ and $\mathfrak{B}$.

The following assertions summarize the preceding:

4.1. *For any two normal algorithms $\mathfrak{A}$ and $\mathfrak{B}$, there exists the normal composition* $(\mathfrak{B} \circ \mathfrak{A})$.

4.2. *The normal composition of normal algorithms $\mathfrak{A}$ and $\mathfrak{B}$ is a normal algorithm over the union of these algorithms' alphabets and is determined uniquely in its essentials.*

4.3. *For any two normal algorithms $\mathfrak{A}$ and $\mathfrak{B}$,*

$$(\mathfrak{B} \circ \mathfrak{A})(P) \simeq \mathfrak{B}(\mathfrak{A}(P)) \quad (P \text{ is a word in } A),$$

where A *denotes the union of the alphabets of the algorithms* $\mathfrak{A}$ *and* $\mathfrak{B}$.

5. The proof given for Theorem 1.1 is a typical example of a *constructive* proof: asserting the existence of a normal algorithm $\mathfrak{C}$, related in a definite way to normal algorithms $\mathfrak{A}$ and $\mathfrak{B}$, we indicated a method[2] which would permit us to potentially realize this algorithm as soon as the initial algorithms $\mathfrak{A}$ and $\mathfrak{B}$ are given.

A typical example of a non-constructive proof of this theorem would be along the following lines: assuming that the non-existence of the desired algorithm $\mathfrak{C}$ has somehow been proven, one reduces this assumption to a contradiction without carrying out a construction of $\mathfrak{C}$ (a "pure existence" apagogical proof).[3]

In many cases, such non-constructive proofs turn out to be simpler *in a technical sense* that the corresponding constructive proofs. And this is understandable—no construction is required of them. However, this should not lead us into temptation. *In a fundamental sense*, a constructive existence proof is, nevertheless, simpler: here one need not ponder over why existence should be understood so indirectly, how an object can turn out to be existing even through no approaches to a search for it are seen. But as for understanding existence as potential feasibility, to us it seems natural and, in a fundamental sense, simple.

Incidentally, the problem of finding a constructive proof can also turn out to be simpler in a technical sense that the corresponding problem of finding a "pure existence proof". For example, we do not see how to refute the assumption that the normal algorithm $\mathfrak{C}$ is Theorem 1.1 does not exist without actually constructing it. We should particularly like to call the reader's attention to this.

6.[4] We shall now prove that the normalization principle for algorithms [§27.6] is equivalent to the following statement:

6.1. *Every algorithm in an alphabet* A *is equivalent with respect to* A *to some normal algorithm over* A.

Let us assume that 6.1 is true and consider any algorithm $\mathfrak{A}$ in A. We shall show that $\mathfrak{A}$ is completely equivalent to some normal algorithm over A.

According to 6.1, there is a normal algorithm $\mathfrak{B}$ over A, equivalent to $\mathfrak{A}$ with respect to A. Denote its alphabet by Б. We have

$$\text{(1)} \qquad \mathrm{A} \subset \text{Б}.$$

Construct the normal algorithm $\mathfrak{C}$ in Б with the scheme

$$\{\xi \rightarrow \xi,$$

[2] For more details on this, see §54.5.

[3] A proof of P by refuting not P. This should be distinguished from a proof by reductio ad absurdum, which is a proof of not P by refuting P. (Translator's note)

[4] This part is a natural continuation of §27. The statement 6.1, just like the normalization principle, can not be regarded as having a precise mathematical meaning. The same may also be said about the assertion of their equivalence. Nevertheless, one may regard the latter as sufficiently understandable.

where ξ runs through the alphabet Б\А.

It is obvious that

$$\mathfrak{C}(P) \eqcirc P \quad (P \text{ is a word in A}) \tag{2}$$

and that $\mathfrak{C}$ is not applicable to any word in Б which is not a word in A.

Construct $\mathfrak{D}$ as the normal composition of the algorithms $\mathfrak{C}$ and $\mathfrak{C}$:

$$\mathfrak{D} \rightleftharpoons (\mathfrak{C} \circ \mathfrak{B}). \tag{3}$$

$\mathfrak{D}$ is a normal algorithm over Б and

$$\mathfrak{D}(P) \simeq \mathfrak{C}(\mathfrak{B}(P)) \quad (P \text{ is a word in Б}) \quad [(3)]. \tag{4}$$

It follows from this that

$$\mathfrak{D}(P) \simeq \mathfrak{B}(P) \tag{5}$$

if P and $\mathfrak{B}(P)$ are words in A [(4), (2)].

Let us show that $\mathfrak{D}$ is the required normal algorithm over A, completely equivalent to $\mathfrak{A}$.

$\mathfrak{D}$ is an algorithm over A in view of (1).

If the algorithm $\mathfrak{A}$ is applicable to a word P in A, then P and $\mathfrak{A}(P)$ are words in A, since $\mathfrak{A}$ is an algorithm in A. Consequently, $\mathfrak{B}$ is then also applicable to P, and we have

$$\mathfrak{A}(P) \eqcirc \mathfrak{B}(P).$$

By virtue of (5), it follows from this that

$$\mathfrak{A}(P) \simeq \mathfrak{D}(P). \tag{6}$$

Therefore, the equality (6) holds whenever the algorithm $\mathfrak{A}$ is applicable to the word P in A.

Let us now assume that the algorithm $\mathfrak{D}$ is applicable to a word P in A. Then the algorithm $\mathfrak{B}$ is also applicable to P [(14)], and the algorithm $\mathfrak{C}$, to $\mathfrak{B}(P)$. Hence $\mathfrak{B}(P)$ is also a word in A. But then the algorithm $\mathfrak{A}$ is also applicable to P is view of the equivalence of the algorithms $\mathfrak{A}$ and $\mathfrak{B}$ with respect to A. Consequently, the algorithm $\mathfrak{A}$ is applicable to a word P in A whenever the algorithm $\mathfrak{D}$ is applicable to it.

We have thus proven that the algorithm $\mathfrak{D}$ is completely equivalent to the algorithm $\mathfrak{A}$ with respect to the alphabet A. By the same token, we proved that the normalization principle follows from the assertion 6.1. The converse is obvious. Consequently, the normalization principle is equivalent to the assertion 6.1. Q.E.D.

§38 Union of Algorithms

1. One often has to simultaneously consider the results of the performances of two or several algorithms on one and the same initial data. In this case, it may prove helpful to construct a system of all these results. It might, for example, itself serve as the initial data for the performance of some other algorithm.

Thus, if we have already been given algorithms for computing the values of the functions $(x+2)$ and $(x+3)$ for natural values of the variable x, an algorithm for computing the values of the function

$$(x+2) \times (x+3) \tag{1}$$

can be constructed as the following prescription: starting with an arbitrarily given natural number N, compute the numbers $(N+2)$ and $(N+3)$ with the aid of the two given algorithms, construct the pair of these numbers and apply a multiplication algorithm to it. The first two stages of this prescription form an algorithm changing every natural number into the pair of numbers $(N+2) * (N+3)$. The composition of this algorithm and a multiplication algorithm constitutes the desired algorithm for computing the values of the function (1).

Therefore, if verbal algorithms $\mathfrak{A}$ and $\mathfrak{B}$ are given, it may turn out worthwhile to construct the following prescription: ≪start with an arbitrary word P in the union A of these algorithms' alphabets and develop the processes of applying the algorithms $\mathfrak{A}$ and $\mathfrak{B}$ to it. If both these processes end, then construct the word $\mathfrak{A}(P)\mathfrak{B}(P)$ and regard it as the result≫. This prescription, as we see, constitutes a certain algorithm in the alphabet A—the ≪union≫ of the given algorithms $\mathfrak{A}$ and $\mathfrak{B}$.

It is natural to ask whether the union of normalizable algorithms is always a normalizable algorithm. The normalization principle suggests an affirmative answer to this question, and in this section, we shall in fact prove the following theorem:

1.1. UNION THEOREM. *Let $\mathfrak{A}$ and $\mathfrak{B}$ be normal algorithms, and* A, *the union of their alphabets. Then a normal algorithm $\mathfrak{C}$ over* A *can be constructed, such that for any word P in* A, *the conditional equality*

$$\mathfrak{C}(P) \simeq \mathfrak{A}(P)\mathfrak{B}(P) \tag{2}$$

holds.

2. The proof of this theorem is the main content of the present section.

We begin with a proof of the following theorem:

2.1. *Given any normal algorithms $\mathfrak{A}$ and $\mathfrak{B}$ in an alphabet* A, *a normal algorithm $\mathfrak{C}$ over* A *can be constructed, such that for any word P in* A, *the conditional equality*

$$\mathfrak{C}(P) \simeq \mathfrak{A}(P)\mathfrak{B}(P) \tag{1}$$

holds.

For the proof, let us introduce the alphabet $\bar{\text{A}}$ of the doubles of A's letters, as in the proof of Theorem §37.2.1. As before, the double of the letter ξ will be denoted by $\bar{\xi}$, the double of the word P by $[P^-$. Set Б$\rightleftharpoons$ A$\bar{\text{A}}$.

Let us introduce the following normal algorithm over A:

the letter-by-letter coding algorithm $\mathfrak{K}_1$ (see §33.1), such that for any letter ξ of A, its ≪code≫ K_ξ is the word $\xi\bar{\xi}$:

$$K_\xi \rightleftharpoons \xi\bar{\xi}; \tag{2}$$

the letter-by-letter coding algorithm $\mathfrak{K}_2$, such that for any letter ξ of A,

$$K_\xi \rightleftharpoons \xi, K_{\bar{\xi}} \rightleftharpoons \xi; \tag{3}$$

the double projecting algorithm $\mathfrak{L}_1$ (see §33.3), such that for any word P in Б,

$$\mathfrak{L}_1(P) \rightleftharpoons [P^{\text{A}}[P^{\bar{\text{A}}}; \tag{4}$$

the double projecting algorithm $\mathfrak{L}_2$, such that for any P in Б,

$$\mathfrak{L}_2(P) \rightleftharpoons [P^{\bar{\text{A}}}[P^{\text{A}}; \tag{5}$$

$\bar{\mathfrak{A}}$—the normal algorithm in $\bar{\text{A}}$, whose scheme is obtained from the scheme of the algorithm $\mathfrak{A}$ by replacing each letter by its double;

$\bar{\mathfrak{A}}_1$—the natural extension of the algorithm $\bar{\mathfrak{A}}$ to the alphabet Б;

$\mathfrak{B}_1$—the natural extension of the algorithm $\mathfrak{B}$ to the alphabet Б.

Let us construct a normal algorithm $\mathfrak{C}$ as the normal composition

$$(\mathfrak{K}_2 \circ (\bar{\mathfrak{A}}_1 \circ (\mathfrak{L}_2 \circ (\mathfrak{B}_1 \circ (\mathfrak{L}_1 \circ \mathfrak{K}_1))))). \tag{6}$$

Since $\mathfrak{K}_2$ already is an algorithm over A, $\mathfrak{C}$ is all the more so a normal algorithm over A [§37.1.1]. We shall show that the conditional equality (1) holds for $\mathfrak{C}$. As a preliminary, we shall prove several lemmas, in which P will denote an arbitrary word in the alphabet A.

2.2. $\mathfrak{L}_1(\mathfrak{K}_1(P)) \rightleftharpoons P[P^-$.

We first show by the method of right induction that for any P in A,

$$[\mathfrak{K}_1(P)^{\text{A}} \rightleftharpoons P, \tag{7}$$

$$[\mathfrak{K}_1(P)^{\bar{\text{A}}} \rightleftharpoons [P^-. \tag{8}$$

Then

$$\begin{aligned} \mathfrak{L}_1(\mathfrak{K}_1(P)) &\rightleftharpoons [\mathfrak{K}_1(P)^{\text{A}}[\mathfrak{K}_1(P)^{\bar{\text{A}}} && [(4)] \\ &\rightleftharpoons P[P^- && [(7), (8)]. \end{aligned}$$

Q.E.D.

2.3. $\mathfrak{B}_1(\mathfrak{L}_1(\mathfrak{K}_1(P))) \simeq \mathfrak{B}(P)[P^-$.

In fact, since $\mathfrak{B}_1$ is the natural extension of $\mathfrak{B}$ to Б, we have

$$\begin{aligned}\mathfrak{B}_1(\mathfrak{L}_1(\mathfrak{K}_1(P))) &\simeq \mathfrak{B}_1(P[P^-) \quad [(2.2)]\\ &\simeq \mathfrak{B}_1(P)[P^- \quad [\S 35.2(1)].\end{aligned}$$

2.4. $\mathfrak{L}_2(\mathfrak{B}_1(\mathfrak{L}_1(\mathfrak{K}_1(P)))) \simeq [P^-\mathfrak{B}(P)$.

Indeed,

$$\begin{aligned}\mathfrak{L}_2(\mathfrak{B}_1(\mathfrak{L}_1(\mathfrak{K}_1(P)))) &\simeq \mathfrak{L}_2(\mathfrak{B}(P)[P^-) \quad [(2.3)]\\ &\simeq [P^-\mathfrak{B}(P) \quad [(5)].\end{aligned}$$

2.5. $\bar{\mathfrak{A}}([P^-) \simeq [\mathfrak{A}(P)^-$.

This immediately follows from the definition of the algorithm $\bar{\mathfrak{A}}$.

2.6. $\bar{\mathfrak{A}}_1(\mathfrak{L}_2(\mathfrak{B}_1(\mathfrak{L}_1(\mathfrak{K}_1(P))))) \simeq [\mathfrak{A}(P)^-\mathfrak{B}(P)$.

In fact, since $\bar{\mathfrak{A}}_1$ is the natural extension of $\bar{\mathfrak{A}}$ to Б, we have

$$\begin{aligned}\bar{\mathfrak{A}}_1(\mathfrak{L}_2(\mathfrak{B}_1(\mathfrak{L}_1(\mathfrak{K}_1(P))))) &\simeq \bar{\mathfrak{A}}_1([P^-\mathfrak{B}(P)) \quad [2.4]\\ &\simeq \bar{\mathfrak{A}}([P^-)\mathfrak{B}(P) \quad [\S 35.2(1)]\\ &\simeq [\mathfrak{A}(P)^-\mathfrak{B}(P) \quad [2.5].\end{aligned}$$

2.7. $\mathfrak{K}_2([Q^-R) = QR$ *(Q, R are words in* A).

This immediately follows from the definition (3) of the algorithm $\mathfrak{K}_2$.

2.8. $\mathfrak{K}_2(\bar{\mathfrak{A}}_1(\mathfrak{L}_2(\mathfrak{B}_1(\mathfrak{L}_1(P))))) \simeq \mathfrak{A}(P)\mathfrak{B}(P)$.

Indeed,

$$\begin{aligned}\mathfrak{K}_2(\bar{\mathfrak{A}}_1(\mathfrak{L}_2(\mathfrak{B}_1(\mathfrak{L}_1(\mathfrak{K}(P)))))) &\simeq \mathfrak{K}_2([\mathfrak{A}(P)^-\mathfrak{B}(P)) \quad [2.6]\\ &\simeq \mathfrak{A}(P)\mathfrak{B}(P) \quad [2.7].\end{aligned}$$

The conditional equality (1) now follows from (6) and 2.8.

3. Theorem 1.1 can be proven in the following way.

In fact, let $\mathfrak{A}$ and $\mathfrak{B}$ be any normal algorithms, and let A be the union of their alphabets. Construct the formal extensions, $\mathfrak{A}_1$ and $\mathfrak{B}_1$, of the algorithms $\mathfrak{A}$ and $\mathfrak{B}$ to the alphabet A. $\mathfrak{A}_1$ and $\mathfrak{B}_1$ are normal algorithms in one and the same alphabet A, and so Theorem 2.1 may be applied to it. According to this theorem, a normal algorithm $\mathfrak{C}$ over A can be constructed, such that for any P in A, the conditional equality

$$\mathfrak{C}(P) \simeq \mathfrak{A}_1(P)\mathfrak{B}_1(P)$$

holds. Bearing in mind that

$$\begin{aligned}\mathfrak{A}_1(P) &\simeq \mathfrak{A}(P),\\ \mathfrak{B}_1(P) &\simeq \mathfrak{B}(P)\end{aligned}$$

[§35.5.4], we obtain the conditional equality 1(2) from this.

4. The following assertion can be proven with the aid of Theorem 1.1:

4.1. *Let $\mathfrak{A}$ and $\mathfrak{B}$ be arbitrary normal algorithms,* A, *the union of their alphabets, and γ, any letter. Then a normal algorithm $\mathfrak{C}$ over the alphabet* $A \cup \{\gamma\}$ *can be constructed, such that for any word P in* A, *the conditional equality*

$$\mathfrak{C}(P) \simeq \mathfrak{A}(P)\gamma\mathfrak{B}(P)$$

will be fulfilled.

Indeed, let $\mathfrak{D}$ denote the normal algorithm in $A \cup \{\gamma\}$, changing every word in this alphabet into the word γ. We know how such an algorithm is constructed [§29.3]. Applying Theorem 1.1 first to the algorithms $\mathfrak{A}$ and $\mathfrak{D}$, and then to the algorithm obtained as a result of this and the algorithm $\mathfrak{B}$, we get a normal algorithm $\mathfrak{C}$ over the alphabet $A \cup \{\gamma\}$, such that for any word P in A, the conditional equality

$$\mathfrak{C}(P) \simeq \mathfrak{A}(P)\mathfrak{D}(P)\mathfrak{B}(P),$$

i.e.,

$$\mathfrak{C}(P) \simeq \mathfrak{A}(P)\gamma\mathfrak{B}(P),$$

is fulfilled. Q.E.D.

5. Let us formulate some corollaries to the theorems just proven.

5.1. *Given any normal algorithm $\mathfrak{A}$ over an alphabet* A, *one can construct normal algorithms $\mathfrak{C}$ and $\mathfrak{D}$ over* A, *such that for any word P in* A, *the conditional equalities*

(1) $$\mathfrak{C}(P) \simeq \mathfrak{A}(P)P,$$

(2) $$\mathfrak{D}(P) \simeq P\mathfrak{A}(P)$$

hold.

For the proof, we make use of the identity normal algorithm in the alphabet A [§28.4]. Applying Theorem 1.1 to the algorithm $\mathfrak{A}$ and the identity algorithm in A (playing the role of $\mathfrak{B}$), we obtain a normal algorithm $\mathfrak{C}$ over A, for which the conditional equality (1) is valid. In a similar manner, with the roles of the algorithm $\mathfrak{A}$ and the identity algorithm exchanged, we obtain a normal algorithm $\mathfrak{D}$ over A, for which (2) holds.

5.2. *Given any normal algorithm $\mathfrak{A}$ over an alphabet* A *and any letter γ, there exist normal algorithms $\mathfrak{C}$ and $\mathfrak{D}$ over the alphabet* $A \cup \{\gamma\}$, *such that for any P in* A, *the conditional equalities*

$$\mathfrak{C}(P) \simeq \mathfrak{A}(P)\gamma P,$$
$$\mathfrak{D}(P) \simeq P\gamma\mathfrak{A}(P)$$

hold.

The proof is analogous to that of 5.1, with 4.1 applied instead of 1.1.

6. The normal algorithm $\mathfrak{C}$ constructed by the method indicated in the proof of Theorem 1.1 is uniquely determined in its essentials by the normal algorithms $\mathfrak{A}$ and $\mathfrak{B}$. We shall denote the normal union of the algorithms $\mathfrak{A}$ and $\mathfrak{B}$ by the symbol

$$(\mathfrak{A} \frown \mathfrak{B}).$$

The following assertions summarize the preceding:

6.1. *For any two normal algorithms* $\mathfrak{A}$ *and* $\mathfrak{B}$, *there exists*[5] *the normal union* $(\mathfrak{A} \frown \mathfrak{B})$.

6.2. *The normal union of normal algorithms* $\mathfrak{A}$ *and* $\mathfrak{B}$ *is a normal algorithm over the union of these algorithm's alphabets.*

6.3. *For any two normal algorithms* $\mathfrak{A}$ *and* $\mathfrak{B}$, *the conditional equality*

$$(\mathfrak{A} \frown \mathfrak{B})(P) \simeq \mathfrak{A}(P)\mathfrak{B}(P)$$

holds, where P *is any word in the union of these algorithm's alphabets.*

§39 Branching of Algorithms

1. One sometimes has to construct a prescription of the following type: ≪apply the algorithm $\mathfrak{A}$ or the algorithm $\mathfrak{B}$ to the initial data, depending on whether or not these data possess a certain property≫. In order that it be possible to regard such a prescription as an algorithm, it is necessary, of course, that there exist a constructive method for recognizing the initial data's property under discussion. This method may, for example, consist in applying to the initial data some specially chosen algorithm, which changes the initial data into the empty word if and only if they possess the property of interest to us. In order to find out whether the initial data possess this property, one must then apply the recognizing algorithm to them and look at the result of this application. The property holds if this result is the empty word, and fails to hold if it is a non-empty word. In the first case, the prescription requires the application of the algorithm $\mathfrak{A}$ to the initial data, in the second, the algorithm $\mathfrak{B}$. Passing from algorithms in general to verbal algorithms, i.e., to algorithms in certain alphabets, we arrive at prescriptions of the following type.

Verbal algorithms $\mathfrak{A}$, $\mathfrak{B}$ and $\mathfrak{C}$ in an alphabet A are given. Starting with an arbitrary word P in this alphabet, the application of the algorithm $\mathfrak{C}$ to it is prescribed. If the empty word is obtained as a result, then the algorithm $\mathfrak{A}$ ought to be applied to P; but if a non-empty word is obtained, then the algorithm $\mathfrak{B}$ ought to be applied to P.

This prescription may be regarded as a certain verbal algorithm in A. For its applicability to a word P in A, it is of course necessary that the algorithm $\mathfrak{C}$ be applicable to P.

We have arrived, therefore, at a certain combination of three verbal algorithms $\mathfrak{A}$, $\mathfrak{B}$ and $\mathfrak{C}$ in A, which is a new verbal algorithm in A. It is natural to call this combination the *branching of the algorithms* $\mathfrak{A}$ *and* $\mathfrak{B}$, *directed by the algorithm* $\mathfrak{C}$.

It is natural to ask whether the branching of two normalizable algorithms, directed by a third normalizable algorithm, is also a normalizable algorithm. A positive answer to this question is suggested by the normalization principle. It is implied by the following theorem:

---

[5]See §54.5

1.1. BRANCHING THEOREM. *Given any normal algorithms $\mathfrak{A}$, $\mathfrak{B}$ and $\mathfrak{C}$, one can construct a normal algorithm $\mathfrak{D}$ over the union A of their alphabets, such that for any word P in A, the following three conditions are fulfilled:*

(1) *if $\mathfrak{C}(P) \eqcirc \Lambda$, then $\mathfrak{D}(P) \simeq \mathfrak{A}(P)$,*
(2) *if $\mathfrak{C}(P) \neq \Lambda$, then $\mathfrak{D}(P) \simeq \mathfrak{B}(P)$,*
(3) *if $\mathfrak{D}(P)$ is defined, then $\mathfrak{C}(P)$ is also defined.*

Here and in the sequel, when the symbol≪$\neq$≫ connects expressions which might turn out to be undefined, it ought to be understood in the following manner: both expressions connected by this symbol make sense, and their values are graphically different words.

The proof of the Branching Theorem makes up the main content of this section.

2. We shall prove several lemmas.

2.1. *Given any normal algorithm $\mathfrak{C}$ in an alphabet A and any letter α, one can construct a normal algorithm $\mathfrak{E}$ over $A \cup \{\alpha\}$, such that*

$$\mathfrak{E}(P) \eqcirc \begin{cases} \alpha \text{ for words } P \text{ in A, such that } \mathfrak{C}(P) \eqcirc \Lambda, \\ \Lambda \text{ for words } P \text{ in A, such that } \mathfrak{C}(P) \neq \Lambda, \end{cases}$$

and $\mathfrak{E}$ is applicable to those and only those words in A to which $\mathfrak{C}$ is applicable.

We shall make use of the branching normal algorithm $\mathfrak{A}_{A\cup\{\alpha\},\Lambda,\alpha,\Lambda}$ over the alphabet $A \cup \{\alpha\}$, which changes Λ into α and changes every non-empty word in the alphabet $A \cup \{\alpha\}$ into Λ [§31.1]. We shall denote it by $\mathfrak{E}_0$ in order to abbreviate our notation. We have

$$(2) \qquad \mathfrak{E}_0(\Lambda) \eqcirc \alpha,$$

$$(3) \qquad \mathfrak{E}_0(P) \eqcirc \Lambda \quad (P \text{ is a word in A}, P \neq \Lambda).$$

Let us construct the desired algorithm $\mathfrak{E}$ as the normal composition of the algorithms $\mathfrak{C}$ and $\mathfrak{E}_0$:

$$(4) \qquad \mathfrak{E} \eqcirc (\mathfrak{E}_0 \circ \mathfrak{C}).$$

The algorithm $\mathfrak{E}$, just like $\mathfrak{E}_0$, is a normal algorithm over $A \cup \{\alpha\}$ and

$$(5) \qquad \mathfrak{E}(P) \simeq \mathfrak{E}_0(\mathfrak{C}(P)) \quad (P \text{ is a word in A}) \quad [(14), §37.4.3].$$

If P is a word in A and $\mathfrak{C}(P) \eqcirc \Lambda$, then

$$\begin{aligned} \mathfrak{E}(P) &\simeq \mathfrak{E}_0(\Lambda) \quad [(5)] \\ &\eqcirc \alpha \qquad [(2)]. \end{aligned}$$

But if P is a word in A and $\mathfrak{C}(P) \neq \Lambda$, then $\mathfrak{C}(P)$ is a word in A and

$$\mathfrak{E}(P) \eqcirc \Lambda \quad [(5), (3)].$$

Therefore, the algorithm $\mathfrak{E}$ satisfies the condition (1). By virtue of (5), the algorithm $\mathfrak{E}$ is applicable only to those words in A to which $\mathfrak{C}$ is applicable. On the other hand, according to what has already been proven, it is applicable to all such words. The lemma is therefore proven.

2.2. *Given any normal algorithm $\mathfrak{C}$ in an alphabet A and any letter α, one can construct a normal algorithm $\mathfrak{F}$ over $\mathrm{A} \cup \{\alpha\}$, such that*

$$(6) \qquad \mathfrak{F}(P) = \begin{cases} \alpha P \text{ for words } P \text{ in A, such that} & \mathfrak{C}(P) = \Lambda, \\ P \text{ for words } P \text{ in A, such that} & \mathfrak{C}(P) \neq \Lambda, \end{cases}$$

and $\mathfrak{F}$ will be applicable to those and only those words in A to which $\mathfrak{C}$ is applicable.

Indeed, let us construct the normal algorithm $\mathfrak{E}$ in accordance with the preceding lemma. Then, by §38.5.1, we shall construct a normal algorithm $\mathfrak{F}$ over $\mathrm{A} \cup \{\alpha\}$, such that

$$(7) \qquad \mathfrak{F}(P) \simeq \mathfrak{E}(P)P \quad (P \text{ is a word in } \mathrm{A} \cup \{\alpha\}).$$

It is obviously applicable to those and only those words in A to which $\mathfrak{E}$ is applicable, i.e., to those and only those words in A to which $\mathfrak{C}$ is applicable. The validity of the condition (6) for $\mathfrak{F}$ immediately follows from the validity of the condition (1) for $\mathfrak{E}$, by virtue of the conditional equality (7).

3. Let us now prove Theorem 1.1.

Let $\mathfrak{A}$, $\mathfrak{B}$ and $\mathfrak{C}$ be normal algorithms; let A be the union of their alphabets. Let $\mathfrak{A}_1$, $\mathfrak{B}_1$ and $\mathfrak{C}_1$ denote the formal extensions of the algorithms $\mathfrak{A}$, $\mathfrak{B}$ and $\mathfrak{C}$, respectively, to the alphabet A. Let the letters α and β, not belonging to A, be held fixed. We construct a normal algorithm $\mathfrak{F}$ over $\mathrm{A}\alpha$ in accordance with Lemma 2.2 in such a way that it be applicable to those and only those words in A to which $\mathfrak{C}_1$ is applicable, and that the following condition be satisfied:

$$(1) \qquad \mathfrak{F}(P) = \begin{cases} \alpha P \text{ for words } P \text{ in A, such that} & \mathfrak{C}_1(P) = \Lambda, \\ P \text{ for words } P \text{ in A, such that} & \mathfrak{C}_1(P) \neq \Lambda. \end{cases}$$

Let us construct a normal algorithm $\mathfrak{A}_2$ in the alphabet $\mathrm{A}\beta$, defining its scheme as follows:

$$(2) \qquad \begin{cases} \beta \rightarrow\cdot \\ \mathfrak{A}_1. \end{cases}$$

We construct the normal algorithm $\mathfrak{B}_1^{\cdot}$, the closure of the algorithm $\mathfrak{B}_1$ [§36.2], then the normal algorithm $\mathfrak{B}_1^{\cdot\beta}$ in the alphabet $\mathrm{A}\beta$, whose scheme is obtained from $\mathfrak{B}_1^{\cdot}$'s scheme by replacing each right side Q of a final substitution formula by the word βQ, and, finally, the normal algorithm $\mathfrak{B}_2$ in the alphabet $\mathrm{A}\alpha\beta$ with the scheme

$$(3) \qquad \begin{cases} \alpha \rightarrow\cdot \\ \mathfrak{B}_1^{\cdot\beta}. \end{cases}$$

We define the desired algorithm $\mathfrak{D}$ as the normal composition

$$(4) \qquad ((\mathfrak{A}_2 \circ \mathfrak{B}_2) \circ \mathfrak{F}).$$

A direct examination of the schemes (2) and (3) permit us to formulate Lemmas 3.1–3.7, cited below. We leave the minor details in the proofs of these lemmas to the reader, trusting that by now she or he has accumulated enough experience for this.

3.1. *If R and S are words in A, then*

$$\mathfrak{A}_2(R\beta S) \eqcirc RS.$$

In fact, $\mathfrak{A}_2 : R\beta S \vdash\cdot RS$ in this case, from which the truth of our assertion follows.

3.2. *If $\mathfrak{A}_1 : P \vdash Q$, then $\mathfrak{A}_2 : P \vdash Q$.*

3.3. *If $\mathfrak{A}_1 : P \vdash\cdot Q$, then $\mathfrak{A}_2 : P \vdash\cdot Q$.*

3.4. *If $\mathfrak{A}_1 : P\rceil$, then $\mathfrak{A}_2 : P\rceil$.*

3.5. *If P is in A, then $\mathfrak{B}_2(\alpha P) \eqcirc P$.*

3.6. *If $\mathfrak{B}_1 : P \vdash Q$, then $\mathfrak{B}_2 : P \vdash Q$.*

3.7. *If $\mathfrak{B}_1 : P \vdash\cdot Q$, then there exist words R and S in A such that $RS \eqcirc Q$ and $\mathfrak{B}_2 : P \vdash\cdot R\beta S$.*

From 3.2–3.4 follows

3.8. *For any P in A, $\mathfrak{A}_2(P) \simeq \mathfrak{A}_1(P)$.*

The next two assertions follow from 3.6 and 3.7.

3.9. *For any P in A, $!\mathfrak{B}_2(P)$ if and only if $!\mathfrak{B}_1(P)$.*

3.10. *If $!\mathfrak{B}_1(P)$, then there exist words R and S, such that $RS \eqcirc \mathfrak{B}_1(P)$ and $\mathfrak{B}_2(P) \eqcirc R\beta S$.*

We shall now show that the conditions 1(1)–1(3) are fulfilled.
Indeed, let

$$\mathfrak{C}(P) \eqcirc \Lambda. \tag{5}$$

then

$$\mathfrak{C}_1(P) \eqcirc \Lambda \qquad [\S35.5.4,\ (5)], \tag{6}$$

$$\mathfrak{F}(P) \eqcirc \alpha P \qquad [(1),\ (6)], \tag{7}$$

$$\begin{aligned}
\mathfrak{D}(P) &\simeq (\mathfrak{A}_2 \circ \mathfrak{B}_2)\mathfrak{F}(P)) && [\S37.4.3,\ (4)]\\
&\simeq (\mathfrak{A}_2 \circ \mathfrak{B}_2)(\alpha P) && [(7)]\\
&\simeq \mathfrak{A}_2(\mathfrak{B}_2(\alpha P)) && [\S37.4.3]\\
&\simeq \mathfrak{A}_2(P) && [3.5]\\
&\simeq \mathfrak{A}_1(P) && [3.8]\\
&\simeq \mathfrak{A}(P) && [\S35.5.4],
\end{aligned}$$

i.e., $\mathfrak{D}(P) \simeq \mathfrak{A}(P)$, so that the condition 1(1) is satisfied.

Let us now verify the condition 1(2). Assume that P is in A and that

$$\mathfrak{C}(P) \not\eqcirc \Lambda. \tag{8}$$

Then

$$(9) \qquad \mathfrak{C}_1(P) \neq \Lambda \qquad [\S 35.5.4,\ (8)],$$

$$(10) \qquad \mathfrak{F}(P) \risingdotseq P \qquad [(1),\ (9)],$$

$$(11) \qquad \begin{aligned} \mathfrak{D}(P) &\simeq (\mathfrak{A}_2 \circ \mathfrak{B}_2)(\mathfrak{F}(P)) && [\S 37.4.3,\ (4)] \\ &\simeq (\mathfrak{A}_2 \circ \mathfrak{B}_2)(P) && [(10)] \\ &\simeq \mathfrak{A}_2(\mathfrak{B}_2(P)) && [\S 37.4.3]. \end{aligned}$$

We shall show that

$$(12) \qquad \mathfrak{A}_2(\mathfrak{B}_2(P)) \simeq \mathfrak{B}(P).$$

Indeed, let the left side of this equality make sense, i.e., let

$$(13) \qquad !\mathfrak{A}_2(\mathfrak{B}_2(P)).$$

Then

$$(14) \qquad !\mathfrak{B}_2(P) \qquad [(13)],$$

$$(15) \qquad !\mathfrak{B}_1(P) \qquad [3.9,\ (14)],$$

$$(16) \qquad !\mathfrak{B}(P) \qquad [\S 35.5.4,\ (15)],$$

i.e., the right side of the equality (12) makes sense. Besides, there exist words R and S in A, such that

$$(17) \qquad RS \risingdotseq \mathfrak{B}_1(P)$$

and

$$(18) \qquad \mathfrak{B}_2(P) \risingdotseq R\beta S \quad [3.10],$$

so that

$$\begin{aligned} \mathfrak{A}_2(\mathfrak{B}_2(P)) &\risingdotseq (R\beta S) && [(18)] \\ &\risingdotseq RS && [8.1] \\ &\risingdotseq \mathfrak{B}_1(P) && [(17)] \\ &\risingdotseq \mathfrak{B}(P) && [\S 35.5.4], \end{aligned}$$

i.e., $\mathfrak{A}_2(\mathfrak{B}_2(P)) \risingdotseq \mathfrak{B}(P)$.

It remains to show that from the meaningfulness of (12)'s right side follows the meaningfulness of its left side. Thus, let

$$(19) \qquad !\mathfrak{B}(P).$$

Then

$$(20) \qquad !\mathfrak{B}_1(P) \qquad [\S 35.5.4],$$

$$(21) \qquad \mathfrak{B}_2(P) \eqsim R\beta S$$

for some words R and S in the alphabet A. But then

$$(22) \qquad \mathfrak{A}_2(R\beta S) \eqsim RS \quad [3.1],$$

i.e., $\mathfrak{A}_2(\mathfrak{B}_2(P))$ is meaningful [(21), (22)]. Therefore, the conditional equality (12) is satisfied, so that we have $\mathfrak{D}(P) \simeq \mathfrak{B}(P)$ [(11)]. Thus, the condition 1(2) is verified.

It remains to verify the condition 1(3). It too is satisfied. In fact, let $!\mathfrak{D}(P)$ for some P in A. Then $!\mathfrak{F}(P)$ [(14)], and so $!\mathfrak{C}_1(P)$ [2.2], but then we also have $!\mathfrak{C}(P)$ [§35.5.4].

Therefore, Theorem 1.1 is proven.

4. We shall call the normal algorithm $\mathfrak{D}$, constructed by the method just described and in essence uniquely determined by the normal algorithms $\mathfrak{A}$, $\mathfrak{B}$ and $\mathfrak{C}$, the *normal branching of the algorithms* $\mathfrak{A}$ *and* $\mathfrak{B}$, *directed by the algorithm* $\mathfrak{C}$. We shall denote this algorithm by the symbol

$$(\mathfrak{A} \curlyvee \mathfrak{B} | \mathfrak{C}).$$

§40 Repetition of an Algorithm

1. It is often necessary to construct prescriptions of the following type in mathematics: ≪apply a given algorithm $\mathfrak{A}$ to the initial data; if the result of the application does not possess a certain property $\mathfrak{P}$, apply the algorithm $\mathfrak{A}$ to this result; if the new result again fails to possess the property $\mathfrak{P}$, apply the algorithm $\mathfrak{A}$ to it, too etc., until an object possessing the property $\mathfrak{P}$ is obtained≫. In order for it to be possible to regard such a prescription as an algorithm, it is necessary, of course, to have a constructive method for recognizing the property $\mathfrak{P}$. This method may, for example, consist in applying to the object under consideration the "recognizing" algorithm, changing this object into the empty word if and only if it possesses the property $\mathfrak{P}$. Turning now from algorithms in general to verbal algorithms, i.e., to algorithms in certain alphabets, we arrive at prescriptions of the following type.

Verbal algorithms $\mathfrak{A}$ and $\mathfrak{C}$ in an alphabet A are given. It is prescribed to start with an arbitrary word P in A, apply the algorithm $\mathfrak{A}$ to it, which (in case of applicability) yields the word P_1; the algorithm $\mathfrak{C}$ ought to be applied to P_1, and if the empty word is obtained as a result, the application of $\mathfrak{A}$ to P_1 is prescribed, which (in case of applicability) yields the word P_2; one should treat it just as P_1 was treated, etc.; this process should be broken off when a word P_n, changed by the algorithm $\mathfrak{C}$ into the empty word, is obtained.

We have arrived, therefore, at a certain combination of the algorithms $\mathfrak{A}$ and $\mathfrak{C}$ in A, a combination which is a verbal algorithm in A. We shall call this combination the *repetition of the algorithm* $\mathfrak{A}$*, directed by the algorithm* $\mathfrak{C}$.

It is natural to ask whether the repetition of a normalizable algorithm, directed by a normalizable algorithm, is normalizable. A positive answer to this question is suggested by the normalization principle. It will in fact follow from Theorem 3.1, to whose formulation we now turn. Before this, however, we must perform a slight preliminary task related to the fact that in the above formulation of the problem, as yet to be made precise, we made use of variable indices and the phrase ≪etc.≫, something which never happened previously. We should like to show the reader that this was merely a tribute to custom and that an exact presentation can be easily and conveniently kept within the bounds of the inductive style adopted by us.

2. Let $\mathfrak{A}$ be a normal algorithm in an alphabet A which does not contain the symbol γ as a letter.

A γ-system S in the alphabet A will be called an *iterative chain of the algorithm* $\mathfrak{A}$ if it satisfies the following condition: whenever a word P is the neighbor of a word Q on the left in S, i.e., whenever the word $\gamma P \gamma Q \gamma$ occurs in S, the equality $\mathfrak{A}(P) \simeq Q$ holds. It is easy to see that

2.1. *Every γ-system in an alphabet A, occurring in an iterative chain of an algorithm $\mathfrak{A}$, is itself an iterative chain of the algorithm $\mathfrak{A}$.*

2.2. *If S is an iterative chain of an algorithm $\mathfrak{A}$, whose last term is P, and $\mathfrak{A}(P) \simeq Q$, then $SQ\gamma$ is an iterative chain of the algorithm $\mathfrak{A}$, whose last term is Q.*

An iterative chain S—like, incidentally, any γ-system— will be called *many-term* if $[S^0 > 1$.

3. The aim of the present section is the proof of the following theorem:

3.1 REPETITION THEOREM. *Given any normal algorithms $\mathfrak{A}$ and $\mathfrak{C}$, one can construct a normal algorithm $\mathfrak{B}$ over the union A of their alphabets, with the following property: $\mathfrak{B}$ changes a word P in the alphabet A into a word Q if and only if $\mathfrak{C}(Q) \simeq \Lambda$ and there exists a many-term iterative chain S of the algorithm $\mathfrak{A}$, connecting P to Q and such that $\mathfrak{C}$ changes each interior*[6] *term of S into a non-empty word.*

Let us begin with the proof of the following lemma.

3.2. *If $\mathfrak{A}$ is a normal algorithm in an alphabet A and β is a letter of this alphabet, then one can construct a normal algorithm $\mathfrak{G}$ over A with the following property: $\mathfrak{G}$ changes a word P in A into a word Q if and only if Q is a word in A, beginning with the letter β, and there exists a many-chain iterative chain S of the algorithm $\mathfrak{A}$, connecting P to Q, and such that none of its interior terms begins with the letter β.*

---

[6]See §24.8.

Let, in fact, $\mathfrak{A}$ be a normal algorithm in an alphabet A, and β, a letter of A. We introduce a letter, α, not belonging to A, and construct the normal algorithm $\mathfrak{G}$ in Aα, whose scheme can be written down in the following abbreviated form:

$$(1) \qquad \begin{cases} \xi\alpha \to \alpha\xi \\ \alpha\beta \to\cdot \beta \\ \alpha \to \\ \mathfrak{A}^{\alpha}, \end{cases}$$

where ξ runs through A and $\mathfrak{A}^{\alpha}$ has the same meaning as in the proof of Theorem §37.2.1.

$\mathfrak{G}$ is a normal algorithm over A. We shall show that it possesses the required property.

For this, let us first of all note that everything said in the proof of Theorem §37.2.1 about the system of formulas $\mathfrak{A}^{\alpha}$ remains valid now, too. In particular, Lemmas §37.2.2–§37.2.6 are valid. Bearing this in mind, it is easy to convince oneself of the validity of Lemmas 3.3–3.6, formulated below. We omit the proofs of these lemmas, since they repeat—with slight changes, yielding simplifications—the proofs of Lemmas §37.2.7 and §37.2.14–§37.2.16, respectively.

Thus we have

3.3. *The algorithm $\mathfrak{G}$ is closed.*

3.4. *If $\mathfrak{A}^{\cdot} : P \vdash Q$, then $\mathfrak{G} : P \vdash Q$.*

3.5. *If $\mathfrak{A}^{\cdot} : P \vdash\cdot Q$, then there exist words R and S such that $Q \eqcirc RS$ and $\mathfrak{G} : P \vdash R\alpha S$.*

3.6. *If R and S are words in* A, *then $\mathfrak{G} : R\alpha S \models \alpha RS$.*

From this, as in the proof of Theorem §37.2.1, we obtain

3.7. *If $\mathfrak{A}^{\cdot} : P \models\cdot Q$, then $\mathfrak{G} : P \models \alpha Q$* (see §37.2.19).

3.8. *If P is a word in* A *and the algorithm $\mathfrak{G}$ is applicable to P, then the algorithm $\mathfrak{A}^{\cdot}$ is also applicable to P* (see §37.2.22).

Examining the scheme (1), we convince ourselves of the truth of the following two lemmas:

3.9. *If Q is a word in* A, *beginning with the letter β, then $\mathfrak{G} : \alpha Q \vdash\cdot Q$.*

3.10. *If Q is a word in* A, *not beginning with the letter β, then $\mathfrak{G} : \alpha Q \vdash Q$.*

Further, let us prove

3.11. *If $\mathfrak{A}(P) \eqcirc Q$ and Q begins with the letter β, then $\mathfrak{G} : P \models\cdot Q$.*

Let, in fact, $\mathfrak{A}(P) \eqcirc Q$, where Q begins with the letter β. Then $\mathfrak{A}^{\cdot}(P) \eqcirc Q$ [§36.2.10], and hence $\mathfrak{A}^{\cdot} : P \models\cdot Q$ [§36.2.1, §36.1.3]. Therefore

$$(2) \qquad \mathfrak{G} : P \models \alpha Q \quad [3.7].$$

Moreover, Q is a word in the alphabet A, since $\mathfrak{A}(P) \coloneqq Q$ and $\mathfrak{A}$ is an algorithm in A. Since Q begins with the letter β, we have

(3) $$\mathfrak{G} : \alpha Q \vdash\cdot Q \quad [3.9],$$
$$\mathfrak{G} : P \models\cdot Q \quad [(2), (3)].$$

Q.E.D.

3.12. *If $\mathfrak{A}(P) \coloneqq Q$ and Q does not begin with the letter β, then $\mathfrak{G} : P \models Q$ and there exists a converting system X of the algorithm $\mathfrak{G}$'s scheme, connecting P to Q, such that $[X^\circ \geqslant 3$.*

Indeed, as in the case of the preceding lemma, $\mathfrak{A}^\cdot : P \models\cdot Q$, and so $\mathfrak{G} : P \models \alpha Q$ [3.7]. In addition, $\mathfrak{G} : \alpha Q \vdash Q$ [3.10]. Consequently, there exists a converting system Y of the algorithm $\mathfrak{G}$'s scheme, connecting P to αQ. Since P is a word in A, but α is not a letter of this alphabet, $[Y^\circ \geqslant 2$. The system $X \rightleftharpoons YQ\gamma$ connects P to Q and, moreover, $[X^\circ = [Y^\circ + 1$, i.e., $[X^\circ \geqslant 3$. Q.E.D.

3.13. *Let S be a many-term iterative chain of the algorithm $\mathfrak{A}$, connecting the word P to the word Q and such that none of its interior terms begins with the letter β. Let Q begin with β. Then $\mathfrak{G}(P) \coloneqq Q$.*

In fact, let the lemma's premises be satisfied. Then P and Q are words in A, where $\gamma P \gamma$ is an initial segment of S and $\gamma Q \gamma$ is a terminal segment of S. There exists a γ-system T in the alphabet A, such that

(4) $$S \coloneqq \gamma PTQ\gamma$$

It follows from 2.1 that T is an iterative chain of the algorithm $\mathfrak{A}$. Every term of T is an interior term of S and, therefore, does not begin with β. Consider the case when $T \not\coloneqq \gamma$. Then T has first and last terms. Let U be T's first term, and V, its last. U and V are words in A. There exist words X and Y in the alphabet Aγ, such that

(5) $$T \coloneqq \gamma U \gamma X$$

and

(6) $$T \coloneqq Y \gamma V \gamma.$$

We have
$$S \coloneqq \gamma P \gamma U \gamma X Q \gamma \quad [(4), (5)]$$
and
$$S \coloneqq \gamma P Y \gamma V \gamma Q \gamma \quad [(4), (6)].$$

Therefore, P is U's neighbor on the left in S, and V is Q's.

We shall show that $\mathfrak{G} : P \models W$ for every term W of the system T. We have $\mathfrak{A}(P) \coloneqq U$, since S is an iterative chain of $\mathfrak{A}$. U does not begin with the letter β, since U is a term of T. Hence $\mathfrak{G} : P \models U$ [3.12]. Now assume it has been established that $\mathfrak{G} : P \models W$ for the term W of the system T, and that W is R's neighbor on the left in T. Then $\mathfrak{A}(W) \coloneqq R$, since T is an iterative chain of $\mathfrak{A}$. R is a term of T and so does not begin with β. Consequently, $\mathfrak{G} : W \models R$ [3.12], and hence $\mathfrak{G} : P \models R$. Applying the method of induction along a system [§24.1.6], we see that each term W of the γ-system T satisfies the condition $\mathfrak{G} : P \models W$. In particular, the last term T satisfies it, i.e., we have

$$\mathfrak{G} : P \models V. \tag{7}$$

On the other hand, $\mathfrak{A}(V) \coloneqq Q$, since S is an iterative chain of $\mathfrak{A}$. But since Q begins with the letter β, we have

$$\mathfrak{G} : V \models\cdot Q \quad [3.11]. \tag{8}$$

Hence

$$\mathfrak{G} : P \models\cdot Q \quad [(7), (8)],$$

and, consequently, $\mathfrak{G}(P) \coloneqq Q$.

Thus, we convinced ourselves that if $T \not\coloneqq \gamma$, then $\mathfrak{G}(P) \coloneqq Q$. But if $T \coloneqq \gamma$, then

$$S \coloneqq \gamma P \gamma Q \gamma \quad [(4)]$$

and P is the neighbor of Q on the left in S. Hence $\mathfrak{A}(P) \coloneqq Q$, and since Q begins with the letter β, we have $\mathfrak{G} : P \models\cdot Q$ [3.11], from which it again follows that $\mathfrak{G}(P) \coloneqq Q$. Lemma 3.13 is proven.

3.14. *Let N be a natural number and let P be a word in an alphabet* A. *Also let*

$$!\mathfrak{G}(P) \tag{9}$$

and

$$(\mathfrak{G} : P) \leqslant N. \tag{10}$$

Then $\mathfrak{G}(P)$ is a word in A, *beginning with the letter β, and there exists a many-term iterative chain S of the algorithm $\mathfrak{A}$, connecting P to $\mathfrak{G}(P)$ and such that none of its interior terms begins with β and its volume does not exceed $N + 1$.*

We shall carry out the proof by the method of arithmetical induction. To this end, let F denote the following one-place arithmetical predicate with the free natural variable N:

> ≪Given any word P in A, if it satisfies the conditions (9) and (10), then $\mathfrak{G}(P)$ is a word in A, beginning with the letter β, and there exists a many-term iterative chain S of the algorithm $\mathfrak{A}$, connecting P to $\mathfrak{G}(P)$, such that [$S^\circ \leqslant N+1$ and no interior term of S begins with β ≫.

Lemma 3.14 will be proven if we prove that every natural number satisfies F.

The number 0 satisfies F trivially. Indeed, if a word P in the alphabet A satisfies the condition (9), then by virtue of $\mathfrak{G}$'s being closed [3.3], we have $\mathfrak{G} : P \models\cdot \mathfrak{G}(P)$ and hence $(\mathfrak{G} : P) > 0$, which contradicts the condition (10) when $N = 0$. Therefore, if $N = 0$ there exist no words P in the alphabet A, satisfying the conditions (9) and (10), and so [§9.2] every universal statement about such words is true. In particular, for every such word P, the following statement is true:

> $\ll \mathfrak{G}(P)$ is a word in A, beginning with the letter β, and there exists a many-term iterative chain S of the algorithm $\mathfrak{A}$, connecting P to $\mathfrak{G}(P)$, such that $[S^{\circ} \leqslant 1$ and no interior term of S begins with $\beta \gg$.

Now suppose that for some N we have convinced ourselves that it satisfies the predicate F. Let us show that $N+1$ also satisfies F. Let P be a word in the alphabet A, satisfying (9) and the condition

(11) $$(\mathfrak{G} : P) \leqslant N + 1.$$

It follows from (9) that $!\mathfrak{A}(P)$ [3.8]. Therefore, there exists a word R in the alphabet A, such that

(12) $$\mathfrak{A}(P) \simeq R.$$

Let us determine whether or not R begins with β. Two cases are possible: a) R begins with the letter β; b) R does not begin with the letter β.

a) *R begins with β.* $\mathfrak{G} : P \models\cdot R$ in this case [3.11], and so $\mathfrak{G}(P) \simeq R$. Consequently, we have

(13) $$\mathfrak{G}(P) \simeq \mathfrak{A}(P) \quad [(12)]$$

and $\mathfrak{G}(P)$ is a word in A, beginning with the letter β. Put

(14) $$S \rightleftharpoons \gamma P \gamma \mathfrak{G}(P) \gamma.$$

Since P and $\mathfrak{G}(P)$ are words in the alphabet A, S is a γ-system in the alphabet A, connecting the word P to the word $\mathfrak{G}(P)$. If some word X is the neighbor of some word Y on the left in S, then $X \simeq P$ and $Y \simeq \mathfrak{G}(P)$ [(14)], from which it follows that $\mathfrak{A}(X) \simeq Y$ [(13)]. Consequently, S is an iterative chain of the algorithm $\mathfrak{A}$. S has no interior terms. Consequently, no interior term of S begins with the letter β. S is many-term, since $[S^{\circ} = 2$. In particular, it follows from this that $[S^{\circ} \leqslant N + 2$. Therefore, S is the desired iterative chain in the case under consideration.

b) *R does not begin with β.* In this case, there exists a converting system U of the scheme of the algorithm $\mathfrak{G}$, connecting P to R and such that $[U^{\circ} \geqslant 3$ [3.12]. We therefore have

(15) $$\mathfrak{G} : P \models R,$$

and since

$$(16) \qquad \mathfrak{S} : P \models\cdot \mathfrak{S}(P)$$

by virtue of $\mathfrak{S}$'s being closed [3.3], we have

$$\mathfrak{S} : R \models\cdot \mathfrak{S}(P) \quad [(16), (15)],$$

and hence

$$(17) \qquad \mathfrak{S}(R) \coloneqq \mathfrak{S}(P),$$

where

$$(18) \qquad (\mathfrak{S} : R) \leqslant (\mathfrak{S} : P) - 2 < N.$$

Let us now make use of the fact that N satisfies the predicate F. Since the word R in the alphabet A satisfies the conditions (17) and (18), $\mathfrak{S}(P)$ is a word in A, beginning with the letter β, and there exists a many-term iterative chain T of the algorithm $\mathfrak{A}$, connecting R to $\mathfrak{S}(P)$, such that none of its interior terms begins with the letter β and

$$(19) \qquad [T^{\circ} \leqslant N + 1.$$

Set

$$(20) \qquad S \rightleftharpoons \gamma PT.$$

We shall show that S is a many-term iterative chain of the algorithm $\mathfrak{A}$, connecting P to $\mathfrak{S}(P)$, such that none of its interior terms begins with β and

$$(21) \qquad [S^{\circ} \leqslant N + 2.$$

This will prove Lemma 3.14.

Since P is a word in the alphabet A and T is a γ-system in A, S is a γ-system in A. P is the first term of S. Since T connects R to $\mathfrak{S}(P)$, the last term of T is $\mathfrak{S}(P)$. Therefore, S connects P to $\mathfrak{S}(P)$.

The number of occurrences of the letter γ in S is obviously greater than two, i.e., S is a many-term system.

Since γ does not occur in the word P in the alphabet A, the number of occurrences of γ in S is one greater than the number of occurrences of γ in T [(20)], and hence

$$(22) \qquad [S^{\circ} = [T^{\circ} + 1.$$

Consequently, (21) holds [(22), (19)].

If X is an interior term of the system S, then X is either the first or an interior term of the system T. The first term of T is the word R, which does not begin with β. No interior term of T begins with the letter β either. Therefore, no interior term of the system S begins with β.

It remains to prove that S is an iterative chain of the algorithm $\mathfrak{A}$. Let the word X be the neighbor of the word Y on the left in the system S. Then we have one of the following two possibilities: b_1) $X \coloneqq P$ and $Y \coloneqq R$; b_2) X is the neighbor of Y on the left in T.

$\mathfrak{A}(X) \coloneqq Y$ in case b_1) [(12)]. We also have $\mathfrak{A}(X) \coloneqq Y$ in case b_2), since T is an iterative chain of the algorithm $\mathfrak{A}$. Therefore, $\mathfrak{A}(X) \coloneqq Y$ whenever X is the neighbor of Y on the left in S, i.e., S is an iterative chain of $\mathfrak{A}$, which is what remained to be proved.

3.15. *Let P be a word in the alphabet A, and let $!\mathfrak{G}(P)$. Then $\mathfrak{G}(P)$ is a word in A, beginning with the letter β, and there exists a many-term iterative chain of the algorithm $\mathfrak{A}$, connecting P to $\mathfrak{G}(P)$ and such that none of its interior terms begins with the letter β.*

This follows from Lemma 3.14.

It follows from Lemmas 3.13 and 3.15 that the normal algorithm $\mathfrak{G}$ satisfies the condition imposed by Lemma 3.2. Therefore, Lemma 3.2 is proven.

We shall now prove Theorem 3.1.

First consider the case when $\mathfrak{A}$ and $\mathfrak{C}$ are normal algorithms in one and the same alphabet A.

Let β be a letter not occurring in A; let $\mathfrak{E}$ be the normal algorithm in the alphabet Aβ with the scheme

$$\{\beta \rightarrow \cdot .$$

We construct the normal algorithm $\mathfrak{F}$ over the alphabet Aβ according to §39.2.2 in such a way that the following conditions hold:

1° The algorithm $\mathfrak{F}$ is applicable to those and only those words in the alphabet A to which the algorithm $\mathfrak{C}$ is applicable.

2° $\mathfrak{F}(X) \coloneqq \beta X$ for every word X in A, such that $\mathfrak{C}(X) \coloneqq \Lambda$.

3° $\mathfrak{F}(X) \coloneqq X$ for every word X in A, such that $\mathfrak{C}$ changes X into a non-empty word.

Let

$$\mathfrak{D} \rightleftharpoons (\mathfrak{F} \circ \mathfrak{A}). \tag{23}$$

Let Д be the alphabet of the algorithm $\mathfrak{D}$. It is clear that β is a letter of Д. We construct the normal algorithm $\mathfrak{G}$ over the alphabet Д according to 3.2 in such a way that the following condition is satisfied: the equality

$$\mathfrak{G}(P) \coloneqq Q, \tag{24}$$

where P is any word in the alphabet Д, holds if and only if Q is a word in the alphabet Д, beginning with the letter β, and there exists a many-term chain S of the algorithm $\mathfrak{D}$, connecting P to Q, such that none of its interior terms begins with the letter β. Let

$$\mathfrak{B} \rightleftharpoons (\mathfrak{E} \circ \mathfrak{G}). \tag{25}$$

$\mathfrak{B}$ is a normal algorithm over the alphabet A. We shall prove that $\mathfrak{B}$ satisfies the condition formulated in Theorem 3.1. As a preliminary, we prove several lemmas.

3.16. *Let P and Q be words in the alphabet A. Let $\mathfrak{C}(\mathfrak{A}) \coloneqq \Lambda$ and let S be a many-term iterative chain of the algorithm $\mathfrak{A}$, connecting P to Q and such that $\mathfrak{C}$ changes each interior term of S into some non-empty word. Then the equality $\mathfrak{B}(P) \coloneqq Q$ holds.*

In fact, since S is an iterative chain of the algorithm $\mathfrak{A}$ in the alphabet A, S is a γ-system in A. Since S is many-term and connects P to Q, there exists a γ-system T in the alphabet A, such that

$$S \coloneqq \gamma PTQ\gamma. \tag{26}$$

T is an iterative chain of the algorithm $\mathfrak{A}$ [2.1]. Q (just as P) is a word in the alphabet A. Set

(27) $$W \rightleftharpoons \gamma P T_\beta Q \gamma.$$

Since γ, P, T and Q are words in the alphabet Aγ, W is a word in the alphabet A$\beta\gamma$. Since W begins and ends with the letter γ, W is a γ-system in the alphabet Aβ. Since T begins and ends with the letter γ, P is the first, and βQ, the last term of W, i.e., W connects P to βQ. Since T contains γ, W is many-term. We shall show that W is an iterative chain of the algorithm $\mathfrak{D}$.

Let X be the neighbor of Y on the left in W. We have one of the following two possibilities:

a) X is the last term of the γ-system γPT and $Y \equiv \beta Q$;

b) X is Y's neighbor on the left in γPT. We consider these two cases.

a) In this case, X is Q's neighbor on the left in S [(26)], and since S is an iterative chain of the algorithm $\mathfrak{A}$, we have

(28) $$\mathfrak{A}(X) \equiv Q.$$

X is a word in the alphabet A. In view of $\mathfrak{C}(Q) \equiv \Lambda$, we have

(29) $$\mathfrak{F}(Q) \equiv \beta Q \quad [2°].$$

Consequently, $\mathfrak{D}(X) \equiv \beta Q$ [(28), (29), (23)], i.e.,

(30) $$\mathfrak{D}(X) \equiv Y.$$

b) In this case, X is Y's neighbor on the left in S and hence

(31) $$\mathfrak{A}(X) \equiv Y.$$

Y is a term of T, and so Y is an interior term of S [(26)]. Consequently, the algorithm $\mathfrak{C}$ changes Y into a non-empty word. Moreover, Y is a word in the alphabet A [(31)]. Hence

(32) $$\mathfrak{F}(Y) \equiv Y \quad [3°];$$

the equality (30) also holds in the given case [(31), (32), (23)].

Therefore, W is an iterative chain of the algorithm $\mathfrak{D}$. Let us show that no interior term of W begins with the letter β.

Let Y be an interior term of W. Then Y is a term of T [(27)], and hence, since T is a γ-system of A, Y is a word in A. Consequently, Y does not begin with the letter β.

Thus, W is a many-term iterative chain of the algorithm $\mathfrak{D}$, connecting the word P in the alphabet A to the word βQ in the alphabet Aβ and such that none of its interior terms begins with the letter β. Here P is a word in the alphabet Д containing the alphabet A [(23)]; the letter β also belongs to Д and βQ is a word in Д, and hence also in the algorithm $\mathfrak{G}$'s alphabet; the word βQ begins with the letter β. According to the construction of the algorithm $\mathfrak{G}$, we can conclude that the following equality holds:

(33) $$\mathfrak{G}(P) \equiv \beta Q.$$

According to the scheme of the algorithm $\mathfrak{E}$, we have

(34) $$\mathfrak{E}(\beta Q) \equiv Q.$$

The equality $\mathfrak{B}(P) \equiv Q$ is therefore satisfied [(33), (34), (25)]. Q.E.D.

3.17. *If X is a word in the alphabet* A, *$\mathfrak{D}(X) \mathrel{=\!\!\!=} Y$ and the word Y does not begin with the letter β, then Y is a word in the alphabet* A, *$\mathfrak{A}(X) \mathrel{=\!\!\!=} Y$ and $\mathfrak{C}$ changes Y into a non-empty word.*

Indeed, if the lemma's hypotheses hold, then there is a word Z in the alphabet A, such that

$$\mathfrak{A}(X) \mathrel{=\!\!\!=} Z \tag{35}$$

and

$$\mathfrak{F}(Z) \mathrel{=\!\!\!=} Y \tag{36}$$

[(23)]. $\mathfrak{F}$ is applicable to Z [(36)], and so $\mathfrak{C}$ is applicable to Z [1°], i.e., there is a word U such that $\mathfrak{C}(Z) \mathrel{=\!\!\!=} U$. If the word U were empty, then we would have

$$\mathfrak{F}(Z) \mathrel{=\!\!\!=} \beta Z \quad [2°], \tag{37}$$

whence $Y \mathrel{=\!\!\!=} \beta Z$ [(36), (37)], which is impossible, since Y does not begin with the letter β. Consequently, $U \not\mathrel{=\!\!\!=} \Lambda$ and so

$$\mathfrak{F}(Z) \mathrel{=\!\!\!=} Z \quad [3°], \tag{38}$$

whence

$$Y \mathrel{=\!\!\!=} Z \tag{39}$$

[(36), (38)]. Therefore, Y is a word in the alphabet A and $\mathfrak{A}(X) \mathrel{=\!\!\!=} Y$ [(35), (39)]. Q.E.D.

3.18. *If X is a word in the alphabet* A *and $\mathfrak{D}(X) \mathrel{=\!\!\!=} \beta Y$, then Y is a word in the alphabet* A, *$\mathfrak{A}(X) \mathrel{=\!\!\!=} Y$ and $\mathfrak{C}(Y) \mathrel{=\!\!\!=} \Lambda$.*

In fact, when this lemma's hypotheses are satisfied, there exists a word Z in the alphabet A, such that (35) holds and

$$\mathfrak{F}(Z) \mathrel{=\!\!\!=} \beta Y \tag{40}$$

[(23)]. $\mathfrak{F}$ is applicable to Z, and so $\mathfrak{C}$ is applicable to Z [1°], i.e., there exists a word U such that $\mathfrak{C}(Z) \mathrel{=\!\!\!=} U$. If U were non-empty, then we would have

$$\mathfrak{F}(Z) \mathrel{=\!\!\!=} Z \quad [3°], \tag{41}$$

whence $Z \mathrel{=\!\!\!=} \beta Y$ [(40), (41)], which is impossible, since Z is a word in an alphabet A, not containing β. Consequently, $U \mathrel{=\!\!\!=} \Lambda$ and $\mathfrak{C}(Z) \mathrel{=\!\!\!=} \Lambda$, by virtue of which we have

$$\mathfrak{F}(Z) \mathrel{=\!\!\!=} \beta Z \quad [2°]. \tag{42}$$

Therefore $\beta Y \mathrel{=\!\!\!=} \beta Z$ [(40), (42)] and so $Y \mathrel{=\!\!\!=} Z$. Thus, Y is a word in the alphabet A, $\mathfrak{A}(X) \mathrel{=\!\!\!=} Y$ [(35)] and $\mathfrak{C}(Y) \mathrel{=\!\!\!=} \Lambda$. Q.E.D.

3.19. *If the equality* $\mathfrak{B}(P) \simeq Q$ *holds, where* P *is in* A, *then* Q *is a word in the alphabet* A, $\mathfrak{C}(Q) \simeq \Lambda$ *and there exists a many-term iterative chain* S *of the algorithm* $\mathfrak{A}$, *connecting* P *to* Q, *and such that* $\mathfrak{C}$ *changes each interior term of* S *into a non-empty word.*

Indeed, let the lemma's hypothesis hold. Then there exists a word R such that

$$(43) \qquad \mathfrak{G}(P) \simeq R$$

and

$$(44) \qquad \mathfrak{E}(R) \simeq Q$$

[(25)]. According to the construction of the normal algorithm $\mathfrak{G}$, R is a word in Д, beginning with the letter β, and there exists a many-term chain W of the algorithm $\mathfrak{D}$, connecting P to R, and such that none of its interior terms begins with the letter β. W is a γ-system in the alphabet Д of the algorithm $\mathfrak{D}$. In view of the fact that W is many-term, there exists a γ-system T in the alphabet Д, such that

$$(45) \qquad W \rightleftharpoons \gamma PTR\gamma.$$

The system T is an iterative chain of the algorithm $\mathfrak{D}$ [2.1]. In view of (44), R is a word in the alphabet Aβ of the algorithm $\mathfrak{E}$. In view of the fact that R begins with the letter β, there exists a word U in the alphabet Aβ, such that

$$(46) \qquad R \rightleftharpoons \beta U.$$

It is clear that every term of the γ-system T is an interior term of W and, consequently, does not begin with the letter β. We shall prove that every term of T is a word in the alphabet A and that the algorithm $\mathfrak{C}$ changes every term of T into a non-empty word. This is true when $T \rightleftharpoons \gamma$, since there exist no terms of T in this case [§9]. Let $T \not\rightleftharpoons \gamma$. Then there is a unique first term V of the system T. V is also the first term of the γ-system $TR\gamma$. P is therefore the neighbor of V on the left in W. Since W is an iterative chain of the algorithm $\mathfrak{D}$,

$$(47) \qquad \mathfrak{D}(P) \simeq V.$$

Here P is a word in the alphabet A and V does not begin with the letter β. Therefore, V is a word in the alphabet A and $\mathfrak{C}$ changes V into a non-empty word [3.17].

Now let X be a word in the alphabet A, let $\mathfrak{C}$ change X into a non-empty word and let X be Y's neighbor on the left in T. Then Y, as a term of T, does not begin with the letter β, and X is Y's neighbor on the left in W, in view of which

$$(48) \qquad \mathfrak{D}(X) \simeq Y.$$

Hence Y is a word in A and $\mathfrak{C}$ changes Y into a non-empty word [3.17].

Applying the method of induction on the system's construction, we conclude that each term of T is a word in the alphabet A and that the algorithm $\mathfrak{C}$ changes each term of T into a non-empty word. Thus, T is a γ-system in the alphabet A. We shall show that T is an iterative chain of $\mathfrak{A}$.

Indeed, let X be the neighbor of Y on the left in T. Then X is Y's neighbor on the left in W [(47)], in view of which we have (48). Here X and Y are terms of T and, therefore, words in A. Y does not begin with β. Hence

$$\mathfrak{A}(X) \simeq Y \quad [3.17]. \tag{49}$$

We now put

$$S \rightleftharpoons \gamma P T Q \gamma. \tag{50}$$

Here P is a word in A, T is a γ-system in A and Q is a word in the alphabet $A\beta$ of the algorithm $\mathfrak{C}$ [(44)]. Consequently S is a γ-system in the alphabet $A\beta$ [(50)]. Let us show that S is an iterative chain of the algorithm $\mathfrak{A}$.

Let X be Y's neighbor on the left in S. Then we have one of the following four possibilities:

1° $X \simeq P$, Y is the first term of T.

2° X is Y's neighbor on the left in T.

3° X is the last term of T, $Y \simeq Q$.

4° $X \simeq P$, $Y \simeq Q$, $T \simeq \gamma$.

When 1° is realized, X is Y's neighbor on the left in W, and so we have (48). Moreover, X is a word in the alphabet A and the word Y, as a term of T, does not begin with the letter β. We therefore have (49) [3.17].

When 2° is realized, we again have (49), since T, as we have seen, is an iterative chain of the algorithm $\mathfrak{A}$.

Suppose that 3° is realized. Then X is R's neighbor on the left in W (45), and so

$$\mathfrak{D}(X) \simeq R, \tag{51}$$

i.e.,

$$\mathfrak{D}(X) \simeq \beta U \quad [(46)], \tag{52}$$

whence

$$\mathfrak{A}(X) \simeq U \quad [3.18], \tag{53}$$

i.e., (49), since by $\mathfrak{C}$'s scheme, (44) and (46), we have $U \simeq Q$.

Finally, let 4° be realized. Then

$$W \simeq \gamma P \gamma R \gamma,$$

where P and R are words in the alphabet $A\beta$. Consequently, P is R's neighbor on the left in W, and so we have (51), i.e., (52). From this we obtain, as above (53), i.e., (49).

This completes the proof that S is an iterative chain of the algorithm $\mathfrak{A}$.

Thus, Q is a word in the alphabet A. We show that $\mathfrak{C}(Q) \simeq \Lambda$.

Two cases are possible: $T \not\simeq \gamma$ and $T \simeq \gamma$. In the first case, let V be the first term of T. Then V is the neighbor of Q on the left in S and, setting $X \rightleftharpoons V$, $Y \rightleftharpoons Q$, we have 3°, considered above, for X and Y. As we have seen, the equality (52) then holds, from which follows $\mathfrak{C}(Q) \simeq \Lambda$ if we take $U \simeq Q$ into account [3.18].

In the second case, putting $X \rightleftharpoons P$ and $Y \rightleftharpoons Q$, we have 4° for X and Y. (52) is also realized then, whence $\mathfrak{C}(Q) \simeq \Lambda$ again.

Thus S is an iterative chain of the algorithm $\mathfrak{A}$. It connects P to Q and is many-term. Every interior term of S is a term of T, and so the algorithm $\mathfrak{C}$ changes each interior term of S into a non-empty word. Finally, we have $\mathfrak{C}(Q) \simeq \Lambda$.

Lemma 3.19 is thus proven.

From Lemmas 3.16 and 3.19 follows the lemma

3.20. *The equality $\mathfrak{B}(P) \simeq Q$, where P is any word in the alphabet* A, *holds if and only if $\mathfrak{C}(Q) \simeq \Lambda$ and there exists a many-term iterative chain S of the algorithm $\mathfrak{A}$, connecting P to Q, and such that $\mathfrak{C}$ changes each interior term of S into a non-empty word.*

Here $\mathfrak{A}$ and $\mathfrak{C}$ are arbitrary normal algorithms in the alphabet A, while $\mathfrak{B}$ is the normal algorithm constructed with the aid of $\mathfrak{A}$ and $\mathfrak{C}$ as described above. By the same token, Theorem 3.1 is proven in case $\mathfrak{A}$ and $\mathfrak{C}$ are normal algorithms in one and the same alphabet. In order to prove the Repetition Theorem in the general case, we can make use of a device already used more than once: passing to formal extensions of the algorithms under consideration. A detailed proof is left to the reader.

We shall call the resulting normal algorithm $\mathfrak{B}$ the *normal repetition of the algorithm $\mathfrak{A}$, directed by the algorithm $\mathfrak{C}$.*

We shall denote this algorithm by means of

$$(\mathfrak{A} \circlearrowleft \mathfrak{C}).$$

We note two simple, but important corollaries of Theorem 3.1.

3.21. *If $\mathfrak{C}(\mathfrak{A}(P)) \simeq \Lambda$, then $(\mathfrak{A} \circlearrowleft \mathfrak{C})(P) \simeq \mathfrak{A}(P)$.*

Indeed, let $\mathfrak{C}(\mathfrak{A}(P)) \simeq \Lambda$. Then $\mathfrak{A}(P)$ makes sense. Put

$$S \rightleftharpoons \gamma P \gamma \mathfrak{A}(P) \gamma.$$

As is easily seen, the γ-system S connects P to $\mathfrak{A}(P)$ and is an iterative chain of the algorithm $\mathfrak{A}$. In fact, P is the first term of S and $\mathfrak{A}(P)$ is its last term. Furthermore, if X is Y's neighbor on the left in S, then $X \simeq P$ and $Y \simeq \mathfrak{A}(P)$, so $\mathfrak{A}(X) \simeq Y$. The iterative chain S is many-term, since [$S^\circ = 2$. The algorithm $\mathfrak{C}$ changes all its interior terms (there simply are not any) into non-empty words [§9]. By hypothesis, the algorithm $\mathfrak{C}$ changes the last term of S into the empty word. Therefore, on the basis of Theorem 3.1 $(\mathfrak{A} \circlearrowleft \mathfrak{C})(P) \simeq \mathfrak{A}(P)$. Q.E.D.

3.22. *If* $\mathfrak{C}(\mathfrak{A}(P)) \neq \Lambda$, *then* $(\mathfrak{A} \circlearrowleft \mathfrak{C})(P) \simeq (\mathfrak{A} \circlearrowleft \mathfrak{C})(\mathfrak{A}(P))$.

Indeed, assuming that $\mathfrak{C}(\mathfrak{A}(P)) \neq \Lambda$ it is not hard to show that:

1°. If the word Q is such that $\mathfrak{C}(Q) \Bumpeq \Lambda$ and S is a many-term iterative chain of the algorithm $\mathfrak{A}$, connecting $\mathfrak{A}(P)$ to Q and such that the algorithm $\mathfrak{C}$ changes each of its interior terms into a non-empty word, then the γ-system γPS is a many-term iterative chain of the algorithm $\mathfrak{A}$, connecting P to Q and possessing the property that the algorithm $\mathfrak{C}$ changes each of its interior terms into a non-empty word. (Note that in comparison with S, the system γPS possesses one additional interior term, $\mathfrak{A}(P)$, but by assumption, the algorithm $\mathfrak{C}$ changes this term into a non-empty word.)

2°. If the word Q is such that $\mathfrak{C}(Q) \Bumpeq \Lambda$ and S is a many-term iterative chain of the algorithm $\mathfrak{A}$, connecting P to Q and such that the algorithm $\mathfrak{C}$ changes each of its interior terms into a non-empty word, then S has the form $\gamma P\gamma\mathfrak{A}(P)TQ\gamma$, where T is a γ-system, and so the γ-system $\gamma\mathfrak{A}(P)TQ\gamma$ is a many-term iterative chain of the algorithm $\mathfrak{A}$, connecting $\mathfrak{A}(P)$ to Q and possessing the property that the algorithm $\mathfrak{C}$ changes each of its interior terms into a non-empty word. (It is easy to see that every interior term of the "reduced" system $\gamma\mathfrak{A}(P)TQ\gamma$ is an interior term of S.)

Starting with 1° and 2°, it is not difficult to prove the conditional equality $(\mathfrak{A} \circlearrowleft \mathfrak{C})(P) \simeq (\mathfrak{A} \circlearrowleft \mathfrak{C})(\mathfrak{A}(P))$ of interest to us.

We leave details of the argument to the reader.

4. Theorem 3.1 deals with a repeated application of a given normal algorithm until we obtain a word with a desired property, where the initial word, itself, is not tested for this property.

It can, however, prove necessary to begin testing for the desired property immediately with the initial word, itself. In this case, the process terminates with the initial word if it possesses this property. But if it does not possess it, the algorithm $\mathfrak{A}$ is applied to it, and then the process proceeds as described above. To this modification of a repetition of the algorithm $\mathfrak{A}$, directed by $\mathfrak{C}$, there corresponds the following modifications of Theorem 3.1:

4.1. *Given any normal algorithms* $\mathfrak{A}$ *and* $\mathfrak{C}$, *one can construct a normal algorithm* $\mathfrak{D}$ *over the union* A *of their alphabets with the following property:* $\mathfrak{D}$ *changes a word* P *in the alphabet* A *into a word* Q *if and only if* $\mathfrak{C}(Q) \Bumpeq \Lambda$ *and there exists an iterative chain* S *of the algorithm* $\mathfrak{A}$, *connecting* P *to* Q *and such that* $\mathfrak{C}$ *changes each term of* S, *except the last, into a non-empty word.*

Theorem 4.1 can be proven in the following manner.

Let $\mathfrak{A}$ and $\mathfrak{C}$ be arbitrary normal algorithms; let A be their alphabets' union. Construct the normal algorithm $\mathfrak{B}$ over A according to Theorem 3.1 in such a way that it possesses the property indicated in this theorem. Construct the identity normal algorithm $\mathfrak{I}_{\text{A}}$ in the alphabet A [§28.4]. Apply the Branching Theorem to the algorithms $\mathfrak{I}_{\text{A}}$, $\mathfrak{B}$ and $\mathfrak{C}$ [§39.1.1]. Construct, according to this theorem, the normal algorithm $\mathfrak{D}$ over the union of these algorithm's alphabets, i.e., over the alphabet Б of the algorithm $\mathfrak{B}$, in such a way that the following conditions are satisfied for any P in Б.

(1) if $\mathfrak{C}(P) \Bumpeq \Lambda$, then $\mathfrak{D}(P) \simeq \mathfrak{I}_{\text{A}}(P)$,

(2) if $\mathfrak{C}(P) \neq \Lambda$, then $\mathfrak{D}(P) \simeq \mathfrak{B}(P)$,

(3) if $\mathfrak{D}$ is applicable to P, then $\mathfrak{C}$ is also applicable to P.

Then $\mathfrak{D}$ is the required algorithm.

In fact, Б is an extension of A, since $\mathfrak{B}$ is an algorithm over A and simultaneously an algorithm in Б. Hence $\mathfrak{D}$, as an algorithm over Б, is an algorithm over A.

Let P be in A and let $\mathfrak{D}(P) \eqcirc Q$. Then $\mathfrak{D}$ is applicable to P, and so $\mathfrak{C}$ is applicable to P [(3)].

Let $\mathfrak{C}(P) \eqcirc \Lambda$. Then $\mathfrak{D}(P) \eqcirc P$ [(1), §28.4]. Therefore $P \eqcirc Q$, and since $\mathfrak{C}(P) \eqcirc \Lambda$, we have

$$\mathfrak{C}(Q) \eqcirc \Lambda.$$

Set

$$S \rightleftharpoons \gamma P \gamma.$$

S is a γ-system in the alphabet A whose first term is P and whose last term is Q. No word of the form $\gamma X \gamma Y \gamma$ occurs in S, by virtue of which S is an iterative chain of the algorithm $\mathfrak{A}$ [§9]. S has no terms which are not its last term. Consequently, the algorithm $\mathfrak{C}$ changes each such term of S into a non-empty word [§9].

Now let $\mathfrak{C}(P) \neq \Lambda$. Then $\mathfrak{D}(P) \simeq \mathfrak{B}(P)$ [(2)], and so $\mathfrak{B}(P) \eqcirc Q$. According to the algorithm $\mathfrak{B}$'s construction, the equality $\mathfrak{C}(Q) \eqcirc \Lambda$ holds and there exists a many-term iterative chain of the algorithm $\mathfrak{A}$, connecting P to Q and such that $\mathfrak{C}$ changes each of its interior terms into a non-empty word. But $\mathfrak{C}$ also changes the first term of S—the word P—into a non-empty word. Consequently, $\mathfrak{C}$ changes all the terms of S, except the last, into non-empty words.

Thus, we have proven that as soon as the equality $\mathfrak{D}(P) \eqcirc Q$ holds for a word P in A, the condition $\mathfrak{C}(Q) \eqcirc \Lambda$ is satisfied and there exists an iterative chain S of the algorithm $\mathfrak{A}$, connecting P to Q and such that $\mathfrak{C}$ changes all the terms of S, except the last, into empty words.

Now suppose that the equality $\mathfrak{C}(Q) \eqcirc \Lambda$ is satisfied for the words P and Q in A and there exists an iterative chain S of the algorithm $\mathfrak{A}$, connecting P to Q, and such that $\mathfrak{C}$ changes all the terms of S, except the last, into non-empty words. We shall prove that then the equality $\mathfrak{D}(P) \eqcirc Q$ holds.

Since no interior term of S is the last term of this system, the algorithm $\mathfrak{C}$ changes each interior term of S into a non-empty word. Let us find the volume $[S^\circ$ of the system S. Since S connects P to Q, $[S^\circ \geqslant 1$. We consider two cases: $[S^\circ = 1$ and $[S^\circ > 1$.

In the first case, we obviously have $P \eqcirc Q$, and so $\mathfrak{C}(P) \eqcirc \Lambda$. Since $\mathfrak{I}_{\mathrm{A}}(P) \eqcirc P$ [§28.4], we have $\mathfrak{D}(P) \eqcirc P$ [(1)], and since $P \eqcirc Q$, $\mathfrak{D}(P) \eqcirc Q$.

In the second case, the system S is many-term. By the construction of the algorithm $\mathfrak{B}$, the equality $\mathfrak{B}(P) \eqcirc Q$ holds. The system S has the first term P, which is not its last term. Therefore, $\mathfrak{C}$ changes P into a non-empty word and we have $\mathfrak{D}(P) \eqcirc Q$ [(12)].

This completes the proof of Theorem 4.1.

5. In theorems 3.1 and 4.1, the question was one of continuing an iterative chain up to a term satisfying a certain algorithmically verifiable condition.

Analogously, the question in Theorem 5.1 is one of continuing an iterative chain up to a given volume. A role in this theorem will be played by words of the form $N * P$, where N is a natural number and P is a word in a given alphabet A (we shall assume that $*$ is not a letter of A).

Words of this kind may be regarded as pairs whose first term is a natural number, and whose second, a word in a given alphabet. The letter C in Theorem 5.1 denotes the two-letter alphabet $\mathsf{I}*$.

5.1. *Given any normal algorithm $\mathfrak{A}$ in an alphabet* A, *one can construct a normal algorithm $\mathfrak{B}$ over the alphabet* $\mathrm{A} \cup \mathrm{C}$, *such that the equality*

$$\mathfrak{B}(N * P) \simeq Q \tag{1}$$

holds for the word P in A *and the natural number N if and only if there exists an iterative chain of volume $N+1$ of the algorithm $\mathfrak{A}$, connecting P to Q.*

Let us prove this theorem.

Let

$$\text{Б} \rightleftharpoons \mathrm{A} \cup \mathrm{C}. \tag{2}$$

Let α be a letter not occurring in Б and let

$$\mathrm{B} \rightleftharpoons \text{Б}\alpha. \tag{3}$$

Construct the normal algorithms $\mathfrak{C}$, $\mathfrak{E}$ and $\mathfrak{F}$ in the alphabet B with the following schemes:

$$\mathfrak{C} : \begin{cases} \alpha\mathsf{I} \rightarrow\cdot \\ \alpha * \xi \rightarrow \alpha * \\ \alpha * \rightarrow\cdot \\ \quad\ \rightarrow \alpha, \end{cases}$$

$$\mathfrak{E} : \{ * \xi \rightarrow *$$

$$\mathfrak{F} : \begin{cases} \mathsf{I}* \rightarrow * \\ * \rightarrow, \end{cases}$$

where the literal variable ξ runs through the alphabet Б. The next four lemmas, in which N denotes an arbitrary natural number, and P, an arbitrary word in the alphabet Б, are easily proven.

5.2. $\mathfrak{C}((N+1) * P) \simeq N * P$.

5.3. $\mathfrak{C}(*P) \simeq \Lambda$.

5.4. $\mathfrak{E}(N * P) \simeq N *$.

5.5. $\mathfrak{F}(N * P) \simeq P$.

We construct the normal compositions $(\mathfrak{E} \circ \mathfrak{C})$ and $(\mathfrak{A} \circ \mathfrak{F})$. These are normal algorithms over the alphabet B. Let

$$\mathfrak{H} \rightleftharpoons ((\mathfrak{E} \circ \mathfrak{C}) \frown (\mathfrak{A} \circ \mathfrak{F}) \quad [\S 38.6]. \tag{4}$$

$\mathfrak{H}$ is a normal algorithm over B [§38.6.2]. We shall prove

5.6. *If*

$$\mathfrak{A}(P) \simeq Q, \tag{5}$$

then for any natural number N we have

$$\mathfrak{H}((N+1) * P) \simeq N * Q. \tag{6}$$

Indeed, assume the equality (5) holds. Then P is a word in the alphabet A and we have

$$(\mathfrak{E} \circ \mathfrak{C})((N+1) * P) \simeq N* \tag{7}$$

for any natural number N [5.2, 5.4]. Further, we have

$$\mathfrak{F}(N+1) * P \simeq P \quad [5.5], \tag{8}$$

$$(\mathfrak{A} \circ \mathfrak{F})((N+1) * P) \simeq Q \quad [(8), (5)] \tag{9}$$

and the equality (6) [(7), (9)].

We also have

5.7. *If* $!\mathfrak{H}(N * R)$, *where* N *is a natural number and* R *is a word in the alphabet* A, *then* $!\mathfrak{A}(R)$ [(4), §38.6.3, §37.4.3, 5.5].

Now let Γ denote the algorithm $\mathfrak{H}$'s alphabet. We construct the normal algorithm $\mathfrak{D}$ in accordance with Theorem 4.1 in such a way that the equality $\mathfrak{D}(P) \simeq Q$, where P is any word in Γ, will hold if and only if

$$\mathfrak{C}(Q) \simeq \Lambda \tag{10}$$

and there exists an iterative chain T of the algorithm $\mathfrak{H}$, connecting P to Q and such that $\mathfrak{C}$ changes all the terms of T except the last, into non-empty words. Construct the normal algorithm $\mathfrak{B}$ as the normal composition of $\mathfrak{D}$ and $\mathfrak{F}$:

$$\mathfrak{B} \rightleftharpoons (\mathfrak{F} \circ \mathfrak{D}). \tag{11}$$

We shall show that $\mathfrak{B}$ satisfies the condition formulated in Theorem 5.1.

For this we prove the lemma

5.8. *If the equality* (1) *holds, where* N *is a natural number and* P *is a word in the alphabet* A, *then there exists an iterative chain of volume* $N+1$ *of the algorithm* $\mathfrak{A}$, *connecting* P *to* Q.

Let the hypothesis of 5.8 be fulfilled for some fixed N, P and Q. In view of (11) and (1), there exists a word R such that

$$\mathfrak{D}(N * P) \simeq R \tag{12}$$

and

$$\mathfrak{F}(R) \simeq Q. \tag{13}$$

According to the construction of the algorithm $\mathfrak{D}$, we have

$$\mathfrak{C}(R) \simeq \Lambda \tag{14}$$

and there exists an iterative chain T of the normal algorithm $\mathfrak{H}$, connecting $N * P$ to R and such that the algorithm $\mathfrak{C}$ changes all the terms of T, except the last, into non-empty words [(12)].

We shall prove that every natural number K less than or equal to N satisfies the following condition F:

> ≪there exists an iterative chain of the algorithm $\mathfrak{A}$, whose first term is P, whose volume is $K+1$ and whose last term X is such that the word $(N-K) * X$ is the $(K+1)$st term of the chain T≫.

The number 0 satisfies the condition F. In fact, the word $\gamma P \gamma$ is trivially an iterative chain of the algorithm $\mathfrak{A}$; P is the first term of this iterative chain whose volume is equal to 1, the last term of the chain $\gamma P \gamma$ is P and $N * P$ is the first term of the chain T.

Now assume that some number K, less than N, satisfies F. Let us show that $K+1$ also satisfies F.

There is an iterative chain S of the normal algorithm $\mathfrak{A}$, whose first term is P, whose volume is $K+1$ and whose last term X is such that $(N-K) * X$ is the $(K+1)$st term of T. Since $K < N$, the number $N-K$ is positive. X is a term of the iterative chain S of the algorithm $\mathfrak{A}$ in the alphabet A, by virtue of which X is a word in A. We have

$$\mathfrak{C}((N-K) * X) \simeq ((N-K)-1) * X \quad [5.2]. \tag{15}$$

Thus, the algorithm $\mathfrak{C}$ changes $(N-K) * X$ into a non-empty word and by virtue of (14) we have

$$(N-K) * X \neq R. \tag{16}$$

Consequently, $(N-K) * X$ is not the last term of the chain T, in view of which $(N-K) * X$ is the neighbor of one of T's terms Z on the left in T. Since T is an iterative chain of the algorithm $\mathfrak{H}$,

$$\mathfrak{H}((N-K) * X) \simeq Z \tag{17}$$

and, consequently

$$!\mathfrak{H}((N-K) * X). \tag{18}$$

Therefore, $!\mathfrak{A}(X)$ [(18), 5.7] and there is a word Y such that

$$\mathfrak{A}(X) \simeq Y. \tag{19}$$

Construct

$$U \rightleftharpoons SY\gamma. \tag{20}$$

U is an iterative chain of the algorithm $\mathfrak{A}$ whose last term is Y. We have

(21) $$\mathfrak{H}((N-K)*X) \eqsim (N-(K+1))*Y \quad [(19), 5.6],$$

by virtue of which

$$Z \eqsim (N-(K+1))*Y \quad [(17), (21)].$$

Thus, $(N-(K+1))*Y$ is a term of the iterative chain T of the algorithm $\mathfrak{H}$. The volume of the chain U is obviously one greater than the volume of the chain S [(20)], i.e., it equals $K+2$. U's first term coincides with the first term of S [(20)], i.e., U's first term is the word P. U's last term is a word Y [(20)] such that $(N-(K+1))*Y$ is a term of T.

We have therefore proven that the number $K+1$ satisfies the predicate F as soon as the number K, less than N, satisfies it. By the same token, it is proven that every natural number not exceeding N satisfies this predicate. In particular, N satisfies it, i.e., there exists an iterative chain W of the normal algorithm $\mathfrak{A}$, whose first term is P, whose volume is $N+1$ and whose last term V is such that the word $*V$ is a term of T. Being a term of an iterative chain of the algorithm $\mathfrak{A}$ in the alphabet A, V is a word in A. Hence

$$\mathfrak{C}(*V) \eqsim \Lambda \quad [5.3].$$

It follows from this that $*V$ is the last term of the chain T. Consequently

(22) $$*V \eqsim R.$$

Hence

(23) $$\mathfrak{F}(R) \eqsim V \quad [5.5, (22)].$$

Therefore

$$V \eqsim Q \quad [(23), (13)].$$

Thus, W is the required iterative chain of the algorithm $\mathfrak{A}$, connecting P to Q and having volume $N+1$. Lemma 5.8 is proven.

Theorem 5.1 will be proven if we prove the lemma

5.9. *If N is a natural number, P and Q are words in the alphabet* A *and there exists an iterative chain of volume $N+1$ of the algorithm $\mathfrak{A}$, connecting P to Q, then the equality* (1) *holds.*

Indeed, let this lemma's hypothesis be fulfilled. Let W be an iterative chain of volume $N+1$ of the algorithm $\mathfrak{A}$, connecting P to Q. We show that every natural number K less than or equal to N satisfies the following condition G:

> ≪there exists an iterative chain of volume $K+1$ of the algorithm $\mathfrak{H}$, connecting the word $N*P$ to the word $(N-K)*X$, where X is the $(K+1)$st term of W≫.

The number 0 satisfies the condition G. In fact, the word $\gamma N * P\gamma$ is an iterative chain of the algorithm $\mathfrak{H}$ of volume 1, connecting the word $N * P$ to itself, where P is the 1st term of W.

Suppose that the number K, less than N, satisfies G. Then there is an iterative chain T of the algorithm $\mathfrak{H}$ of volume $K+1$, connecting the word $N*P$ to the word $(N-K)*X$, where X is the $(K+1)$st term of W. Since $K < N$, $K+2 \leqslant N+1$, i.e., $K+2 \leqslant [W^{\circ}$. Hence there exists a term Y of the chain W, whose number is $K+2$. Since X is W's term with number $K+1$, X is the neighbor of Y on the left in W. Since W is an iterative chain of the algorithm $\mathfrak{A}$, we have

$$\mathfrak{A}(X) \simeq Y,$$

by virtue of which

$$\mathfrak{H}((N-K)*X) \simeq (N-(K+1))*Y \quad [5.6]. \tag{24}$$

Since T is an iterative chain of the algorithm $\mathfrak{H}$, connecting $N*P$ to $(N-K)*X$, the word $(N-K)*X$ is the last term of T. Therefore

$$T(N-(K+1))*Y\gamma$$

is an iterative chain of $\mathfrak{H}$ of volume $K+2$ with the last term $(N-(K+1))*Y$ [(24)]. But the first term of the iterative chain $T(N-(K+1))*Y\gamma$ coincides with the first term of T, i.e., with $N*P$. Thus, $T(N-(K+1))*Y\gamma$ is an iterative chain of the algorithm $\mathfrak{H}$ of volume $K+1$, connecting $N*P$ to $(N-(K+1))*Y$, where Y is a term of the chain W with the number $K+2$. We have proven that as soon as a natural number K, less than N, satisfies the condition G, the natural number $K+1$ also satisfies it. Consequently, every natural number less than or equal to N satisfies the condition G. In particular, N satisfies the condition G. This means that there exists an iterative chain S of the algorithm $\mathfrak{H}$ of volume $N+1$, connecting $N*P$ to the word $*X$, where X is the term of W whose number is $N+1$. Since the volume of W is equal to $N+1$, X is the last term of W. Nevertheless, the last term of the chain W is the word Q, since W connects P to Q. We therefore have $X \simeq Q$. Thus, the iterative chain S of the algorithm $\mathfrak{H}$ connects the word $N*P$ to the word $*Q$.

Let us prove that whatever be the natural number K less than or equal to N, the term of the chain S with the number $K+1$ has the form $(N-K)*X$, where X is the term of the chain W with the number $K+1$.

This is so when $K=0$, in view of the fact that the first term of the chain S is the word $N*P$, where P is the first term of the chain W.

Now let $K < N$ and let it be known that the term of the chain S with the number $K+1$ has the form $(N-K)*X$, where X is the term of the chain W with the number $K+1$. We have $K+2 \leqslant N+1$, and hence there is a term of the chain W with the number $K+2$ and a term of the chain S with the same number. Let Y be W's term with the number $K+2$, Z, the term of S with the number $K+2$. Then X is Y's neighbor on the left in W, and so

$$\mathfrak{A}(X) \simeq Y; \tag{25}$$

$(N - K) * X$ is Z's neighbor on the left in S, and so

(26) $$\mathfrak{H}((N - K) * X) \coloneqq Z.$$

In view of (25), we have

(27) $$\mathfrak{H}((N - K) * X) \coloneqq (N - (K + 1)) * Y \quad [5.6].$$

Consequently, $Z \coloneqq (N - (K + 1)) * Y$ [(26),(27)], where Z is the term of S with the number $K + 2$ and Y is W's term with the same number. Using the method of arithmetical induction, we discover that when $K \leqslant N$, S's term with the number $K + 1$ has the form $(N - K) * X$, where X is W's term with the number $K + 1$. Therefore, when $K < N$, the algorithm $\mathfrak{C}$ changes the term of the chain S with the number $K + 1$ into a non-empty word [5.2].

Now let U be a term of S other than the last. There exists a number M of the term U in the chain S, which is less than the volume of S, i.e., less than $N + 1$. It is positive and, therefore, has the form $K + 1$, where K is a natural number. We have $K + 1 < N + 1$, and so $K < N$. Thus, U is the term of S with the number $K + 1$, where $K < N$. As we have just seen, it follows from this that the algorithm $\mathfrak{C}$ changes U into a non-empty word. Consequently, this algorithm changes each term, other than the last, of the chain S connecting the word $N * P$ to the word $*Q$, into a non-empty word. Moreover, Q is a word in the alphabet A and

$$\mathfrak{C}(*Q) \coloneqq \Lambda \quad [5.3].$$

According to the construction of the algorithm $\mathfrak{D}$, we perceive that

(28) $$\mathfrak{D}(N * P) \coloneqq *Q.$$

Furthermore

(29) $$\mathfrak{F}(*Q) \coloneqq Q \quad [5.5].$$

Consequently,

$$\mathfrak{B}(N * P) \coloneqq Q \quad [(11), (28), (29)].$$

Q.E.D.

Lemma 5.9 has been proven. Together with it, Theorem 5.1 is also proven.

§41 Translation of an Algorithm

When constructing normal algorithms which fulfill, in a given alphabet, one or another task of interest to us, we are often forced to introduce "additional" letters in the course of the construction. These letters are absent from the initial data and the result of the algorithm's performance. They appear temporarily in the "current" word of a process and, having fulfilled a definite auxiliary function, disappear before the process breaks off. This was the case, for example, with the letter α in the right joining algorithm [§28.3(2)], the letters α, β and γ in the doubling algorithm [§31.1(1)], etc. The number of additional letters changes from algorithm to algorithm. It is impossible to manage entirely without them, as Theorem §32.3.1, for example, shows. It is natural, however, to ask whether it is possible to manage with some fixed (preferably a small) number of additional letters. When this question is made precise in a natural manner, the answer to it turns out to be affirmative (Markov [2], III. §7.4.2; Nagorny [1]).

The present section is devoted to a consideration of this question. Its main content is contained in Part 6. A general theory of the translation of words is presented in the preceding parts.

1. Let A be an alphabet not containing the letters α, β and $*$. We shall call letters of A and words of the form aBa, where B is a non-empty word in the one-letter alphabet b, A-*literoids*.

It is obvious that

1.1. *Every* A-*literoid is a non-empty word in the alphabet* Aab.

It is also easy to show that

1.2. *If the* A-*literoid* X *occurs in the* A-*literoid* Y, *then* $X \leftrightharpoons Y$.

In particular,

1.3. *If the* A-*literoid* X *begins or ends with the* A-*literoid* Y, *then* $Y \leftrightharpoons Y$.

1.4. *No* A-*literoid can be a proper initial or terminal segment or some other* A-*literoid.*

It is also clear that

1.5. *If* L *is an* A-*literoid and* P *is a word in* Aa, *such that* $[P^{\delta} \geqslant 2$, *then* P *does not occur in* L.

Let us prove

1.6. *Every occurrence of the* A-*literoid* L *in the word* RMS, *where* M *is an* A-*literoid and* R, S *are words in* Aab *such that* R *does not end, while* S *does not begin with the letter* b, *has one of the following three forms:*
KMS, *where* K *is an occurrence of* L *in* R;
RMK, *where* K *is an occurrence of* L *in* S;
$R * L * S$, *where* $M \leftrightharpoons L$.

Indeed, let

$$(1) \qquad X * L * Y$$

be an occurrence of the A-literoid L in the word RMS, where M is an A-literoid and R, S are words in Aab satisfying the requirements listed in the lemma's formulation. Then X and Y are words in Aab and the following equality is fulfilled:

$$(2) \qquad XLY \leftrightharpoons RMS.$$

By virtue of (2), the words X and R are initial segments of one and the same word. We therefore have one of the following three possibilities:

1°. X is a proper initial segment of R.

2°. R is a proper initial segment of X.

3°. $R \asymp X$.

Let 1° hold. Then X is a word in Aab and there exists a non-empty word W such that

$$R \asymp XW. \tag{3}$$

But then

$$XLY \asymp XWMS \quad [(2), (3)] \tag{4}$$

and, therefore,

$$LY \asymp WMS \quad [(4)]. \tag{5}$$

By virtue of (5), L and M are initial segments of one and the same word. Therefore, either L is an initial segment of W or W is a proper initial segment of L.

Assume that W is a proper initial segment of L. Then there exists a non-empty word V such that $L \asymp WV$. But then $WVY \asymp WMS$ [(5)] and so $VY \asymp MS$. Denoting the first letter of V by η, we see from this that it is the first letter of the A-literoid M and is, consequently, different from b. We discover from (3) that W's last letter also differs from b. Denote it by means of ξ. Then we shall have $L \asymp W'\xi\eta V'$ for some V' and W', which is obviously impossible [1.5].

Hence, L is an initial segment of W and so there exists a word Z such that

$$W \asymp LZ. \tag{6}$$

But then

$$LY \asymp LZMS \quad [(5), (6)] \tag{7}$$

and, therefore,

$$Y \asymp ZMS \quad [(7)]. \tag{8}$$

Thus,

$$X * L * Y \asymp X * L * ZMS \quad [(1), (8)]. \tag{9}$$

But

$$XLZ \asymp R \quad [(3), (6)], \tag{10}$$

so that, setting

$$K \rightleftharpoons X * L * Z,$$

we see that in Case 1°, the occurrence (1) has the form KMS [(9)], where K is an occurrence of L in R [(10)].

Now consider Case 2°. In this case, there exists a non-empty word W such that

(11) $$X \leftrightharpoons RW.$$

But then

(12) $$RWLY \leftrightharpoons RWS \quad [(2), (11)],$$

and so

(13) $$WLY \leftrightharpoons MS \quad [(12)].$$

By virtue of (13), W and M are initial segments of one and the same word. Therefore, either M is an initial segment of W or W is a proper initial segment of M. In the first case,

(14) $$W \leftrightharpoons MZ$$

for some Z, and so

(15) $$MZLY \leftrightharpoons MS \quad [(13), (14)],$$

whence

(16) $$ZLY \leftrightharpoons S \quad [(15)].$$

Thus,

(17) $$X * L * Y \leftrightharpoons RMZ * L * Y \quad [(11), (14)]$$

and, putting

(18) $$K \rightleftharpoons Z * L * Y,$$

we see that under our assumption,

$$X * L * Y \leftrightharpoons RMK \quad [(17), (18)],$$

where K is an occurrence of L in S [(16)].

We shall show that W can not be a proper initial segment of M. In fact, suppose there exists a non-empty word T such that

(19) $$M \leftrightharpoons WT.$$

Then

$$(20) \qquad WLY \eqcirc WTS \quad [(13), (19)],$$

and hence

$$(21) \qquad LY \eqcirc TS \quad [(20)].$$

L can not be an initial segment of T, since there would then exist a word U such that

$$(22) \qquad T \eqcirc LU$$

and so we would have $M \eqcirc WLU$ [(19), (22)], which would contradict 1.2 in view of the non-emptiness of W. Thus it remains to assume that T is a proper initial segment of L, i.e., that there exists a non-empty word V such that

$$(23) \qquad L \eqcirc TV.$$

But then

$$(24) \qquad TVY \eqcirc TS \quad [(21), (23)]$$

and, consequently,

$$(25) \qquad VY \eqcirc S \quad [(24)].$$

The words T and V are non-empty, and so L, being an A-literoid, has the form aBa. The equality (23) shows that the last letter of V is a, i.e., $V \eqcirc V'a$ for some V'. Analogously, we conclude from (19) that $T \eqcirc T'a$ for some T'. But then $L \eqcirc T'aV'a$ [(23)] and so $T' \eqcirc \Lambda$ and V' is a non-empty word in the alphabet b. Consequently, S begins with the letter b [(25)], which contradicts the lemma's hypothesis.

This completes the examination of Case 2°.

In Case 3°, when $X \eqcirc R$, we have

$$(26) \qquad XLY \eqcirc XMS \quad [(2)],$$

and therefore

$$(27) \qquad LY \eqcirc MS \quad [(26)].$$

Consequently, L begins with M or M begins with L. From this we get $L \eqcirc M$ by 1.3, and so

$$(28) \qquad Y \eqcirc S \quad [(27)],$$

and hence

$$X * L * Y \eqcirc R * M * S \quad [(1), (28)].$$

Thus, all the cases 1°–3° have been considered. The assertion 1.6 is proven.

From this easily follows

1.7. *Every occurrence of the* A-*literoid* L *in the word* RM, *where* M *is an* A-*literoid and* R *is a word in* Aab *not ending with the letter* b, *has one of the following forms:*

KM, *where* K *is an occurrence of* L *in* R;

$R * L*$, *where* $M \rightleftharpoons L$.

1.8. *Every occurrence of the* A-*literoid* L *in the word* MS, *where* M *is an* A-*literoid and* S *is a word in* Aab *not beginning with the letter* b, *has one of the following forms:*

MK, *where* K *is an occurrence of* L *in* S;

$*L * S$, *where* $M \rightleftharpoons L$.

2. We define a certain "grammatical category" of words in the alphabet Aab, called ≪A-verboids≫. These words will be defined by the two following construction rules:

Vd 1. The empty word is regarded as an A-*verboid.*

Vd 2. Whenever an A-*verboid* X has been constructed, every word constructed as XL, where L is any A-literoid, is regarded as an A-verboid.

A word X will be regarded as an A-verboid if and only if the truth of the statement

≪ X is regarded as an A-verboid ≫

can be inferred on the basis of the rules Vd 1 and Vd 2, which we shall call the A-*verboid construction rules.* According to the definition, the following words, for example, are A-verboids:

$$\Lambda, abba, abbbba, abbbaabbba.$$

The process of constructing A-verboids splits up naturally into steps; a certain A-verboid is constructed at each of them, where one of the rules, Vd 1 or Vd 2 is applied.

In view of the ≪inductive≫ nature of our definition of the concept of an A-verboid—a definition by means of describing a process for constructing one—in order to prove general statements about A-verboids, we can apply the method of induction on the construction, consisting in the following:

Suppose we want to prove that every A-verboid satisfies a certain one-place verbal predicate F. For this we prove:

firstly, that Λ satisfies F;

secondly, that whenever some A-verboid X satisfies F, every A-verboid XL, where L is any A-literoid, also satisfies F.

We then have the right to conclude that every A-verboid satisfies F. In particular, this pertains to the last step of the process, when the A-verboid of interest to us is obtained.

The following statement is easily proven by the method of induction on the construction:

2.1. *Every* A-*verboid is either empty or else begins and ends with an* A-*literoid.*

In other words, we have

2.2. *Every non-empty* A-*verboid begins and ends with an* A-*literoid.*

By the method of induction on the construction, we can also easily prove

2.3. *Every word of the form XY, where X and Y are* A-*verboids is an* A-*verboid.*

We shall prove

2.4. *Both wings of every occurrence of an* A-*literoid in an* A-*verboid are* A-*verboids.*

Let us agree to call an A-verboid X *proper* during the course of the proof if both wings of every occurrence of an A-literoid in X are A-verboids. It is required to prove that every A-verboid is proper. The A-verboid Λ is proper, since there exist no occurrence of A-literoids in Λ. Let X be a proper A-verboid, and M, an A-literoid. We show that the A-verboid MX is proper.

Let H be any occurrence of an A-literoid L in XM. According to 1.7, we have one of the following two possibilities:

1°. there is an occurrence K of the literoid L in X such that

$$H \asymp KM. \tag{1}$$

2°. $H \asymp X * L *$
and

$$L \asymp M.$$

In Case 1°, let P be K's left wing, and Q, K's right wing. We have

$$H \asymp P * L * QM \quad [(1)],$$

and so P and QM are the wings of H. Here P and Q are A-verboids by virtue of X's being proper. QM is also an A-verboid according to Vd 1. Therefore, both wings of H are A-verboids.

In Case 2°, X and Λ are H's wings. Both these words are A-verboids.

Thus, the A-verboid XM is proper. Q.E.D.

We shall prove

2.5. *Both wings of an occurrence of a non-empty* A-*verboid in an* A-*verboid are* A-*verboids.*

Let H be an occurrence of a non-empty A-verboid Y in an A-verboid X. According to 2.2, Y ends with an A-literoid, i.e., there exists a word Z and an A-literoid L such that

$$Y \asymp ZL. \tag{2}$$

Denoting H's left and right wings by Q and P, respectively, we have

$$\begin{aligned} X &\asymp PYQ \\ &\asymp PZLQ \quad [(2)], \end{aligned} \tag{3}$$

from which it follows that $PZ * L * Q$ is an occurrence of the A-literoid L in the A-verboid X. Q is the right wing of this occurrence, and so by 2.4, Q is an A-verboid.

We have therefore proven that the right wing of H is an A-verboid. Analogously, it can be perceived that the left wing of H is an A-verboid. Thus, Theorem 2.5 is proven.

2.6. *If X is an* A*-verboid and Y is a word for which XY is an* A*-verboid, then Y is an* A*-verboid.*

In fact, $*X*Y$ is an occurrence of X in XY and Y is this occurrence's right wing. If $X \neq \Lambda$, then Y is an A-verboid by 2.5. But if $X = \Lambda$, then $Y = XY$ and, consequently, is an A-verboid, since XY is an A-verboid.

Analogously, one proves

2.7. *If Y is an* A*-verboid and X is a word for which XY is an* A*-verboid, then X is an* A*-verboid.*

The following statement is proven by the method of induction on the construction of a word, with the aid of 1.1:

2.8. *Every* A*-verboid is a word in the alphabet* Aab.

3. Let Б be any alphabet which has no letters in common with the alphabet A. We number its letters from left to right, beginning with the first, to which we assign the number 1. In view of the fact that each letter of the alphabet Б has a unique occurrence in this alphabet, every letter of Б will receive a unique number. We shall denote the number of the letter ξ by $\nu(\xi)$. $\nu(\xi)$ may be characterized as the natural number which is one greater than the length of the left wing of ξ's occurrence in Б. To each letter ξ of the alphabet Б, we assign the word

(1) $$t_{\text{Б}}(\xi) \rightleftharpoons ab^{\nu(\xi)}a$$

in the two-letter alphabet ab. We shall call this word the Б-*translation of the letter* ξ.

Furthermore, the Б-*translation of each letter* η of the alphabet A will mean this same letter η:

$$t_{\text{Б}}(\eta) \rightleftharpoons \eta,$$

where $t_{\text{Б}}(\eta)$ denotes the Б-translation of the letter η of A.

We have thus defined the Б-translation of every letter of the alphabet A Б. We obviously have the following true statements:

3.1. *The* Б*-translation of every letter of the alphabet* AБ *is an* A*-literoid.*

3.2. *The* Б*-translations of two different letters of the alphabet* AБ *are different.*

Therefore, we have the true statement

3.3. *If ς and χ are letters of the alphabet* AБ*, then $\varsigma = \chi$ if and only if $t_{\text{Б}}(\varsigma) = t_{\text{Б}}(\chi)$.*

The notation $t_{\text{Б}}$ may be regarded as a symbol of a certain operation on the letters of the alphabet AБ.

4. We now extend the operation $t_{\text{Б}}$, defined so far for the letters of the alphabet AБ, to the words of this alphabet, , setting

(1) $$\mathfrak{T}_{\text{A,Б}}(\Lambda) \rightleftharpoons \Lambda$$

(2) $$\mathfrak{T}_{\text{A,Б}}(P\xi) \rightleftharpoons \mathfrak{T}_{\text{A,Б}}(P)t_{\text{Б}}(\xi)$$

for any word P in the alphabet АБ and for any letter ξ of this alphabet.

It is obvious that

$$\mathfrak{T}_{\text{А,Б}}(\xi) = t_{\text{Б}}(\xi) \tag{3}$$

for any letter ξ of the alphabet АБ;

$$\mathfrak{T}_{\text{А,Б}}(PQ) = \mathfrak{T}_{\text{А,Б}}(P)\mathfrak{T}_{\text{А,Б}}(Q) \tag{4}$$

for any words P and Q in the alphabet АБ.

We shall call the result of applying the operation $\mathfrak{T}$ to the word P in the alphabet АБ the (А,Б)-*translation* of the word P. By virtue of (3), this definition is compatible with our definition of the Б-*translation* of a letter of the alphabet Б (see Part 3). According to §33, there is a normal algorithm changing each word in the alphabet АБ into this word's (А, Б)-translation. Where this will not give rise to misunderstandings, we shall simply write ≪translation≫ instead of ≪(А,Б)-translation≫. We shall also omit the subscripts А and Б to the letters $\mathfrak{T}$ and t.

4.1. *The translation of each word in the alphabet* АБ *is an* А-*verboid.*

This is proven by induction on the construction of a word in the alphabet АБ, with the aid of (1), (2) and 3.1.

4.2. *The translations of any two distinct words in the alphabet* АБ *are distinct.*

For any natural number N, let B_N denote the statement

> ≪given any words P and Q in the alphabet АБ, such that $[P^\delta \leqslant N$ and $[Q^\delta \leqslant N$, if $\mathfrak{T}(P) = \mathfrak{T}(Q)$, then $P = Q$≫.

Let $B_{P,Q}$ denote the implication

$$\ll \text{if } \mathfrak{T}(P) = \mathfrak{T}(Q), \text{ then } P = Q \gg .$$

Furthermore, let us introduce the condition

$$\ll [P^\delta \leqslant N \text{ and } [Q^\delta \leqslant N \gg . \tag{5}$$

The statement B_0 is true, since for $[P^\delta \leqslant 0$ and $[Q^\delta \leqslant 0$, we have $P = \Lambda$ and $Q = \Lambda$, and hence $P = Q$.

Assume we have already convinced ourselves of the truth of the statement B_N for some natural number N. Let us prove the truth of the statement B_{N+1}.

Let the words P and Q in the alphabet АБ satisfy the condition

$$\ll [P^\delta \leqslant N+1 \text{ and } [Q^\delta \leqslant N+1 \gg . \tag{6}$$

If these words satisfy the condition (5), then by virtue of B_N's truth, the statement $B_{P,Q}$ is true. Assume that the condition (5) is not satisfied. Then

$[P^\delta = N+1$ or $[Q^\delta = N+1$. With no loss of generality in our reasoning, let us assume that

$$(7) \qquad [P^\delta = N+1.$$

Then $P \neq \Lambda$, and so there exists a word R in АБ and a letter ξ of АБ, such that

$$(8) \qquad P \mathrel{=\!\!=} R\xi.$$

We have

$$(9) \qquad \mathfrak{T}(P) \mathrel{=\!\!=} \mathfrak{T}(R)t(\xi) \quad [(8), (2)],$$
$$(10) \qquad [R^\delta = N \quad [(8), (7)],$$
$$(11) \qquad \mathfrak{T}(P) \neq \Lambda \quad [(9)].$$

Let us now assume that

$$(12) \qquad \mathfrak{T}(P) \mathrel{=\!\!=} \mathfrak{T}(Q).$$

Then

$$\mathfrak{T}(Q) \neq \Lambda \quad [(11), (12)]$$

and hence

$$Q \neq \Lambda \quad [(1)].$$

Therefore, there exists a word S in the alphabet АБ and a letter η of АБ, such that

$$(13) \qquad Q \mathrel{=\!\!=} S\eta.$$

We have

$$(14) \qquad \mathfrak{T}(Q) \mathrel{=\!\!=} \mathfrak{T}(S)t(\eta) \quad [(13), (2)].$$

By virtue of (9), (14) and (12), $t(\xi)$ and $t(\eta)$ are terminal segments of one and the same word.

One of the words $t(\xi)$ and $t(\eta)$ is therefore a terminal segment of the other. But since these words are A-literoids [3.1], they coincide [1.3]. Hence

$$(15) \qquad \xi \mathrel{=\!\!=} \eta \quad [3.2].$$

We have

$$(16) \qquad \mathfrak{T}(R)t(\xi) \mathrel{=\!\!=} \mathfrak{T}(S)t(\eta) \quad [(9), (14), (12)],$$

and since $t(\xi) \mathrel{=\!\!=} t(\eta)$, we have

$$(17) \qquad \mathfrak{T}(R) \mathrel{=\!\!=} \mathfrak{T}(S) \quad [(16)].$$

In view of (13) and (6),

(18) $$[S^\delta \leqslant N.$$

Thus, the condition (5) with P and Q replaced by R and S, respectively, is fulfilled. In view of the truth of B_N, it follows from this that the statement $B_{R,S}$ is true. By virtue of (17), we have, finally,

(19) $$R \equiv S$$

and

(20) $$P \equiv Q \quad [(8), (13), (19), (15)].$$

The last equality holds as soon as the equality (12) is fulfilled. This means that the implication $B_{P,Q}$ is true. Here P and Q are any words in the alphabet АБ, satisfying the condition (12) and the condition (6). Consequently, the statement B_{N+1} holds. It is true as soon as B_N is true. The method of arithmetical induction permits us to conclude that every statement B_N is true. Therefore, the equality (20) holds for all the words P and Q in the alphabet АБ for which the equality (12) holds. Consequently, in case (20) is not satisfied, (12) is not satisfied, i.e., the words $\mathfrak{T}(P)$ and $\mathfrak{T}(Q)$ are different. Q.E.D.

An equality analogous to (4) holds for three words:

$$\mathfrak{T}(PQR) \equiv \mathfrak{T}(P)\mathfrak{T}(Q)\mathfrak{T}(R),$$

where P, Q and R are any words in the alphabet АБ. This equality can be proved in the obvious way with the aid of (4).

5. As above, we shall assume that the asterisk $*$ is not a letter of the alphabet АБ. Occurrences in the alphabet АБ will be treated as usual like words in the alphabet АБ$*$ of the form $R * P * S$, where R, P and S are words in АБ.

We shall call the word

$$\mathfrak{T}(R) * \mathfrak{T}(P) * \mathfrak{T}(S)$$

the *translation of the occurrence* $R * P * S$.

5.1. *The translation of an occurrence of the word P in the word Q is an occurrence of the translation of the word P in the translation of the word Q.*

Indeed, let $R * P * S$, where R, P and S are words in the alphabet АБ, be an occurrence of the word P in the word Q. Then

$$RPS \equiv Q$$

and so

(1) $$\mathfrak{T}(R)\mathfrak{T}(P)\mathfrak{T}(S) \equiv \mathfrak{T}(Q),$$

where $\mathfrak{T}(R)$, $\mathfrak{T}(P)$, $\mathfrak{T}(S)$ and $\mathfrak{T}(Q)$ are words in the alphabet Аab, not containing the asterisk. Hence $\mathfrak{T}(R) * \mathfrak{T}(P) * \mathfrak{T}(S)$ is an occurrence in the alphabet Аab with the base $\mathfrak{T}(P)$, i.e., an occurrence, in view of (1), in the word $\mathfrak{T}(Q)$.

Let us prove the statement

5.2. *Every occurrence of an A-literoid in the translation of a word Q in the alphabet* АБ *is the translation of an occurrence of some letter of the alphabet* АБ *in the word Q.*

We carry out the proof by the method of induction on the construction of a word in the alphabet АБ. Let us agree to call a word Q in this alphabet *proper* during the course of the proof if every occurrence of an A-literoid in $\mathfrak{T}(Q)$ is the translation of an occurrence of some letter of the alphabet АБ in Q. It is required to prove that every word in the alphabet АБ is proper.

The empty word is proper, since there are no occurrences of A-literoids in $\mathfrak{T}(\Lambda)$.

Assume it to have been made clear that a certain word Q in the alphabet АБ is proper, and let η be a letter of the alphabet АБ. We prove that the word $Q\eta$ is also proper.

Let K be any occurrence of an A-literoid L in $\mathfrak{T}(Q\eta)$. We have

$$\mathfrak{T}(Q\eta) \leftrightharpoons \mathfrak{T}(Q)t(\eta) \quad [4(2)],$$

where $\mathfrak{T}(Q)$ is an A-verboid [4.1] and $t(\eta)$ is an A-literoid [3.1]. According to 1.7, one of the following two possibilities holds:

1°. There is an occurrence H of the A-literoid L in the word

$$\mathfrak{T}(Q), \text{ such that } K \leftrightharpoons Ht(\eta). \tag{2}$$

2°. $K \leftrightharpoons \mathfrak{T}(Q) * L*$, where $L \leftrightharpoons t(\eta)$.

In Case 1°, H is the translation of an occurrence of some letter ξ of the alphabet АБ in the word Q, since Q is proper. In this case, therefore,

$$H \leftrightharpoons \mathfrak{T}(R) * \mathfrak{T}(\xi) * \mathfrak{T}(S), \tag{3}$$

where the words R and S are such that $R * \xi * S$ is an occurrence of the letter ξ in the word Q. We have

$$R\xi S \leftrightharpoons Q,$$
$$R\xi S\eta \leftrightharpoons Q\eta$$

and so $R * \xi * S\eta$ is an occurrence of ξ in the word $Q\eta$. The translation of this occurrence is the word $\mathfrak{T}(R) * \mathfrak{T}(\xi) * \mathfrak{T}(S\eta)$, i.e., the word K [4(2), (3), (2)]. Thus, in Case 1°, K is the translation of an occurrence of some letter of the alphabet АБ in the word $Q\eta$.

In Case 2°, K is the translation of the occurrence $Q * \eta*$ of the letter η of the alphabet АБ in the word $Q\eta$.

We have therefore proven that the word $Q\eta$ is proper.

Thus, as soon as a word Q is proper, every word $Q\eta$ is also proper, where η is any letter of the alphabet АБ. By the same token, the statement 5.2 is proven.

From 5.2 follows

5.3. *Every A-literoid occurring in the translation of a word Q is the translation of some letter of the word Q.*

We shall prove

5.4. *If every* A*-literoid occurring in an* A*-verboid Q is the translation of a letter of the alphabet* AБ, *then Q is the translation of a word in the alphabet* AБ.

We shall say that the A-verboid Q is *proper* if the following implication is true:

> ≪if each A-literoid occurring in Q is the translation of a letter of the alphabet AБ, then Q is the translation of a word in the alphabet AБ≫.

It is required to prove that every A-verboid is proper.

The A-verboid Λ is proper by virtue of 4(1).

Let Q be a proper A-verboid and let L be an A-literoid. We prove that the A-verboid QL is proper.

Suppose that each A-literoid occurring in QL is the translation of some letter of the alphabet AБ. Then, in particular, each A-literoid occurring in Q is the translation of a letter of AБ. Since Q is proper, Q is the translation of some word P in AБ:

$$Q \leftrightharpoons \mathfrak{T}(P). \tag{4}$$

Since the A-literoid L occurs in QL, L is the translation of some letter ξ of the alphabet AБ:

$$L \leftrightharpoons \mathfrak{T}(\xi). \tag{5}$$

Consequently,

$$QL \leftrightharpoons \mathfrak{T}(P\xi) \quad [(4), (5), 4(4)],$$

and we see that QL is the translation of some word in the alphabet AБ. This proves that the word QL is proper.

Applying the method of induction, we conclude that every word in the alphabet AБ is proper. Q.E.D.

It follows from the statements 5.3 and 5.4 that

5.5. *In order for an* A*-verboid Q to be the translation of a word in the alphabet* AБ, *it is necessary and sufficient that every* A*-literoid occurring in Q be the translation of some letter of* AБ.

Let us prove

5.6. *Every* A*-verboid occurring in the translation of any word in the alphabet* AБ *is itself the translation of some word in* AБ.

Indeed, let the A-verboid Q occur in $\mathfrak{T}(P)$, where P is a word in the alphabet AБ. Every A-literoid occurring in Q then occurs in $\mathfrak{T}(P)$ and is therefore the translation of some letter of the alphabet AБ [5.3]. It follows from this that Q is the translation of a word in the alphabet AБ [5.5].

5.7. *If Q and S are words in the alphabet АБ and $Q \not\asymp \Lambda$ then every occurrence of the translation of Q in the translation of S is the translation of some occurrence of Q in S.*

Let the hypotheses of Theorem 5.7 be fulfilled. Consider any occurrence K of the translation of Q in $\mathfrak{T}(S)$. Let T and U be K's left and right wings, respectively. We have

$$K \asymp T * \mathfrak{T}(Q) * U. \tag{6}$$

$\mathfrak{T}(S)$ and $\mathfrak{T}(Q)$ are A-verboids [4.1], where $\mathfrak{T}(Q) \not\asymp \Lambda$, since $Q \not\asymp \Lambda$ [4.2, 4(1)]. Hence T and U are also A-verboids [2.5]. In view of (6), we have

$$\mathfrak{T}(S) \asymp T\mathfrak{T}(Q)U.$$

Consequently, T and U occur in the A-verboid $\mathfrak{T}(S)$ and are therefore translations of some words in the alphabet АБ [5.6]. Let

$$T \asymp \mathfrak{T}(P), \tag{7}$$
$$U \asymp \mathfrak{T}(R), \tag{8}$$

where P and R are words in the alphabet АБ. We have

$$K \asymp \mathfrak{T}(P) * \mathfrak{T}(Q) * \mathfrak{T}(R) \quad [(6), (7), (8)],$$

i.e., K is the translation of the occurrence $P * Q * R$, and the theorem is proven.
We prove

5.8. *If Q and S are words in the alphabet АБ, then $\mathfrak{T}(Q)$ occurs in $\mathfrak{T}(S)$ if and only if Q occurs in S.*

Note that when $Q \asymp \Lambda$, the statement of Theorem 5.8 becomes trivial, since $\mathfrak{T}(Q) \asymp \Lambda$ [4(1)], Q occurs in S and $\mathfrak{T}(Q)$ occurs in $\mathfrak{T}(S)$. Now let $Q \not\asymp \Lambda$.

If Q occurs in S, then there is an occurrence K of Q in S, and the translation of K is an occurrence of the translation of Q in the translation of S [5.1]. Consequently, $\mathfrak{T}(Q)$ then occurs in $\mathfrak{T}(S)$.

Conversely, if $\mathfrak{T}(Q)$ occurs in $\mathfrak{T}(S)$, then there is an occurrence H of the word $\mathfrak{T}(Q)$ in the word $\mathfrak{T}(S)$. H is the translation of some occurrence of Q in S [5.7]. Consequently, Q then occurs in S.

Theorem 5.8 is therefore proven. We shall prove the following theorem:

5.9. *If Q and S are words in АБ, then the translation of S begins with the translation of Q if and only if the word S begins with the word Q.*

This statement is also trivial for $Q \asymp \Lambda$. Let $Q \not\asymp \Lambda$.

If S begins with Q, then there is a word R in АБ, such that

$$S \asymp QR. \tag{9}$$

We then have

$$\mathfrak{T}(S) \asymp \mathfrak{T}(Q)\mathfrak{T}(R) \quad [4(4)]$$

and, consequently, the word $\mathfrak{T}(S)$ begins with the word $\mathfrak{T}(Q)$.

On the other hand, let the word $\mathfrak{T}(S)$ begin with the word $\mathfrak{T}(Q)$.Then there is a word T in the alphabet Aab for which

$$\mathfrak{T}(S) \rightleftharpoons \mathfrak{T}(Q)T.$$

Therefore

$$*\mathfrak{T}(Q) * T$$

is an occurrence of the word $\mathfrak{T}(Q)$ in the word $\mathfrak{T}(S)$. Since $Q \neq \Lambda$, this occurrence is the translation of some occurrence K of the word Q in the word S [5.7]. The translation of K's left wing is the left wing of the occurrence $*\mathfrak{T}(Q) * T$, i.e., Λ. It follows from this that K's left wing is empty [4(2)], i.e., that K has the form $*Q*R$. Since K is an occurrence of Q in S, the equality (9) holds. Consequently, S begins with the word Q. Q.E.D.

Let us prove the theorem

5.10. *Let Q and S be words in the alphabet АБ and let K and H be occurrences of Q in S. If K precedes H, then $\mathfrak{T}(K)$ precedes $\mathfrak{T}(H)$.*

Let P and T be the left wings of the occurrences K and H, respectively. Then P is a proper initial segment of T, since K precedes H. Hence there is a non-empty word R such that

$$T \rightleftharpoons PR. \tag{10}$$

We have

$$\mathfrak{T}(T) \rightleftharpoons \mathfrak{T}(P)\mathfrak{T}(R) \quad [(10),\ 4(4)].$$

Here $\mathfrak{T}(P) \neq \Lambda$, since $R \neq \Lambda$ [4.2, 4(1)]. Consequently, $\mathfrak{T}(P)$ is a proper initial segment of $\mathfrak{T}(T)$. But by the definition of an occurrence's translation, $\mathfrak{T}(P)$ is the left wing of the translation of K. However, the translations of K and H are occurrences of the translation of Q in the translation of S [5.1]. Thus, the translation of K precedes the translation of H. Q.E.D.

From this follows

5.11. *If Q and S are words in the alphabet АБ, then different occurrences of Q in S have different translations.*

We now prove

5.12. *If Q, S are words in the alphabet АБ and K, H are occurrences of Q in S, then the translation of K precedes the translation of H if and only if K precedes H.*

Indeed, if K precedes H, then the translation of K precedes the translation of H [5.10]. Conversely, let $\mathfrak{T}(K)$ precede $\mathfrak{T}(H)$. Then $\mathfrak{T}(K) \neq \mathfrak{T}(H)$, and so $K \neq H$. Consequently, K precedes H or H precedes K. However, the second possibility can be eliminated, since then $\mathfrak{T}(H)$ would precede $\mathfrak{T}(K)$ [5.10], contrary to the assumption that $\mathfrak{T}(K)$ precedes $\mathfrak{T}(H)$. Consequently, L precedes H, which is what remained to be proven.

Theorems 5.1, 5.11 and 5.12 show that for $Q \neq \Lambda$, there is a one-to-one order-preserving correspondence between the occurrences of Q in S and the occurrences of $\mathfrak{T}(Q)$ in $\mathfrak{T}(S)$. To each occurrence of Q in S, it assigns the translation of this occurrence.

5.13. *If S is a word in the alphabet* АБ *and Q occurs in S, then the first occurrence of the translation of Q in the translation of S is the translation of the first occurrence of Q in S.*

In fact, let this theorem's hypothesis be fulfilled. Then there is a first occurrence of Q in S. Let us denote it by the letter K and show that the translation of K is the first occurrence of $\mathfrak{T}(Q)$ in $\mathfrak{T}(S)$.

The translation of K is an occurrence of $\mathfrak{T}(Q)$ in $\mathfrak{T}(S)$ [5.1]. If $Q \rightleftharpoons \Lambda$, then $\mathfrak{T}(Q) \rightleftharpoons \Lambda$ [4.1], $K \rightleftharpoons * * S$ and the translation of K is the occurrence $* * \mathfrak{T}(S)$. Consequently, the translation of K is the first occurrence of $\mathfrak{T}(Q)$ in $\mathfrak{T}(S)$.

Now let $Q \not\rightleftharpoons \Lambda$. Let H be any occurrence of the translation of Q in the translation of S, distinct from the translation of K. H is the translation of some occurrence G of the word Q in the word S [5.7]. G is distinct from K, since the translation of G is distinct from the translation of K. Therefore, since K is the first occurrence of Q in S, K precedes G, while the translation of K precedes the translation of G [5.10], i.e., H. Thus, the translation of K precedes every occurrence, distinct from the translation of K, of the translation of Q in the translation of S i.e., the translation of K is the first occurrence of the translation of Q in the translation of S. Q.E.D.

5.14. *Let S, Q and U be words in the alphabet* АБ. *If Q occurs in S, then the translation of the result of substituting U for the first occurrence of Q in S coincides with the result of substituting the translation of U for the first occurrence of the translation of Q in the translation of S.*

Indeed, let Q occur in S and let K be the first occurrence of Q in S. The translation of K is then the first occurrence of $\mathfrak{T}(Q)$ in $\mathfrak{T}(S)$ [5.13].Let $K \rightleftharpoons P * Q * R$. Then the translation of K is the occurrence

$$\mathfrak{T}(P) * \mathfrak{T}(Q) * \mathfrak{T}(R),$$

the result of substituting U for K is the word PUR and the translation of the result of substituting U for the first occurrence of Q in S is the word $\mathfrak{T}(P)\mathfrak{T}(U)\mathfrak{T}(R)$, coinciding with the result of substituting $\mathfrak{T}(U)$ for the translation of K.

5.15. $\mathfrak{T}(P) \rightleftharpoons P$ *for every word P in the alphabet* А.

This is proven by the method of induction on the construction of a word, with the aid of 4(2) and 3(2)..

5.16. *The translation of each word in the alphabet* АБ *is a word in the alphabet* А*ab* [4.1, 2.8].

6. Let А and Б be alphabets without any letters in common, containing neither an asterisk, an arrow nor a dot. Assume, moreover, that А contains neither the letter a nor the letter b. Then, as shown above, one can define the translation operation $\mathfrak{T}$ on the words in the alphabet АБ. The result of its application to a word P in the alphabet АБ is the ≪translation≫ $\mathfrak{T}(P)$ of the word P, which is a certain А-verboid [4.1].

We shall call the word $\mathfrak{T}(P) \to \mathfrak{T}(Q)$, where P and Q are words in the alphabet АБ, the *translation of the simple substitution formula* $P \to Q$ in the alphabet АБ. We shall call the word $\mathfrak{T}(P) \to\cdot \mathfrak{T}(Q)$, where P and Q are words in the alphabet АБ, the *translation of the final substitution formula* $P \to\cdot Q$. The translation of a substitution formula F will be denoted by $\mathfrak{T}(F)$, thus extending the scope of the operation $\mathfrak{T}$ to substitution formulas.

6.1. *The translation of a simple substitution formula in the alphabet* АБ *is a simple substitution formula in the alphabet* Аab; *the translation of a final substitution formula in the alphabet* АБ *is a final substitution formula in the alphabet* Аab [5.16].

6.2. *The substitution formula F in the alphabet* АБ *acts on the word P in the alphabet* АБ *if and only if its translation acts on P's translation* [5.8].

6.3. *If the substitution formula F in the alphabet* АБ *acts on the word P in this alphabet, then the translation of the result of F's action on P is the result of the action of $\mathfrak{T}(F)$ on $\mathfrak{T}(P)$* [6.2, 5.14].

Replacing each substitution formula in the scheme S of a normal algorithm $\mathfrak{A}$ in the alphabet АБ by its translation, we obtain a configuration which is the scheme of some normal algorithm in the alphabet Аab [6.1]. We shall call this uniquely determined normal algorithm the *translation of the normal algorithm* $\mathfrak{A}$.

6.4. *The translation of the normal algorithm $\mathfrak{A}$ is a normal algorithm in the alphabet* Аab.

We shall denote the translation of the algorithm $\mathfrak{A}$ by the letter $\mathfrak{B}$, and $\mathfrak{B}$'s scheme, by the letter T.

An obvious one-to-one, order-preserving correspondence holds between the lines of the scheme S and the lines of the scheme T. We shall denote the line of T, corresponding to a given line C of S, by the symbol $\widetilde{C}$.

We have, by the definition of the translation of a normal algorithm, the following true statements:

6.5. *Whenever a substitution formula F is written down in a line C of a scheme S, the translation of F is written down in the line $\widetilde{C}$ of the scheme T.*

6.6. *The line D of the scheme S is situated higher than the line C of this scheme if and only if the line $\widetilde{D}$ of the scheme T is situated higher than the line $\widetilde{C}$ of the scheme T.*

We prove the statement

6.7. *Whenever a substitution formula F of a scheme S is active on a word P in the scheme S, the substitution formula $\mathfrak{T}(F)$ is active on the word $\mathfrak{T}(P)$ in the scheme T.*

In fact, let the substitution formula F of the scheme S be active on the word P in the scheme S. Then F acts on P and there is a line C in the scheme S, such that the substitution formula F is written down in it, and in no higher line of C is there written down a substitution formula acting on P. In the line

$\widetilde{C}$ of the scheme T, there is written down the substitution formula $\mathfrak{T}(F)$ of the scheme T [6.5], acting on $\mathfrak{T}(P)$ [6.2]. Let E be any line of the scheme T, situated higher than $\widetilde{C}$ in T. E corresponds to some line D of the scheme S, i.e., E is $\widetilde{D}$. A certain substitution formula G of the scheme S is written down in D. The substitution formula $\mathfrak{T}(G)$ of the scheme T is written down in $\widetilde{D}$, i.e., in E [6.5]. D is situated higher than C in S, since E is situated higher than $\widetilde{C}$ in T [6.6]. The substitution formula G of the scheme S, written down in the line D, does not act on the word P, since the line D, in which the formula G is written down, is situated higher than C. Hence the formula $\mathfrak{T}(G)$ does not act on $\mathfrak{T}(P)$ [6.2]. We have therefore proven that the substitution formula written down in any line situated higher than $\widetilde{C}$ in the scheme T does not act on $\mathfrak{T}(P)$. This completes the proof that the substitution formula $\mathfrak{T}(F)$ is active on the word $\mathfrak{T}(P)$ in the scheme T.

We shall prove

6.8. *Whenever a substitution formula F of a scheme S is such that the formula $\mathfrak{T}(F)$ is active on the word $\mathfrak{T}(P)$ in the scheme T, F is active on the word P in the scheme S. (Here P is any word in* AБ.*)*

Indeed, let P be a word in the alphabet AБ, let F be a substitution formula of a scheme S and let the formula $\mathfrak{T}(F)$ be active on the word $\mathfrak{T}(P)$ in the scheme T. We show that F is active on P in S.

Since $\mathfrak{T}(F)$ is active on $\mathfrak{T}(P)$ in the scheme T, $\mathfrak{T}(F)$ acts on $\mathfrak{T}(P)$, and so F acts on P. There is a line E of the scheme T, such that the formula $\mathfrak{T}(F)$ is written down in it and the formula written down in any line of T, situated higher than E, does not act on $\mathfrak{T}(P)$. The line E corresponds to some line C of the scheme S. For this line C the line E is $\widetilde{C}$. Let the substitution formula G be written down in some line D situated higher than C in the scheme S. Then the formula $\mathfrak{T}(G)$ is written down in $\widetilde{D}$ [6.5]. $\widetilde{D}$ is situated higher then E in T, since D is situated higher than C in S [6.6]. Therefore, the formula written down in $\widetilde{D}$ does not act on $\mathfrak{T}(P)$, and so the formula written down in D, i.e., G, does not act on P [6.2]. We have thus proven that in every line situated higher than C in the scheme S, there is written down a substitution formula which does not act on P. This completes the proof that the formula F is active on the word P in the scheme S. By the same token, Theorem 6.8 is proven.

6.9. *If $\mathfrak{A} : P \vdash Q$, then $\mathfrak{B} : \mathfrak{T}(P) \vdash \mathfrak{T}(Q)$.*

Indeed, let $\mathfrak{A} : P \vdash Q$. Then there exists a simple substitution formula F of the scheme S, active on P in S and such that Q is the result of the action of F on P. $\mathfrak{T}(F)$ is a simple substitution formula of the scheme T [6.1], active on $\mathfrak{T}(P)$ in T [6.7]. $\mathfrak{T}(Q)$ is the result of the action of $\mathfrak{T}(F)$ on $\mathfrak{T}(P)$ [6.3]. Hence $\mathfrak{B} : \mathfrak{T}(P) \vdash (Q)$.

6.10. *If $\mathfrak{A} : P \vdash\cdot Q$, then $\mathfrak{B} : \mathfrak{T}(P) \vdash\cdot \mathfrak{T}(Q)$.*

This is proven analogously to the way in which the preceding statement was proved.

6.11. *If* $\mathfrak{B} : \mathfrak{T}(P) \vdash \mathfrak{T}(Q)$, *where* P *and* Q *are words in the alphabet* АБ, *then* $\mathfrak{A} : P \vdash Q$.

In fact, let $\mathfrak{B} : \mathfrak{T}(P) \vdash \mathfrak{T}(Q)$, where P and Q are words in the alphabet АБ. Then there exists a simple substitution formula G of the scheme T, active on $\mathfrak{T}(P)$ in T and such that $\mathfrak{T}(Q)$ is the result of the action of G on $\mathfrak{T}(P)$. According to the construction of the scheme T, the formula G is the translation of some substitution formula F of the scheme S. F is also a simple substitution formula [6.1]. Since $\mathfrak{T}(F)$, i.e., G, is active on $\mathfrak{T}(P)$ in T, F is active on P in S [6.8]. F acts on P. Let H be the result of the action of F on P. $\mathfrak{T}(H)$ is the result of the action of $\mathfrak{T}(F)$, i.e., G, on $\mathfrak{T}(P)$. Consequently, $\mathfrak{T}(H) \equiv \mathfrak{T}(Q)$ and $H \equiv Q$ [4.2]. Thus, Q is the result of the action of the formula F on P. Moreover, as we have seen, F is active on P in S. consequently, $\mathfrak{A} : P \vdash Q$. Q.E.D.

6.12. *If* $\mathfrak{B} : \mathfrak{T}(P) \vdash\cdot \mathfrak{T}(Q)$, *where* P *and* Q *are words in the alphabet* АБ, *then* $\mathfrak{A} : P \vdash\cdot Q$.

The proof is analogous to the preceding proof.

6.13. $\mathfrak{A} : P\rceil$ *if and only if* $\mathfrak{B} : \mathfrak{T}(P)\rceil$ [6.2].

6.14. *If* $\mathfrak{A} : P \models Q$, *then* $\mathfrak{B} : \mathfrak{T}(P) \models \mathfrak{T}(Q)$.

This is proven by the method of induction on the steps of the performance of the algorithm $\mathfrak{A}$, with the aid of 6.9.

6.15. *If* $\mathfrak{B} : \mathfrak{T}(P) \vdash X$, *then there exists a word* Q *in the alphabet* АБ *such that* $X \equiv \mathfrak{T}(Q)$ *and* $\mathfrak{A} : P \vdash Q$.

Let, in fact, $\mathfrak{B} : \mathfrak{T}(P) \vdash Q$. Here P is a word in the alphabet АБ. The scheme T acts on the word $\mathfrak{T}(P)$. Hence the scheme S acts on P [6.2]. Denote the result of S's action by the letter Q. We have $\mathfrak{A} : P \vdash Q$ or $\mathfrak{A} : P \vdash\cdot Q$. However, in case $\mathfrak{A} : P \vdash\cdot Q$, we would have $\mathfrak{B} : \mathfrak{T}(P) \vdash\cdot \mathfrak{T}(Q)$ [6.10], which is incompatible with our initial assumption. Therefore, $\mathfrak{A} : P \vdash Q$, and so $\mathfrak{B} : \mathfrak{T}(P) \vdash \mathfrak{T}(Q)$ [6.9]. Since, moreover, $\mathfrak{B} : \mathfrak{T}(P) \vdash X$, we have $X \equiv \mathfrak{T}(Q)$, which is just what remained to be proven.

6.16. *If* $\mathfrak{B} : \mathfrak{T}(P) \vdash\cdot X$, *then there exists a word* Q *in the alphabet* АБ, *such that* $X \equiv \mathfrak{T}(Q)$ *and* $\mathfrak{A} : P \vdash\cdot Q$.

This is proven analogously to the way in which we proved the preceding statement.

6.17. *If* $\mathfrak{B} : \mathfrak{T}(P) \models X$, *then there exists a word* Q *in the alphabet* АБ *such that* $X \equiv \mathfrak{T}(Q)$ *and* $\mathfrak{A} : P \models Q$.

This is proven with the aid of the statement 6.15 by the method of induction on the steps of the performance of the algorithm $\mathfrak{B}$.

6.18. *If* $\mathfrak{B} : \mathfrak{T}(P) \models \mathfrak{T}(Q)$, *then* $\mathfrak{A} : P \models Q$.

This follows from the statements 6.17 and 4.2.

6.19. $\mathfrak{A} : P \models Q$ *if and only if* $\mathfrak{B} : \mathfrak{T}(P) \models \mathfrak{T}(Q)$ [6.14, 6.18].

6.20. $\mathfrak{A} : P \models Q\urcorner$ *if and only if* $\mathfrak{B} : \mathfrak{T}(P) \models \mathfrak{T}(Q)\urcorner$ [6.19, 6.13].

6.21. $\mathfrak{A} : P \models\cdot Q$ *if and only if* $\mathfrak{B} : \mathfrak{T}(P) \models\cdot \mathfrak{T}(Q)$ [6.19, 6.10, 6.12].

6.22. $\mathfrak{A}(P) \simeq Q$ *if and only if* $\mathfrak{B}(\mathfrak{T}(P)) \simeq \mathfrak{T}(Q)$ [6.20, 6.21].

6.23. *If* P *and* Q *are words in the alphabet* A, *then* $\mathfrak{A}(P) \simeq Q$ *if and only if* $\mathfrak{B}(P) \simeq Q$ [6.22, 5.15].

6.24. *If* $\mathfrak{B} : \mathfrak{T}(P) \models\cdot X$, *then there exists a word* Q *in the alphabet* AБ *such that* $X \simeq \mathfrak{T}(Q)$ *and* $\mathfrak{A} : P \models\cdot Q$.

Indeed, in this case there exists a word R in the alphabet Aab, such that $\mathfrak{B} : \mathfrak{T}(P) \models R$ and $\mathfrak{B} : R \vdash\cdot X$. There exists a word W in the alphabet AБ, such that $R \simeq \mathfrak{T}(W)$ and $\mathfrak{A} : P \models W$ [6.17]. We have $\mathfrak{B} : \mathfrak{T}(W) \vdash\cdot X$. Therefore, there exists a word Q in the alphabet AБ, such that $X \simeq \mathfrak{T}(Q)$ and $\mathfrak{A} : W \vdash\cdot Q$ [6.16]. We have $\mathfrak{A} : P \models\cdot Q$, which is just what remained to be proven.

6.25. *If* $\mathfrak{B} : \mathfrak{T}(P) \models X\urcorner$, *then there exists a word* Q *in the alphabet* AБ *such that* $X \simeq \mathfrak{T}(Q)$ *and* $\mathfrak{A} : P \models Q$ [6.17, 6.13].

6.26. *If* $\mathfrak{B}(\mathfrak{T}(P)) \simeq X$, *then there exists a word* Q *in the alphabet* AБ *such that* $X \simeq \mathfrak{T}(Q)$ *and* $\mathfrak{A}(P) \simeq Q$ [6.24, 6.25].

6.27. *The algorithm* $\mathfrak{A}$ *is applicable to a word* P *in the alphabet* A *if and only if the algorithm* $\mathfrak{B}$ *is applicable to this word* [6.22, 5.15].

6.28. *The algorithms* $\mathfrak{A}$ *and* $\mathfrak{B}$ *are strongly equivalent with respect to the alphabet* A [§26.6, 6.23, 6.27].

7. In the statement 6.28, $\mathfrak{A}$ and $\mathfrak{B}$ are normal algorithms; moreover, $\mathfrak{A}$ is an algorithm in the alphabet AБ and $\mathfrak{B}$ is constructed in a definite way, starting with A, Б and $\mathfrak{A}$, as a normal algorithm in the alphabet Aab. As for the letters a and b, they are assumed in this context not to occur in the alphabet A. The algorithm $\mathfrak{B}$ turns out strongly equivalent to the algorithm $\mathfrak{A}$ with respect to the alphabet A.

It is obvious that in this construction, the letters a and b may be replaced by any two distinct letter. This yields the following *Reduction Theorem.*

7.1. *Given any normal algorithm* $\mathfrak{A}$ *over the alphabet* A *and any letters* α *and* β, *distinct from each other not occurring in* A, *one can construct a normal algorithm* $\mathfrak{B}$ *in the alphabet* A$\alpha\beta$, *strongly equivalent to* $\mathfrak{A}$ *with respect to the alphabet* A.

8. By complicating the operation of translating an algorithm, one of the supplementary letters in Theorem 7.1 can be eliminated. In fact, the following *Strengthened Reduction Theorem* holds (see Nagorny [1] [2][7]):

[7]In these papers, the assertion corresponding to our Theorem 8.1 is formulated for the relation of *equivalence with respect to* A. However, the construction presented there is actually suitable for our purposes also. We reproduce it with the necessary modifications and with slight simplifications.

8.1. *Let* A *be a non-empty alphabet and let* α *be a letter not belonging to* A. *Every normal algorithm* $\mathfrak{A}$ *over the alphabet* A *is strongly equivalent with respect to* A *to some normal algorithm* $\mathfrak{B}$ *in* Aα.

Indeed, let $\mathfrak{A}$ be a normal algorithm in the alphabet AБ, where A and Б have no letters in common. Let α be a letter not belonging to A.

Let us denote the ith letters of the alphabets A and Б by a_i and b_i respectively, and define the following operation of translating the alphabet AБ· into the alphabet Aα:

$$\begin{aligned} [a_i^\tau &\rightleftharpoons \alpha a_i \alpha \quad (1 \leqslant i \leqslant [\mathrm{A}^\delta), \\ [b_i^\tau &\rightleftharpoons \alpha a_1^{i+4} \alpha \quad (1 \leqslant i \leqslant [\mathrm{Б}^\delta), \\ [\cdot^\tau &\rightleftharpoons \alpha a_1^2 \alpha, \\ [\Lambda^\tau &\rightleftharpoons \Lambda, \\ [P\xi^\tau &\rightleftharpoons [P^\tau[\xi^\tau \end{aligned}$$

(here P denotes an arbitrary word in AБ·, and ξ, an arbitrary letter of this alphabet).

Let $F \rightleftharpoons P \to \varepsilon Q$ be a substitution formula in the alphabet AБ (here $\varepsilon \leftrightharpoons \cdot$ or $\varepsilon \leftrightharpoons \Lambda$). We define the translation F as follows:

$$[F^\tau \rightleftharpoons \begin{cases} [P^\tau \to [\varepsilon Q^\tau & \text{if } P \not\leftrightharpoons \Lambda, \\ \alpha a_1^3 \alpha \to \alpha a_1^3 \alpha [\varepsilon Q^\tau, & \text{if } P \leftrightharpoons \Lambda. \end{cases}$$

We construct $\mathfrak{A}^\cdot$—the closure of the algorithm $\mathfrak{A}$. Let $\mathfrak{C}$ denote the scheme obtained as a result of a formula-by-formula translation of the scheme of the algorithm $\mathfrak{A}^\cdot$.

We now define $\mathfrak{B}$ as the normal algorithm in the alphabet Aα, whose scheme can be written down in the following abbreviated form:

$$\left\{ \begin{aligned} \alpha a_4^1 \alpha \xi &\to [\xi^\tau \alpha a_1^4 \alpha \\ \alpha a_1^4 \alpha &\to \\ \alpha a_1^2 \alpha [\varsigma^\tau &\to [\varsigma^\tau \alpha a_1^2 \alpha \\ [\xi^\tau \alpha a_1^2 \alpha &\to \alpha a_1^2 \alpha \xi \\ [\eta^\tau \alpha a_1^2 \alpha &\to \cdot [\eta^\tau \alpha a_1^2 \alpha \\ \alpha a_1^3 \alpha \alpha a_1^2 \alpha &\to \cdot \\ &\mathfrak{C} \\ &\to \alpha a_1^3 \alpha \alpha a_1^4 \alpha, \end{aligned} \right.$$

where the variables ξ, η and ς run through the alphabets A, Б and AБ, respectively.

We shall show that the algorithm $\mathfrak{A}$ is strongly equivalent to $\mathfrak{B}$ with respect to A. Assuming that the reader has accumulated the necessary experience, we shall limit ourselves, on the whole, to stating the lemmas required here, and shall not go into the details of the proofs.

8.2. *The translation of any non-empty word P in АБ is a juxtaposition of words of the form $\alpha a_i \alpha$ $(1 \leqslant i \leqslant [\mathrm{A}^\delta)$ and $\alpha a_1^{j+4}\alpha$ $(1 \leqslant j \leqslant [\text{Б}^\delta)$, and so the words $\alpha a_1^2\alpha$, $\alpha a_1^3\alpha$ and $\alpha a_1^4\alpha$ do not occur in $[P^\tau$.*

8.3. *The first occurrence of $\alpha a_1^4\alpha\xi$ in the word $\alpha a_1^3\alpha[Q^\tau \alpha a_1^4\alpha\xi R$ is the occurrence $\alpha a_1^1\alpha[Q^\tau * \alpha a_1^4\alpha\xi * R$ (Q, R are words in A, ξ is a letter of this alphabet).*

8.4. *The first occurrence of $\alpha a_1^4\alpha$ in the word $\alpha a_1^3\alpha[Q^\tau \alpha a_1^4\alpha$ is the occurrence $\alpha a_1^3\alpha[Q^\tau * \alpha a_1^4\alpha *$ (Q is a word in A).*

8.5. *The first occurrence of $\alpha a_1^2\alpha[\varsigma^\tau$ in the word $\alpha a_1^3\alpha[Q^\tau \alpha a_1^2\alpha[\varsigma R^\tau$ is the occurrence $\alpha a_1^3\alpha[Q^\tau * \alpha a_1^2\alpha[\varsigma^\tau * [R^\tau$ (Q, R are words in АБ, ς is a letter of this alphabet).*

8.6. *The first occurrence of $[\varsigma^\tau \alpha a_1^2\alpha$ in the word $\alpha a_1^3\alpha[Q\varsigma^\tau \alpha a_1^2\alpha R$ is the occurrence $\alpha a_1^3\alpha[Q^\tau * [\varsigma^\tau \alpha a_1^2\alpha * R$ (Q, R are words in АБ, ς is a letter of this alphabet).*

Let R, T and S be words in АБ and let $T \neq \Lambda$. We shall call the occurrence

$$a_1^3\alpha[R^\tau * [T^\tau * [S^\tau$$

the *image of the occurrence* $R * T * S$.

8.7. *A non-empty word T in the alphabet АБ occurs in a word P in this alphabet if and only if $[T^\tau$ occurs in $\alpha a_1^3\alpha[P^\tau$.*

8.8. *If a non-empty word T occurs in a word P in the alphabet АБ, then the image of the first occurrence of T in P is the first occurrence of $[T^\tau$ in $\alpha a_1^3\alpha[P^\tau$.*

It follows from an examination of the scheme of the algorithm $\mathfrak{B}$ that

8.9. *The algorithm $\mathfrak{B}$ is closed.*

8.10. *α occurs in the left side of every substitution formula, except the last, of the algorithm $\mathfrak{B}$'s scheme.*

Further, let us formulate the following assertions:

8.11. *If P is a word in A, then*

$$\mathfrak{B} : P \vdash \alpha a_1^3\alpha\alpha a_1^4\alpha P.$$

8.12. *If P is a word in A, then*

$$\mathfrak{B} : \alpha a_1^3\alpha\alpha a_1^4\alpha P \models \alpha a_1^3\alpha[P^\tau \alpha a_1^4\alpha.$$

8.13. *If P is a word in A, then*

$$\mathfrak{B} : \alpha a_1^3\alpha[P^\tau \alpha a_1^4\alpha \vdash \alpha a_1^3\alpha[P^\tau \quad [8.4].$$

8.14. *If P is a word in A, then*

$$\mathfrak{B} : P \models \alpha a_1^3\alpha[P^\tau \quad [8.11, 8.12, 8.13].$$

8.15. *If R and S are words in* АБ, *then*

$$\mathfrak{B} : \alpha a_1^3 \alpha [R^\tau \alpha a_1^2 \alpha [S^\tau \models \alpha a_1^3 \alpha [RS^\tau \alpha a_1^2 \alpha.$$

8.16. *If Q is a word in* АБ *and R is a word in* А, *then*

$$\mathfrak{B} : \alpha a_1^3 \alpha [Q^\tau [R^\tau \alpha a_1^2 \alpha \models \alpha a_1^3 \alpha [Q^\tau \alpha a_1^2 \alpha R.$$

8.17. *If Q, R are words in* АБ, *and η is a letter of* Б, *then*

$$\mathfrak{B} : \alpha a_1^3 \alpha [Q^\tau [\eta^\tau \alpha a_1^2 \alpha R \vdash\cdot \alpha a_1^3 \alpha [Q^\tau [\eta^\tau \alpha a_1^2 \alpha R.$$

8.18. *If Q is a word in* АБ *and* $[Q^{\text{Б}} \neq \Lambda$,[8] *then*

$$\mathfrak{B} : (\alpha a_1^3 \alpha [Q^\tau \alpha a_1^2 \alpha)$$

is defined and is not a word in А.

Indeed, let the lemma's hypothesis be fulfilled. Then $Q \eqcirc R\eta S$, where η is a letter of Б and S is a word in А. But then

$$\begin{aligned} \alpha a_1^3 \alpha [Q^\tau \alpha a_1^2 \alpha &\eqcirc \alpha a_1^3 \alpha [R^\tau [\eta^\tau [S^\tau \alpha a_1^2 \alpha, \\ \mathfrak{B} : \alpha a_1^3 \alpha [R^\tau [\eta^\tau [S^\tau \alpha a_1^2 \alpha &\models \alpha a_1^3 \alpha [R^\tau [\eta^\tau \alpha a_1^2 \alpha S \qquad [8.16], \\ &\vdash\cdot \alpha a_1^3 \alpha [R^\tau [\eta^\tau \alpha a_1^2 \alpha S \qquad [8.17], \end{aligned}$$

from which the lemma's conclusion follows.

8.19. *If Q is a word in* А, *then* $\mathfrak{B} : \alpha a_1^3 \alpha \alpha a_1^2 \alpha Q \vdash\cdot Q$.

We now formulate several lemmas relating the performance of the algorithm $\mathfrak{B}$ to that of the algorithm $\mathfrak{A}$.

8.20. *If* $\mathfrak{A}^\cdot : P \vdash Q$, *then* $\mathfrak{B} : \alpha a_1^3 \alpha [P^\tau \vdash \alpha a_1^3 \alpha [Q^\tau$.

8.21. *If* $\mathfrak{A}^\cdot : P \models Q$, *then* $\mathfrak{B} : \alpha a_1^3 \alpha [P^\tau \models \alpha a_1^3 \alpha [Q^\tau$ [8.20].

8.22. *If* $\mathfrak{A}^\cdot : P \vdash\cdot Q$, *then there exist words R and S such that*

$$RS \eqcirc Q$$

and

$$\mathfrak{B} : \alpha a_1^3 \alpha [P^\tau \vdash \alpha a_1^3 \alpha [R^\tau \alpha a_1^2 \alpha [S^\tau.$$

8.23. *If* $\mathfrak{A}^\cdot : P \vdash\cdot Q$, *then*

$$\mathfrak{B} : \alpha a_1^3 \alpha [P^\tau \models \alpha a_1^3 \alpha [Q^\tau \alpha a_1^2 \alpha \quad [8.22,\ 8.15].$$

---

[8] We recall that $[Q^{\text{Б}}$ denotes the projection of the word Q on the alphabet Б.

8.24. *If* $\mathfrak{A}^{\cdot} : P \models\cdot Q$, *then*

$$\mathfrak{B} : \alpha a_1^3 \alpha [P^\tau \models \alpha a_1^3 \alpha [Q^\tau \alpha a_1^2 \alpha \quad [8.21,\ 8.23].$$

8.25. *If* P *is a word in* A *and* $\mathfrak{A}^{\cdot} : P \models\cdot Q$, *then*

$$\mathfrak{B} : P \models \alpha a_1^3 \alpha [Q^\tau \alpha a_1^2 \alpha \quad [8.14,\ 8.24].$$

8.26. *If* P, Q *are words in* A *and* $\mathfrak{A}(P) \eqsim Q$, *then* $\mathfrak{B}(P) \eqsim Q$.

In fact, let the lemma's hypothesis be fulfilled. Then $\mathfrak{A}^{\cdot} : P \models\cdot Q$ and, consequently,

$$\begin{aligned} \mathfrak{B} : P &\models \alpha a_1^3 \alpha [Q^\tau \alpha a_1^2 \alpha && [8.25] \\ &\models \alpha a_1^3 \alpha \alpha a_1^2 \alpha Q && [8.16] \\ &\vdash\cdot Q && [8.19], \end{aligned}$$

from which the lemma's conclusion follows.

8.27. *If* P *is a word in* A *and the algorithm* $\mathfrak{A}$ *is applicable to* P, *then* $\mathfrak{B}$ *is also applicable to* P.

Indeed, let the lemma's hypothesis be fulfilled. Then $\mathfrak{A}(P) \eqsim Q$ for some Q in AБ. Suppose that Q is a word in A. Then the lemma's conclusion follows from 8.26. Now assume that Q is not a word in A. Then $[Q^{\text{Б}} \not\eqsim \Lambda$, and since $\mathfrak{A}^{\cdot} : P \models\cdot Q$, we have

$$\mathfrak{B} : P \models \alpha a_1^3 \alpha [Q^\tau \alpha a_1^2 \alpha \quad [8.25].$$

But $\mathfrak{B}(\alpha a_1^3 \alpha [Q^\tau \alpha a_1^2 \alpha)$ is defined [8.18], and this means that $\mathfrak{B}$ is applicable to P in this case too.

By the same token, Lemma 8.27 is proven.

8.28. *If* P *is a word in* A *and* $\mathfrak{B}$ *is applicable to* P, *then* $\mathfrak{A}$ *is also applicable to* P.

In fact, let the lemma's hypothesis be fulfilled. Since

$$\mathfrak{B} : P \models \alpha a_1^3 \alpha [P^\tau \quad [8.14],$$

we have

$$\mathfrak{B}(\alpha a_1^3 \alpha [P^\tau) \eqsim \mathfrak{B}(P).$$

Let N denote the number of steps in which $\mathfrak{B}$ changes $\alpha a_1^3 \alpha [P^\tau$ into $\mathfrak{B}(P)$. By virtue of Lemma 8.20, $\mathfrak{A}^{\cdot}$ can not complete, starting with P, more than N simple steps. Therefore, $\mathfrak{A}^{\cdot}$, and hence $\mathfrak{A}$ too, is applicable to P. Q.E.D.

8.29. *If P, Q are words in* A *and* $\mathfrak{B}(P) \eqsim Q$, *then* $\mathfrak{A}(P) \eqsim Q$.

Indeed, let P, Q be words in A and let $\mathfrak{B}(P) \eqsim Q$. Then $\mathfrak{A}$ is applicable to P. Let $\mathfrak{A}(P) \eqsim R$. Then $\mathfrak{A}^{\cdot} : P \models^{\cdot} R$. But then

$$\mathfrak{B} : \alpha a_1^3 \alpha [P^\tau \models \alpha a_1^3 \alpha [R^\tau \alpha a_1^2 \alpha \quad [8.24].$$

Let us now assume that $[R^{\text{Б}} \eqsim \Lambda$. Then by virtue of 8.18,

$$\mathfrak{B}(\alpha a_1^3 \alpha [R^\tau \alpha a_1^2 \alpha)$$

is defined and is not a word in A. But

$$\begin{aligned} Q &\eqsim \mathfrak{B}(P) \\ &\eqsim \mathfrak{B}(\alpha a_1^3 \alpha [P^\tau) \\ &\eqsim \mathfrak{B}(\alpha a_1^3 \alpha [R^\tau \alpha a_1^2 \alpha), \end{aligned}$$

and so Q is not a word in A, which contradicts the lemma's hypothesis. Thus, $[R^{\text{Б}} \eqsim \Lambda$, i.e., R is a word in A. But then by virtue of 8.26, $\mathfrak{B}(P) \eqsim R$. Therefore, $\mathfrak{A}(P) \eqsim \mathfrak{B}(P)$, which is just what remained to be proved.

8.30. $\mathfrak{A}$ *is strongly equivalent to* $\mathfrak{B}$ *with respect to* A [8.26, 8.29, 8.27, 8.28].

Thus, Theorem 8.1 is proven.

As was already mentioned in the beginning of the section, it is impossible to further strengthen the Reduction Theorem by further decreasing the number of supplementary letters.

9. Theorem 6.22 was proven by us for a specific word-translation operation $\mathfrak{T}$. Syrkin [1] found necessary and sufficient conditions, to be imposed on an arbitrary two-letter word-coding operation $\mathfrak{T}$ in the alphabet AБ, for this theorem's conclusion to retain its validity.

CHAPTER V

UNIVERSAL ALGORITHM

In this chapter, we shall construct a normal algorithm over a given arbitrary alphabet A, which in a certain sense is capable of doing the work of any algorithm in A—a ≪universal algorithm≫ for the alphabet A.

A universal algorithm, like any other normal algorithm, will admit arbitrary words in its alphabet as initial data. However, we will only be interested in how it performs on words of a special type—on combinations of words in A with schemes of normal algorithms in A, written out linearly [§25.4]. The very same method is applied here as in modern digital computers: not only the initial data, but also the prescription establishing the sequence of operations to be carried out, are written down in the form of words in an appropriate alphabet and introduced into the computer, which is capable of "understanding" written prescriptions and fulfilling them. The standardization of prescriptions, which we achieve by means of schemes of normal algorithms, ensures the possibility of writing them down in such a manner.

The existence of a universal algorithm is a central fact of the general theory of algorithms—the reader will soon be able to evaluate its importance. Of course, this fact is not a specific feature of the theory of *normal* algorithms: theorems analogous to our Theorem §42.1.1 were known for Turing machines and for recursive functions almost as soon as the corresponding concepts were developed.

§42 Statement of the Universal Algorithm Theorem

1. Let us now turn to a precise statement of the Universal Algorithm Theorem.

We shall regard as fixed some non-empty[1] alphabet A such that the pairwise distinct letters α, β, γ and δ do not occur in A. When speaking of schemes of normal algorithms in A, we shall mean γ-schemes in the alphabet $A\alpha\beta$ [§25.4] with the usual representation of substitution formulas. We shall call the scheme

[1]The case when the alphabet A is empty is of little interest, and we do not consider it. There are only two non-equivalent normal algorithms in the empty alphabet, and they are both trivial.

of a normal algorithm $\mathfrak{A}$ in the alphabet A the *representation* of the algorithm $\mathfrak{A}$. We shall denote it by the symbol $\mathfrak{A}^{\mathrm{Rp}}$.

A word in the alphabet $\mathrm{A}\alpha\beta\gamma\delta$ of the form

$$\mathfrak{A}^{\mathrm{Rp}}\delta P, \tag{1}$$

where $\mathfrak{A}$ is a normal algorithm in A and P is a word in A, represents the pair ≪the algorithm $\mathfrak{A}$ and the word P≫ in the sense that it uniquely determines this pair and is itself uniquely determined by it. The former is obvious, while the latter is clear from the fact that P is the right wing of the unique occurrence of δ in $\mathfrak{A}^{\mathrm{Rp}}\delta P$, $\mathfrak{A}^{\mathrm{Rp}}$ is the left wing of this occurrence and the algorithm $\mathfrak{A}$ is uniquely determined by $\mathfrak{A}^{\mathrm{Rp}}$.

It is natural to apply the following prescription to pairs of the form (1): ≪start with a pair $\mathfrak{A}^{\mathrm{Rp}}\delta P$, construct the word P and the algorithm $\mathfrak{A}$, then apply $\mathfrak{A}$ to P≫. This prescription itself constitutes a certain algorithm changing $\mathfrak{A}^{\mathrm{Rp}}\delta P$ into $\mathfrak{A}(P)$, if the algorithm $\mathfrak{A}$ is applicable to P, and inapplicable to this word if $\mathfrak{A}$ is inapplicable to P. There arises the question as to the possibility of presenting such an algorithm in the form of a normal algorithm. A positive answer to this question is given by the following theorem.

1.1. UNIVERSAL ALGORITHM THEOREM. *One can construct a normal algorithm $\mathfrak{U}$ over $\mathrm{A}\alpha\beta\gamma\delta$, such that the conditional equality*

$$\mathfrak{U}(\mathfrak{A}^{\mathrm{Rp}}\delta P) \simeq \mathfrak{A}(P) \tag{2}$$

holds for each word P in A *and each normal algorithm $\mathfrak{A}$ in* A.

The proof of this theorem forms the main content of the present chapter.

At the present time, we can offer several different proofs of this theorem, all of them based on prior proofs of its analogues, in which the concept of a representation is defined in a slightly different manner. Chronologically, the first proof is due to Markov (1951). A detailed presentation of it constitutes Chapter IV of his monograph [2]. This proof indicates a transparent method, based on combination theorems, for constructing the scheme of a universal algorithm, but the scheme itself is not explicitly written down and its complexity (see §31.1 of this book) is not estimated. The second proof chronologically (1958) is due to Orlovskii [1]. It contains an *explicit construction* of a universal algorithm's scheme whose complexity turns out to be quadratically dependent on n, where n is the number of letters of the alphabet A. Later (1973), Ostroukhov [2] constructed a universal algorithm with complexity $10n+$ const, and Zharov [2], with complexity $5n+$ const (1974). In the last two proofs, based on the original definition of a representation, the additive constants are calculated: they are equal to 426 and 296, respectively. In passing to the definition of a representation employed in this book, the constants would be slightly changed, but the leading coefficients would be unaffected.

The construction and proof given below—excepting details for which the book's second author assumes responsibility—go back to Zharov. An auxiliary algorithm $\mathfrak{B}$ is first constructed in §43. It is a slight modification of a universal algorithm for the two-letter alphabet ab. With its aid, a universal algorithm for an arbitrary alphabet is then constructed in §44 by means of a suitable coding.

§43 Case of a Two-Letter Alphabet

1. We introduce the letters

$$a, b, \kappa, \lambda, \mu, \nu, \pi, \sigma, \varphi, \psi, \omega,$$

pairwise distinct and different from the letters of the alphabet $\mathrm{A}\alpha\beta\gamma\delta$. We also introduce the ≪doubles≫[2]

$$\bar{a}, \bar{b}, \bar{\beta}, \bar{\lambda}, \bar{\mu}, \bar{\pi}, \bar{\varphi},$$

which are likewise distinct and different from all letters mentioned so far. Furthermore, it will be convenient for us to set $\bar{\alpha} \rightleftharpoons \alpha$.

We shall denote the alphabet ab by the letter A_0, and the alphabet $\mathrm{A}_0\beta\lambda\mu\varphi$ by the letter Б.

By means of $\mathfrak{B}$, we shall denote the normal algorithm in the alphabet Б $\bar{\text{Б}}\alpha\gamma\ \nu\sigma\pi\psi\omega\kappa$, whose scheme can be written down in the following abbreviated form:

$$\left\{\begin{array}{rll}
\bar{\xi}\iota \to & \iota\bar{\xi} & \langle 1\rangle \\
\bar{\xi}\psi\xi \to & \xi\psi & \langle 2\rangle \\
\bar{\xi}\psi \to & \nu & \langle 3\rangle \\
\bar{\xi}\kappa \to & \xi\kappa & \langle 4\rangle \\
\theta\nu \to & \nu\theta & \langle 5\rangle \\
\varphi\nu \to & \nu\bar{\varphi} & \langle 6\rangle \\
\lambda\mu\nu \to & \varphi & \langle 7\rangle \\
\lambda\nu \to & \lambda\mu & \langle 8\rangle \\
\bar{\lambda}\nu\gamma \to & \gamma\lambda\mu & \langle 9\rangle \\
\mu\nu \to & & \langle 10\rangle \\
\nu \to & \lambda\mu & \langle 11\rangle \\
\bar{\varphi}\xi \to & \nu\xi\varphi\psi & \langle 12\rangle \\
\bar{\varphi} \to & & \langle 13\rangle \\
\sigma \to & \nu & \langle 14\rangle \\
\bar{\lambda}\bar{\mu} \to & \eta\bar{\lambda} & \langle 15\rangle \\
\bar{\lambda}\mu\chi \to & \bar{\chi}\bar{\mu} & \langle 16\rangle \\
\lambda\mu\gamma \to & \gamma\lambda\mu & \langle 17\rangle \\
\mu\xi \to & \xi\mu\bar{\xi} & \langle 18\rangle \\
\lambda \to & \bar{\lambda} & \langle 19\rangle \\
\varphi\varsigma \to & \varphi & \langle 20\rangle \\
\varphi\psi \to & \kappa & \langle 21\rangle \\
\varphi\omega \to & \pi & \langle 22\rangle \\
\bar{\mu} \to & \mu & \langle 23\rangle \\
\kappa \to & \nu & \langle 24\rangle \\
\bar{\beta} \to & \nu\sigma & \langle 25\rangle \\
\omega \to & \nu\omega\varphi\psi & \langle 26\rangle,
\end{array}\right.$$

[2]cf. §37.2

where ξ runs through the alphabet A_0, η—the alphabet $A_0\alpha\beta$, ς—the alphabet $A_0\alpha\beta\gamma$, θ—the alphabet $A\alpha\beta\gamma\omega$, ι—the alphabet $A_0\alpha\beta\gamma\varphi\omega$ and χ—the alphabet $\alpha\beta$.

We number the lines of the scheme of the algorithm $\mathfrak{B}$ for convenience in referring to them, and in what follows, the subscript i in the notation $\mathfrak{B} : X \vdash_i Y$ will give the number of the line representing the substitution formula active on the word X.

2. We are trying to show that the following theorem holds.

2.1. *Given any normal algorithm $\mathfrak{A}$ in the alphabet A_0 and any word Q in this alphabet, the conditional equality*

$$\mathfrak{B}(\mathfrak{A}^{\cdot \mathrm{Rp}}\omega Q) \simeq \pi\mathfrak{A}(Q)$$

holds.(Here $\mathfrak{A}^{\cdot}$ denotes the closure of the algorithm $\mathfrak{A}$ [§36.2].)

The next lemmas below, 2.2–2.5, clarify the performance of the algorithm $\mathfrak{B}$ in general.

2.2. *If $\mathfrak{A}$ is a normal algorithm in A_0 and Q is a word in A_0, then*

$$\mathfrak{B} : \mathfrak{A}^{\cdot \mathrm{Rp}}\omega Q \models \lambda\mu\mathfrak{A}^{\cdot \mathrm{Rp}}\omega\varphi\psi Q.$$

2.3. *If U is a word in $A_0\alpha\beta\gamma\varphi\psi\omega$, then*

$$\mathfrak{B} : \lambda\mu\gamma U \vdash_{17} \gamma\lambda\mu U.$$

2.4. *If L and N are words in the alphabet $A_0\alpha\beta\gamma$, M is a word in $A_0\alpha\beta$ and Q, S are words in A_0, such that Q does not occur in S, then*

$$\mathfrak{B} : L\lambda\mu QM\gamma N\omega\varphi\psi S \models LQM\gamma\lambda\mu N\omega\varphi\psi S.$$

2.5. *If L and N are words in $A_0\alpha\beta\gamma$, if Q, R and S are words in A_0 and if Q occurs in S, then*

a) $$\mathfrak{B} : L\lambda\mu Q\alpha R\gamma N\omega\varphi\psi S \models LQ\alpha R\gamma N\omega\Sigma(R, Q, S);$$

b) $$\mathfrak{B} : L\lambda\mu Q\beta R\gamma N\omega\varphi\psi S \models \pi\Sigma(R, Q, S).$$

Theorem 2.1 follows from Lemmas 2.2–2.5.

Indeed, let $\mathfrak{A}$ be a normal algorithm in the alphabet A_0. Then $\mathfrak{A}^{\cdot \mathrm{Rp}}$ has the form γFW, where F is the first substitution formula of the algorithm $\mathfrak{A}^{\cdot}$ and W is the rest of $\mathfrak{A}^{\cdot}$'s scheme. Let the words P and Q be such that $\mathfrak{A}^{\cdot} : P \vdash Q$. Then

$$\begin{aligned} \mathfrak{B} : \mathfrak{A}^{\cdot \mathrm{Rp}}\omega P &\models \lambda\mu\mathfrak{A}^{\cdot \mathrm{Rp}}\omega\varphi\psi P && [2.2] \\ &\vdash \gamma\lambda\mu FW\omega\varphi\psi P && [2.3] \\ &\models \mathfrak{A}^{\cdot \mathrm{Rp}}\omega Q && [2.4,\ 2.5\mathrm{a})]; \end{aligned}$$

moreover, at least three simple steps of $\mathfrak{B}$'s performance on $\mathfrak{A}^{\cdot \mathrm{Rp}}\omega P$ correspond to one simple step of $\mathfrak{A}$'s performance on P, so that if $!\mathfrak{B}(\mathfrak{A}^{\cdot \mathrm{Rp}}\omega P)$, then $!\mathfrak{A}(P)$.

But if $\mathfrak{A}^{\cdot} : P \vdash\cdot Q$, then

$$\mathfrak{B}(\mathfrak{A}^{\cdot \mathrm{Rp}}\omega P) \risingdotseq \pi Q \quad [2.2, 2.3, 2.4, 2.5b].$$

Therefore, if $!\mathfrak{A}(P)$, then $\mathfrak{A}^{\cdot} : P \models \mathfrak{A}(P)$, and hence

$$\mathfrak{B}(\mathfrak{A}^{\cdot \mathrm{Rp}}\omega P) \risingdotseq \pi \mathfrak{A}(P).$$

3. Thus, we proved that Theorem 2.1 follows from Lemmas 2.2–2.5. Let us now turn to the proof of these lemmas. As a preliminary, we shall establish the auxiliary assertions 3.1–3.8.

3.1. *Let W be a word in the alphabet $\mathrm{A}_0\alpha\beta\gamma\varphi\omega$, let B, C be words in the alphabet $\mathrm{A}_0\alpha\beta\gamma\kappa\lambda\mu\sigma\varphi\psi\omega\bar{\beta}\bar{\lambda}\bar{\mu}\bar{\varphi}$ and let ξ be a letter of the alphabet A_0. Then*

$$\mathfrak{B} : B\bar{\xi}WC \models BW\bar{\xi}C. \tag{1}$$

This is proven by right induction on W.

In fact, the assertion (1) is trivial for $W \risingdotseq \Lambda$. Now suppose that for some word W in the alphabet $\mathrm{A}_0\alpha\beta\gamma\varphi\omega$, the assertion (1) is valid for any B, C and ξ satisfying the hypothesis listed in the statement of the lemma. Let ι be any letter of the alphabet $\mathrm{A}_0\alpha\beta\gamma\varphi\omega$. Then

$$\begin{aligned}\mathfrak{B} : B\bar{\xi}W\iota C &\models BW\bar{\xi}\iota C\\ &\vdash_1 BW\iota\bar{\xi}C,\end{aligned}$$

i.e.,

$$\mathfrak{B} : B\bar{\xi}W\iota C \models BW\iota\bar{\xi}C,$$

which is just what remained to be proven.

3.2. *Let T be a word in the alphabet $\mathrm{A}_0\alpha\beta\gamma\omega$ and let B, C be words in the alphabet $\mathrm{A}_0\alpha\beta\gamma\kappa\lambda\mu\sigma\varphi\psi\omega\bar{\beta}\bar{\lambda}\bar{\mu}\bar{\varphi}$. Then*

$$\mathfrak{B} : BT\nu C \models B\nu TC. \tag{2}$$

This is proven by right induction on T.

In fact, the assertion (2) is trivial for $T \risingdotseq \Lambda$. Now suppose that for some T in the alphabet $\mathrm{A}_0\alpha\beta\gamma\omega$, the assertion (2) is valid for any B and C satisfying the hypotheses given in the lemma. Let θ be any letter of the alphabet $\mathrm{A}_0\alpha\beta\gamma\omega$. Then

$$\begin{aligned}\mathfrak{B} : BT\theta\nu C &\vdash_5 BT\nu\theta C\\ &\models B\nu T\theta C,\end{aligned}$$

which is just what remained to be proven.

3.3. *Let M be a word in the alphabet $A_0\alpha\beta$ and let D, E be words in the alphabet $A_0\alpha\beta\gamma\mu\varphi\psi\omega$. Then*

$$\mathfrak{B} : D\bar{\lambda}ME \models DM\bar{\lambda}E. \tag{3}$$

This is proven by right induction on M. The substitution formula represented by the fifteenth line of the algorithm $\mathfrak{B}$'s scheme should be used for the inductive step.

3.4. *Let N be a word in the alphabet $A_0\alpha\beta\gamma$ and let F, G be words in the alphabet $A_0\alpha\beta\gamma\psi\omega\bar{\beta}\bar{\mu}$. Then*

$$\mathfrak{B} : F\varphi NG \models F\varphi G. \tag{4}$$

This is proven by right induction on N. The substitution formula represented by the twentieth line of the algorithm $\mathfrak{B}$'s scheme should be used for the inductive step.

3.5. *Let L be a word in the alphabet $A_0\alpha\beta\gamma$, let Q, S be words in the alphabet A_0 and let T be a word in the alphabet $A_0\alpha\beta\gamma\omega$. Then*

$$\mathfrak{B} : L\lambda\mu QT\varphi\psi QS \models L\lambda Q\mu T\varphi Q\psi S. \tag{5}$$

This is proven by right induction on Q.

The assertion (5) is trivial for empty Q. Now suppose that for some Q in the alphabet A_0, the assertion (5) is true for all L, T and S satisfying the hypotheses listed in the statement of the lemma. Then for any letter ξ of the alphabet A_0, we have

$$\begin{aligned} \mathfrak{B} : L\lambda\mu Q\xi T\varphi\psi Q\xi S &\models L\lambda Q\mu\xi T\varphi Q\psi\xi S \\ &\vdash_{18} L\lambda Q\xi\mu\bar{\xi}T\varphi Q\psi\xi S \\ &\models L\lambda Q\xi\mu T\bar{\varphi}Q\xi\psi\xi S \qquad [3.1] \\ &\vdash_{2} L\lambda Q\xi\mu T\varphi Q\xi\psi S, \end{aligned}$$

i.e.,

$$\mathfrak{B} : L\lambda\mu Q\xi T\varphi\psi Q\xi S \models L\lambda Q\xi\mu T\varphi Q\xi\psi S,$$

which is just what remained to be proven.

3.6. *Let L be a word in the alphabet $A_0\alpha\beta\gamma$, let Q, S be words in the alphabet A_0 and let T be a word in the alphabet $A_0\alpha\beta\gamma\omega$. If Q is not an initial segment of S, then*

$$\mathfrak{B} : L\lambda\mu QT\varphi\psi S \models L\lambda QT\bar{\varphi}S. \tag{6}$$

Indeed, let the lemma's hypothesis be fulfilled. Then there exist words R, Q_1 and S_1 in the alphabet A_0 and a letter ξ of this alphabet, such that $Q \rightleftharpoons R\xi Q_1$, $S \rightleftharpoons RS_1$ and the word S_1 does not begin with the letter ξ. We have

$$L\lambda\mu QT\varphi\psi S \rightleftharpoons L\lambda\mu R\xi Q_1T\varphi\psi RS_1$$

and

$$\mathfrak{B} : L\lambda\mu R\xi Q_1 T\varphi\psi RS_1 \models L\lambda R\mu\xi Q_1 T\varphi R\psi S_1 \quad [3.5]$$
$$\vdash_{18} L\lambda R\xi\mu\bar{\xi} Q_1 T\varphi R\psi S_1$$
$$\models L\lambda R\xi\mu Q_1 T\varphi R\bar{\xi}\psi S_1 \quad [3.1]$$
$$\vdash_3 L\lambda R\xi\mu Q_1 T\varphi R\nu S_1$$
$$\models L\lambda R\xi\mu Q_1 T\varphi\nu RS_1 \quad [3.2]$$
$$\vdash_6 L\lambda R\xi\mu Q_1 T\nu\bar{\varphi} RS_1$$
$$\models L\lambda R\xi\mu\nu Q_1 T\bar{\varphi} RS_1 \quad [3.2]$$
$$\vdash_{10} L\lambda R\xi Q_1 T\bar{\varphi} RS_1.$$

But

$$L\lambda R\xi Q_1 T\bar{\varphi} RS_1 = L\lambda QT\bar{\varphi}S,$$

and hence

$$\mathfrak{B} : L\lambda\mu QT\varphi\psi S \models L\lambda QT\bar{\varphi}S.$$

Q.E.D.

For arbitrary words Q and S in the alphabet A_0, put[3]

$$\Pi(Q,S \rightleftharpoons \begin{cases} \Sigma(\varphi\psi Q, Q, S) & \text{if } Q \text{ occurs in } S, \\ S\varphi\psi & \text{if } Q \text{ does not occur in } S. \end{cases}$$

3.7. *Let L be a word in the alphabet $A_0\alpha\beta\gamma$, let Q and S be words in A_0 and let T be a word in $A_0\alpha\beta\gamma\omega$. Then*

(7) $$\mathfrak{B} : L\lambda\mu QT\varphi\psi S \models L\lambda\mu QT\Pi(Q,S).$$

This is proven by left induction on S.

The assertion (7) is trivial for empty S, since $\Pi(Q,\Lambda) = \varphi\psi$ for any Q.

Suppose that for some S in A_0, the assertion (7) is valid for all L, T and Q satisfying the hypotheses listed in the statement of the lemma. We shall show that it will also be valid when S is replaced by ξS, where ξ is any letter of the alphabet A_0.

If ξS begins with the word Q, then $\Pi(Q,\xi S) = \varphi\psi\xi S$, in which case the assertion

$$\mathfrak{B} : L\lambda\mu QT\varphi\psi\xi S \models L\lambda\mu QT\Pi(Q,\xi S)$$

is trivial.

Now consider the case when ξS does not begin with the word Q. If Q does not occur in ξS, then neither does Q occur in S, and so

$$\Pi(Q,\xi S) = \xi S\varphi\psi, \quad \Pi(Q,S) = S\varphi\psi.$$

[3]The result of substituting $\varphi\psi Q$ in place of the first occurrence of Q in S (see §23.8, p.75).

But if Q occurs in ξS, then the first occurrence of Q is ξS has the form $\xi S_1 * Q * S_2$, where S_1 and S_2 are words in the alphabet A_0. But then $S_1 * Q * S_2$ is the first occurrence of Q in S, so that

$$\Pi(Q, \xi S) \rightleftharpoons \xi S_1 \varphi\psi Q S_2,$$
$$\Pi(Q, S) \rightleftharpoons S_1 \varphi\psi Q S_2.$$

Consequently, if ξS does not begin with Q, then

$$\Pi(Q, \xi S) \rightleftharpoons \xi \Pi(Q, S)$$

and

$$\begin{aligned} \mathfrak{B} : L\lambda\mu QT\varphi\psi\xi S &\models L\lambda QT\bar{\varphi}\xi S && [3.6] \\ &\vdash_{12} L\lambda QT\nu\xi\varphi\psi S \\ &\models L\lambda\nu QT\xi\varphi\psi S && [3.2] \\ &\vdash_{8} L\lambda\mu QT\xi\varphi\psi S \\ &\models L\lambda\mu QT\xi\Pi(Q, S). \end{aligned}$$

But $\xi\Pi(Q, S) \rightleftharpoons \Pi(Q, \xi S)$ and, therefore,

$$\mathfrak{B} : L\lambda\mu QT\varphi\psi\xi S \models L\lambda\mu QT\Pi(Q, \xi S),$$

which is just what remained to be proven.

3.8. *Let H be a word in the alphabet $\mathrm{A}_0\alpha\beta\bar{\beta}\gamma$, let R, S be words in A_0 and let T be a word in $\mathrm{A}_0\alpha\beta\gamma\omega$. Then*

(8) $$\mathfrak{B} : H\mu RT\kappa S \models HR\mu TR\kappa S.$$

This is proven by right induction on R.

The assertion (8) is trivial for empty R. Suppose that for some R in the alphabet A_0, the assertion (8) is valid for all H, T and S satisfying the hypotheses listed in the statement of the lemma. Then for any letter ξ of the alphabet A_0, we have

$$\begin{aligned} \mathfrak{B} : H\mu R\xi T\kappa S &\models HR\mu\xi TR\kappa S \\ &\vdash_{18} HR\xi\mu\bar{\xi}TR\kappa S \\ &\models HR\xi\mu TR\bar{\xi}\kappa S && [3.1] \\ &\vdash_{4} HR\xi\mu TR\xi\kappa S, \end{aligned}$$

i.e.,

$$\mathfrak{B} : H\mu R\xi T\kappa S \models HR\xi\mu TR\xi\kappa S,$$

which is what remained to be proven.

We now prove Lemmas 2.2–2.5.

PROOF OF LEMMA 2.2:

Let $\mathfrak{A}$ be a normal algorithm in the alphabet A_0 and let Q be a word in this alphabet. Then

$$\begin{aligned}\mathfrak{B} : \mathfrak{A}^{\cdot \mathrm{Rp}}\omega Q &\vdash_{26} \mathfrak{A}^{\cdot \mathrm{Rp}}\nu\omega\varphi\psi Q \\ &\models \nu\mathfrak{A}^{\cdot \mathrm{Rp}}\omega\varphi\psi Q \qquad [3.2] \\ &\vdash_{11} \lambda\mu\mathfrak{A}^{\cdot \mathrm{Rp}}\omega\varphi\psi Q,\end{aligned}$$

i.e.,

$$\mathfrak{B} : \mathfrak{A}^{\cdot \mathrm{Rp}}\omega Q \models \lambda\mu\mathfrak{A}^{\cdot \mathrm{Rp}}\omega\varphi\psi Q.$$

Q.E.D.

THE PROOF OF LEMMA 2.3: is obtained directly from an examination of the scheme of the algorithm $\mathfrak{B}$.

PROOF OF LEMMA 2.4:

Let the lemma's hypotheses be fulfilled. Then

$$Q \eqcirc \Lambda \tag{9}$$

and

$$\begin{aligned}\mathfrak{B} : L\lambda\mu QM\gamma N\omega\varphi\psi S &\models L\lambda\mu QM\gamma N\omega\Pi(Q, S) \qquad [3.7] \\ &\eqcirc L\lambda\mu QM\gamma N\omega S\varphi\psi \\ &\models L\lambda QM\gamma N\omega S\bar{\varphi} \qquad [3.6, (9)] \\ &\vdash_{13} L\lambda QM\gamma N\omega S \\ &\vdash_{19} L\bar{\lambda} QM\gamma N\omega S \\ &\models LQM\bar{\lambda}\gamma N\omega S \qquad [3.3] \\ &\vdash_{26} LQM\bar{\lambda}\gamma N\nu\omega\varphi\psi S \\ &\models LQM\bar{\lambda}\nu\gamma N\omega\varphi\psi S \qquad [3.2] \\ &\vdash_{9} LQM\gamma\lambda\mu N\omega\varphi\psi S,\end{aligned}$$

i.e.,

$$\mathfrak{B} : L\lambda\mu QM\gamma N\omega\varphi\psi S \models LQM\gamma\lambda\mu N\omega\varphi\psi S.$$

Q.E.D.

PROOF OF LEMMA 2.5:

Let χ be a letter of the alphabet $\alpha\beta$, let L, N be words in the alphabet $A_0\alpha\beta\gamma$, let Q, R, S be words in the alphabet A_0 and let Q occur in S. Let $S_1 * Q * S_2$, where S_1 and S_2 are words in the alphabet A_0, be the first occurrence of Q in S. Then

$$\begin{aligned}\Pi(Q, S) &\eqcirc \Sigma(\varphi\psi Q, Q, S) \\ &\eqcirc S_1\varphi\psi QS_2\end{aligned}$$

and

$$
\begin{aligned}
\mathfrak{B} : L\lambda\mu Q\chi R\gamma N\omega\varphi\psi S &\models L\lambda\mu Q\chi R\gamma N\omega\Pi(Q, S) && [3.7]\\
&\rightleftharpoons L\lambda\mu Q\chi R\gamma N\omega S_1\varphi\psi Q S_2\\
&\models L\lambda Q\mu\chi R\gamma N\omega S_1\varphi Q\psi S_2 && [3.5]\\
&\vdash_{19} L\bar{\lambda}Q\mu\chi R\gamma N\omega S_1\varphi Q\psi S_2\\
&\models LQ\bar{\lambda}\mu\chi R\gamma N\omega S_1\varphi Q\psi S_2 && [3.3]\\
&\vdash_{16} LQ\bar{\chi}\bar{\mu}R\gamma N\omega S_1\varphi Q\psi S_2\\
&\models LQ\bar{\chi}\bar{\mu}R\gamma N\omega S_1\varphi\psi S_2 && [3.4]\\
&\vdash_{21} LQ\bar{\chi}\bar{\mu}R\gamma N\omega S_1\kappa S_2\\
&\vdash_{23} LQ\bar{\chi}\mu R\gamma N\omega S_1\kappa S_2\\
&\models LQ\bar{\chi}R\mu\gamma N\omega S_1 R\kappa S_2 && [3.8]\\
&\vdash_{24} LQ\bar{\chi}R\mu\gamma N\omega S_1 R\nu S_2\\
&\models LQ\bar{\chi}R\mu\nu\gamma N\omega S_1 R S_2 && [3.2]\\
&\vdash_{10} LQ\bar{\chi}R\gamma N\omega S_1 R S_2\\
&\rightleftharpoons LQ\bar{\chi}R\gamma N\omega\Sigma(R, Q, S).
\end{aligned}
$$

Since $\bar{\alpha} \rightleftharpoons \alpha$, the assertion a) of Lemma 2.5 is proven (it is sufficient to take α for χ in the preceding argument). We now complete the proof of the assertion b) (here $\chi \rightleftharpoons \beta$).

$$
\begin{aligned}
\mathfrak{B} : LQ\bar{\beta}R\gamma N\omega\Sigma(R, Q, S) &\vdash_{25} LQ\nu\sigma R\gamma N\omega\Sigma(R, Q.S)\\
&\models \nu LQ\sigma R\gamma N\omega\Sigma(R, Q, S) && [3.2]\\
&\vdash_{11} \lambda\mu LQ\sigma R\gamma N\omega\Sigma(R, Q, S)\\
&\vdash_{14} \lambda\mu LQ\nu R\gamma N\omega\Sigma(R, Q, S)\\
&\models \lambda\mu\nu LQR\gamma N\omega\Sigma(R, Q, S) && [3.2]\\
&\vdash_{7} \varphi LQR\gamma N\omega\Sigma(R, Q, S)\\
&\models \varphi\omega\Sigma(R, Q, S) && [3.4]\\
&\vdash_{22} \pi\Sigma(R, Q, S)
\end{aligned}
$$

and the algorithm $\mathfrak{B}$ does not act on the word $\pi\Sigma(R, Q, S)$.

Thus, the proof of Theorem 2.1 is completed.

§44 Proof of the Universal Algorithm Theorem

1. Let us now turn to the proof of Theorem §42.1.1.

Let the length of the alphabet A we are considering be equal to n. We agree to denote its ith letter by the symbol a_i ($1 \leqslant i \leqslant n$). We introduce the translation operation $[P^\tau$ for words in the alphabet A_0, defining it by the following equalities:

$$
\begin{aligned}
&[\Lambda^\tau \rightleftharpoons \Lambda,\\
&[Pa_i^\tau \rightleftharpoons [P^\tau ab^i a
\end{aligned}
$$

(here P means an arbitrary word in the alphabet A, i satisfies the condition $1 \leqslant i \leqslant n$ and a, b are letters of the alphabet A_0, introduced by us at the beginning of §43). Let us further agree that for any normal algorithm $\mathfrak{A}$ in the alphabet A, $\mathfrak{A}^{\tau}$ denotes its translation [§41.6] into the two-letter alphabet A_0 in accordance with the translation operation for words just described. We introduce two new letters, ε and ρ, different from letters of A and those intoroduced in Part 1 of §43. Consider the normal algorithm $\mathfrak{C}$ in the alphabet $AA_0\alpha\beta\gamma\delta\varepsilon\rho\pi\bar{\pi}$, whose scheme can be written down in the following abbreviated form:

$$
\left\{
\begin{array}{rll}
ba_{i+1} \to & a_i & \langle 1 \rangle \\
\bar{\pi}a \to & a_n\pi & \langle 2 \rangle \\
\bar{\pi}b \to & b\bar{\pi} & \langle 3 \rangle \\
\rho a_1\pi \to & a_n\pi & \langle 4 \rangle \\
\rho a_1 \to & baap & \langle 5 \rangle \\
ab\rho b \to & & \langle 6 \rangle \\
a\rho\varsigma \to & \varsigma a\rho & \langle 7 \rangle \\
a\rho\delta \to & \beta\gamma\varepsilon a\rho & \langle 8 \rangle \\
\rho \to & b\rho b & \langle 9 \rangle \\
\pi ab \to & \rho\bar{\pi} & \langle 10 \rangle
\end{array}
\right.
$$

where $1 \leqslant i < n$ and ς runs through the alphabet $\alpha\beta\gamma$.

We are aiming at a proof of the following assertion:

1.1. *Let* $\mathfrak{A}$ *be a normal algorithm in the alphabet* A *and let* P *be a word in* A. *Then:*

a) $$\mathfrak{C} : a\rho\mathfrak{A}^{\mathrm{Rp}}\delta P \models \mathfrak{A}^{\cdot\tau\mathrm{Rp}}\varepsilon[P^{\tau}$$

b) $$\mathfrak{C} : \pi[P^{\tau} \models P\pi.$$

2. As a preliminary, we prove the following lemmas, 2.1–2.6.

2.1. *Let each of the words* X, Y *be a word in the one of the alphabets* $A\alpha\beta\gamma\delta$, $A_0\alpha\beta\gamma\varepsilon\pi$. *Let* $1 \leqslant i \leqslant \eta$. *Then*

(1) $$\mathfrak{C} : Xb^j\rho a_i Y \models Xb^{j+i-1}\rho a_1 Y.$$

This is proven by arithmetical induction on i.

In fact, the assertion (1) is trivial for $i = 1$. Now suppose that for some i $(1 \leqslant i < n)$, the assertion (1) is true for all j and for all X, Y satisfying the hypotheses listed in the statement of the lemma. Then

$$
\begin{aligned}
\mathfrak{C} : Xb^j\rho a_{i+1}Y &\vdash_9 Xb^{j+1}\rho ba_{i+1}Y \\
&\vdash_1 Xb^{j+1}\rho a_i Y \\
&\models Xb^{j+1}\rho a_1 Y \quad \text{[ind. ass.]},
\end{aligned}
$$

i.e.,

$$\mathfrak{C} : Xb^j\rho a_{i+1}Y \models Xb^{j+i}\rho a_1 Y,$$

which is what remained to be proven.

2.2. *Let P be a word in the alphabet* A, *let U be a word in the alphabet* $A_0\alpha\beta\gamma\varepsilon$ *and let V be a word in the alphabet* $A\alpha\beta\gamma\delta$. *Then*

(2) $$\mathfrak{C} : Ua\rho PV \models U[P^\tau a\rho V.$$

This is proven by right induction on P.

Indeed, the assertion (2) is trivial for empty P. Now suppose that for some P, the assertion (2) is true for all U and V satisfying the hypotheses listed in the statement of the lemma. Let $1 \leqslant i \leqslant n$. Then

$$\begin{aligned}\mathfrak{C} : Ua\rho Pa_iV &\models U[P^\tau a\rho a_iV && \text{[ind. ass.]}\\ &\models U[P^\tau ab^{i-1}\rho a_1V && [2.1]\\ &\vdash_5 U[P^\tau ab^i aa\rho V\\ &\eqcirc U[Pa_i^\tau a\rho V,\end{aligned}$$

i.e.,

$$\mathfrak{C} : Ua\rho Pa_iV \models U[Pa_i^\tau a\rho V,$$

which is just what remained to be proven.

2.3. *Let* $\mathfrak{A}$ *be a normal algorithm in the alphabet* A *and let V be a word in the alphabet* $A\alpha\beta\gamma\delta$. *Then*

$$\mathfrak{C} : a\rho\mathfrak{A}^{\mathrm{Rp}}V \models \mathfrak{A}^{\tau\mathrm{Rp}}a\rho V.$$

This is proven by induction along the scheme[4] of the algorithm $\mathfrak{A}$.

In fact, for an algorithm $\mathfrak{A}$ with an ≪empty≫ scheme [§24.3], the assertion (3) is true:

$$\mathfrak{C} : a\rho\gamma V \vdash_7 \gamma a\rho V.$$

Now let the assertion (3) be true for an algorithm $\mathfrak{A}$ and all V satisfying the hypothesis given in the statement of the lemma. We show that it will also be true for a normal algorithm $\mathfrak{A}_1$ in the alphabet A, whose scheme is obtained from the scheme of the algorithm A by the addition of one more substitution formula. Denote the left and right sides of this formula by P_1 and P_2, respectively. Furthermore, set $\omega \eqcirc \alpha$ if this formula is simple and $\omega \eqcirc \beta$ if it is final. Then $\mathfrak{A}_1^{\mathrm{Rp}} \eqcirc \mathfrak{A}^{\mathrm{Rp}}P_1\omega P_2\gamma$ and, consequently

$$\begin{aligned}\mathfrak{C} : a\rho\mathfrak{A}_1^{\mathrm{Rp}}V &\eqcirc a\rho\mathfrak{A}^{\mathrm{Rp}}P_1\omega P_2V\\ &\models \mathfrak{A}^{\tau\mathrm{Rp}}a\rho P_1\omega P_2\gamma V && \text{[ind. ass.]}\\ &\models \mathfrak{A}^{\tau\mathrm{Rp}}[P_1^\tau a\rho\omega P_2\gamma V && [2.2]\\ &\vdash_7 \mathfrak{A}^{\tau\mathrm{Rp}}[P_1^\tau\omega a\rho P_2\gamma V\\ &\models \mathfrak{A}^{\tau\mathrm{Rp}}[P_1^\tau\omega[P_2^\tau a\rho\gamma V && [2.2]\\ &\vdash_7 \mathfrak{A}^{\tau\mathrm{Rp}}[P_1^\tau\omega[P_2^\tau\gamma a\rho V\\ &\eqcirc \mathfrak{A}_1^{\tau\mathrm{Rp}}a\rho V,\end{aligned}$$

i.e.,

$$\mathfrak{C} : a\rho\mathfrak{A}_1^{\mathrm{Rp}}V \models \mathfrak{A}_1^{\tau\mathrm{Rp}}a\rho V,$$

which is what remained to be proven.

---

[4]That is, on the construction of the scheme.

2.4. *Let W be a word in the alphabet $\mathrm{A}\rho$ and let Q be a word in the alphabet A_0. Then*

$$\mathfrak{C}: W\bar{\pi}b^jQ \models Wb^j\bar{\pi}Q.$$

This is proved by induction on j.

In fact, the assertion (4) is trivial when $j = 0$. Now let the assertion (4) be true for some j and for all W and Q satisfying the hypotheses listed in the statement of the lemma. Then

$$\begin{aligned}\mathfrak{C}: W\bar{\pi}b^{j+1}Q &\models Wb^j\bar{\pi}bQ \qquad \text{[ind. ass.]}\\ &\vdash_3 Wb^{j+1}\bar{\pi}Q\end{aligned}$$

which is just what remained to be proven.

2.5. *Let W be a word in the alphabet $\mathrm{A}\rho$, let Z be a word in the alphabet $\mathrm{A}_0\pi$ and let $j < n$. Then*

$$\mathfrak{C}: Wb^{l+j}a_nZ \models Wb^la_{n-j}Z. \tag{5}$$

This is proven by induction on j.

Indeed, the assertion (5) is trivial when $j = 0$. Now let the assertion (5) be true for some j ($j < n - 1$) and for all W and Z satisfying the hypotheses listed in the statement of the lemma. Then

$$\begin{aligned}\mathfrak{C}: Wb^{l+j+1}a_nZ &\models Wb^{l+1}a_{n-j}Z \qquad \text{[ind. ass.]}\\ &\vdash_1 Wb^la_{n-j-1}Z,\end{aligned}$$

which is just what remained to be proven.

2.6. *Let P be a word in the alphabet A and let Q be a word in the alphabet A_0. Then*

$$\mathfrak{C}: \pi[P^{\tau}Q \models P\pi Q.$$

This is proven by right induction on P.

In fact, the assertion (6) is trivial when $P \eqcirc \Lambda$. Now let the assertion (6) be true for some P and for all Q in A_0. Let $1 \leqslant j \leqslant n$. Then

$$\begin{aligned}\mathfrak{C}: \pi[Pa_j^{\tau}Q &\eqcirc \pi[P^{\tau}ab^jaQ\\ &\models P\pi ab^jaQ \qquad \text{[ind. ass.]}\\ &\vdash_{10} P\rho\bar{\pi}b^{j-1}aQ\\ &\models P\rho b^{j-1}\bar{\pi}aQ \qquad [2.4]\\ &\vdash_2 P\rho b^{j-1}a_n\pi Q\\ &\models P\rho a_{n-j+1}\pi Q \qquad [2.5]\\ &\models Pb^{n-j}\rho a_1\pi Q \qquad [2.1]\\ &\vdash_4 Pb^{n-j}a_n\pi Q\\ &\models Pa_j\pi Q \qquad [2.5],\end{aligned}$$

which is just what remained to be proven.

3. Let us now finish the proof of Theorem 1.1. The assertion b) of this theorem follows immediately from Lemma 2.6. Thus, it remains to prove the assertion a).

If $\mathfrak{A}$ is a normal algorithm in the alphabet A and P is a word in the alphabet A, then

$$\begin{aligned} \mathfrak{C} : a\rho\mathfrak{A}^{\mathrm{Rp}}\delta P &\models \mathfrak{A}^{\tau\mathrm{Rp}} a\rho\delta P && [2.3] \\ &\vdash_8 \mathfrak{A}^{\tau\mathrm{Rp}} \beta\gamma\varepsilon a\rho P \\ &\fallingdotseq \mathfrak{A}^{\cdot\tau\mathrm{Rp}} \varepsilon \dot{a}\rho P \\ &\models \mathfrak{A}^{\cdot\tau\mathrm{Rp}} \varepsilon[P^{\tau} a\rho && [2.2] \\ &\vdash_9 \mathfrak{A}^{\cdot\tau\mathrm{Rp}} \varepsilon[P^{\tau} ab\rho b \\ &\vdash_6 \mathfrak{A}^{\cdot\tau\mathrm{Rp}} \varepsilon[P^{\tau}. \end{aligned}$$

4. We shall now complete the proof of the Universal Algorithm Theorem. We shall show that the hypothesis of this theorem is satisfied by the normal algorithm $\mathfrak{U}$ in the alphabet

$$\text{АБ}\bar{\text{Б}}\alpha\gamma\delta\varepsilon\nu\rho\sigma\pi\bar{\pi}\psi\omega \quad [\S 43.1]$$

with the scheme

$$\left\{\begin{array}{ll} \mathfrak{B} & \langle 1\rangle \\ \mathfrak{C} & \langle 2\rangle \\ \varepsilon \rightarrow \omega & \langle 3\rangle \\ \pi \rightarrow \cdot & \langle 4\rangle \\ \rightarrow a\rho. & \langle 5\rangle \end{array}\right.$$

Let $\mathfrak{A}$ be a normal algorithm in the alphabet A, let P be a word in the alphabet A and let V_1, V_2 be words in the alphabet $\mathrm{AA}_0\alpha\beta\gamma\delta\varepsilon\rho\pi\bar{\pi}$ (i.e., in the alphabet of the algorithm $\mathfrak{C}$). No left side of a substitution formula of the algorithm $\mathfrak{B}$'s scheme is a word in the alphabet $\mathrm{AA}_0\alpha\beta\gamma\delta\varepsilon\rho\pi\bar{\pi}$ [see §43.1]; therefore, no such left side occurs in V_1. Hence, if $\mathfrak{C} : V_1 \vdash V_2$, then $\mathfrak{U} : V_1 \vdash V_2$ and, consequently, if $\mathfrak{C} : V_1 \models V_2$, then $\mathfrak{U} : V_1 \models V_2$.

Thus

$$\begin{aligned} \mathfrak{U} : \mathfrak{A}^{\mathrm{Rp}}\delta P &\vdash_5 a\rho\mathfrak{A}^{\mathrm{Rp}}\delta P \\ &\models \mathfrak{A}^{\cdot\tau\mathrm{Rp}} \varepsilon[P^{\tau} && [1.1\mathrm{a})] \\ &\vdash_3 \mathfrak{A}^{\cdot\tau\mathrm{Rp}} \omega[P^{\tau}, \end{aligned}$$

i.e.,

$$\mathfrak{U} : \mathfrak{A}^{\mathrm{Rp}}\delta P \models \mathfrak{A}^{\cdot\tau\mathrm{Rp}}\omega[P^{\tau},$$

and so

$$\mathfrak{U} : (\mathfrak{A}^{\mathrm{Rp}}\delta P) \simeq \mathfrak{U}(\mathfrak{A}^{\cdot\tau\mathrm{Rp}}\omega[P^{\tau}).$$

Besides,

$$\mathfrak{U} : \pi[P^\tau \models P\pi \quad [1.1b)]$$
$$\vdash\cdot_4 P,$$

i.e.,

$$\mathfrak{U}(\pi[P^\tau) \risingdotseq P,$$

so that

$$\mathfrak{U}(\pi[\mathfrak{A}(P)^\tau) \simeq \mathfrak{A}(P).$$

Therefore, for the proof of the Universal Algorithm Theorem, it now remains to show that

$$\mathfrak{U}(\mathfrak{A}^{\cdot\tau\mathrm{Rp}}\omega[P^\tau) \simeq \mathfrak{U}(\pi[\mathfrak{A}(P)^\tau).$$

But since there are no final substitution formulas in the scheme of the algorithm $\mathfrak{B}$, we have

$$\mathfrak{U}(\mathfrak{A}^{\cdot\mathrm{Rp}}\omega[P^\tau) \simeq \mathfrak{U}(\mathfrak{B}(\mathfrak{A}^{\cdot\tau\mathrm{Rp}}\omega[P^\tau)).$$

From this we obtain, according to Theorem §43.2.1,

$$\mathfrak{U}(\mathfrak{A}^{\cdot\tau\mathrm{Rp}}\omega[P^\tau) \simeq \mathfrak{U}(\pi\mathfrak{A}^\tau([P^\tau)).$$

This completes the proof of the Universal Algorithm Theorem, since according to the Translation Theorem [§42.6.22], we have

$$\mathfrak{A}^\tau([P^\tau) \simeq [\mathfrak{A}(P)^\tau.$$

5. The proof of the Universal Algorithm Theorem presented here is based on Zharov's techniques, already mentioned in Part 3 of §31. The basic tendency of the proof is a striving towards economy: the complexity of the algorithm $\mathfrak{U}$'s scheme, which is not hard to calculate, is equal to $5n + 349$; the substantiation of the scheme's correctness also seems sufficiently compact to us.

§45 Modification of the Universal Algorithm Theorem

1. The method of representing normal algorithms adopted by us is simple and convenient, but requires the introduction of new letters. In what follows, we shall need another method of representing normal algorithms, a somewhat more complex method, but one which works with only two letters. In those cases when the initial alphabet contains more than one letter, this method will make it possible for us to write down schemes of normal algorithms in the initial alphabet in the form of words in this same alphabet, which is of considerable significance, as we shall see below.

2. Along with the alphabet $\mathrm{A}\alpha\beta\gamma$, used by us for constructing representations of normal algorithms in A, we shall consider the two-letter alphabet

$A_0 \rightleftharpoons ab$ and define, as in §41.3–4, a translation operation for words in the alphabet $A\alpha\beta\gamma$ in such a way that *all* the letters of this alphabet are translated into Λ-literoids, i.e., words of the form aBa, where B is a non-empty word in the one-letter alphabet b. (Consequently, the role of the alphabet A in §41.3 will here be played by the empty alphabet Λ, and the role of the alphabet Б, by the alphabet $A\alpha\beta\gamma$.) We do not assume in this connection that the letters of the alphabet A_0 are different from the letters of the alphabet $A\alpha\beta\gamma$.

One can construct an algorithm $\mathfrak{T}$ over the alphabet $A \cup A\alpha\beta\gamma$, changing each word in the alphabet $A\alpha\beta\gamma$ into the translation of this word [§33]. It is also easy to construct a normal algorithm $\mathfrak{T}_1$ over $A_0 \cup A\alpha\beta\gamma$, changing the translation of each word in $A\alpha\beta\gamma$ into this word, itself. We shall make use of these algorithms in the sequel.

3. We shall now consider normal algorithms in the alphabet A and their representations. The translation of the representation of a normal algorithm in the alphabet A will be called the *record* of this algorithm. We shall denote the record of the algorithm $\mathfrak{A}$ by the symbol $\mathfrak{A}^{\mathrm{Rc}}$.

By definition of a record,

$$\mathfrak{A}^{\mathrm{Rc}} \rightleftharpoons \mathfrak{T}(\mathfrak{A}^{\mathrm{Rp}}) \tag{1}$$
$$\mathfrak{A}^{\mathrm{Rp}} \rightleftharpoons \mathfrak{T}_1(\mathfrak{A}^{\mathrm{Rc}})$$

for every normal algorithm $\mathfrak{A}$ in the alphabet A.

3.1. *The record of every normal algorithm in the alphabet* A *is a word in the alphabet* A_0.

3.2. *Every normal algorithm in the alphabet* A *is completely determined by its record.*

In fact, if we are given the record $\mathfrak{A}^{\mathrm{Rc}}$ of a normal algorithm $\mathfrak{A}$ in the alphabet A, then the representation $\mathfrak{A}^{\mathrm{Rp}}$ of this algorithm can be obtained as the result of applying the algorithm $\mathfrak{T}_1$ to $\mathfrak{A}^{\mathrm{Rc}}$, but the algorithm $\mathfrak{A}$ is uniquely determined by its representation.

4. The case when the letters a and b belong to the alphabet A is especially interesting. Then the record of each normal algorithm in the alphabet A is a word in the very same alphabet. This possibility of writing down the schemes of normal algorithms in A in the form of words in this same alphabet is related not to A's containing precisely the letters a and b, but only to this alphabet's containing at least two letters. Then one can choose any two of A's letters in an arbitrary way and define the record of a normal algorithm in A so that these letters are forced to play the roles of a and b.

5. We now prove a theorem analogous to Theorem §42.1.1, in which the role of the representation $\mathfrak{A}^{\mathrm{Rp}}$ of the algorithm $\mathfrak{A}$ will be played by its record $\mathfrak{A}^{\mathrm{Rc}}$. This ≪modification of the Universal Algorithm Theorem≫ will be needed by us in the sequel.

5.1. *Let the letter δ fail to belong to the alphabet* $\mathrm{A} \cup \mathrm{A}_0$. *Then one can construct a normal algorithm $\mathfrak{W}$ over the alphabet* $\mathrm{A} \cup \mathrm{A}_0 \cup \delta$, *such that*

$$\text{(1)} \qquad \mathfrak{W}(\mathfrak{A}^{\mathrm{Rc}}\delta P) \simeq \mathfrak{A}(P)$$

for each word P in A *and each normal algorithm $\mathfrak{A}$ in* A.

Indeed, using any letters α, β and γ not belonging to the alphabet $\mathrm{A}\delta$, we construct a universal algorithm $\mathfrak{U}$ over the alphabet $\mathrm{A}\alpha\beta\gamma\delta$ according to §42.1.1 in such a way that the conditional equality §42.1(2) holds. We construct the severing normal algorithms $\mathfrak{I}_{\Gamma,\delta}$ and $\mathfrak{G}_{\Gamma,\delta}$ [§29.4], where $\Gamma \rightleftharpoons \mathrm{A}_0 \cup \mathrm{A}\alpha\beta\gamma$.

We construct the normal algorithm $\mathfrak{R}$ as the normal composition $(\mathfrak{T}_1 \circ \mathfrak{I}_{\Gamma,\delta})$ of the algorithms $\mathfrak{I}_{\Gamma,\delta}$ and $\mathfrak{T}_1$. $\mathfrak{R}$ is a normal algorithm over the alphabet $\Gamma\delta$. We construct the normal algorithm $\mathfrak{S}$ over the alphabet $\Gamma\delta$ according to the Union Theorem [§38.1.1] in such a way that for any word Q in $\Gamma\delta$, the following conditional equality is fulfilled:

$$\text{(2)} \qquad \mathfrak{S}(Q) \simeq \mathfrak{R}(Q)\delta\mathfrak{G}_{\Gamma,\delta}(Q),$$

We construct the desired algorithm $\mathfrak{W}$ as the normal composition $(\mathfrak{U} \circ \mathfrak{S})$ of the algorithms $\mathfrak{S}$ and $\mathfrak{U}$.

$\mathfrak{W}$ is a normal algorithm over the alphabet $\Gamma\delta$ and, therefore, over $\mathrm{A}_0 \cup \mathrm{A}\delta$. We show that the conditional equality (1) holds for it.

Indeed, let $\mathfrak{A}$ be a normal algorithm in A and let P be a word in this alphabet. Then

$$\begin{array}{lrll}
(3) & \mathfrak{I}_{\Gamma,\delta}(\mathfrak{A}^{\mathrm{Rc}}\delta P) & \doteq \mathfrak{A}^{\mathrm{Rc}} & [§29.4.6], \\
(4) & \mathfrak{G}_{\Gamma,\delta}(\mathfrak{A}^{\mathrm{Rc}}\delta P) & \doteq P & [§29.4.7], \\
 & \mathfrak{R}(\mathfrak{A}^{\mathrm{Rc}}\delta P) & \simeq \mathfrak{T}_1(\mathfrak{I}_{\Gamma,\delta}(\mathfrak{A}^{\mathrm{Rc}}\delta P)) & [§37.4.3] \\
(5) & & \doteq \mathfrak{A}^{\mathrm{Rp}} & [(3),\ 3(1)], \\
(6) & \mathfrak{S}(\mathfrak{A}^{\mathrm{Rc}}\delta P) & \doteq \mathfrak{A}^{\mathrm{Rp}}\delta P & [(2),\ (5),\ (4)], \\
 & \mathfrak{W}(\mathfrak{A}^{\mathrm{Rc}}\delta P) & \simeq \mathfrak{U}(\mathfrak{S}(\mathfrak{A}^{\mathrm{Rc}}\delta P)) & [§37.4.3] \\
 & & \simeq \mathfrak{U}(\mathfrak{A}^{\mathrm{Rp}}\delta P) & [(6)] \\
 & & \simeq \mathfrak{A}(P) & [§42.1(2)].
\end{array}$$

Q.E.D.

CHAPTER VI

BASIC THEOREMS ON THE NON-EXISTENCE OF ALGORITHMS

§46 The Concept of a Mass Algorithmic Problem

1. Let A be any alphabet and let F be a one-place verbal predicate in it. It is natural to tie up with each separate word X in A the following problem:

$$\ll \text{determine whether the given } X \text{ satisfies the predicate } F \gg,$$

and with the predicate F itself—the one-parameter family of such problems.

Each of these singular problems requires an affirmative or negative answer to a formulated question, so that there arises the natural problem of seeking a single general method (an algorithm) allowing one to find the correct solution to any of the singular problems in the family under consideration. The latter is just what is customary to call a *mass algorithmic problem* in mathematics.

2. For example, one may take as A the alphabet consisting of the single letter **I**, and take as F, the predicate

$$\ll \text{the number } X \text{ is even and perfect} \gg .$$

Then the corresponding mathematical problem will consist in seeking a single general method which, starting with any natural number, would allow one to say whether this number is even and perfect. It is easy to see that in the given case, the desired algorithm can be found, with the aid of the decomposition of X into prime factors. It is also easy to figure out that it yields an affirmative answer in at least 24 cases (24 even perfect numbers are known at the present time).

If we take the new predicate

$$\ll \text{the number } X \text{ is odd and perfect} \gg$$

as F, then the algorithm required in this case will also exist. True, at the present time we cannot say whether it will answer the formulated question affirmatively at least once, because not a single odd perfect number is known. But this is not required of us.

Having agreed to write out pairs of natural numbers in the form of words in the alphabet $\mathsf{I}*$, one can also become interested in the singular problems of the relative primality of any two given natural numbers. Each of these problems is formulated as a requirement: ≪determine whether the given natural numbers X and Y are relatively prime≫. The mass algorithmic problem corresponding to the family of these singular problems will consist in seeking a general method allowing one to find out, for two given natural numbers X and Y, whether or not they are relatively prime. As is well known, this mass problem is solvable; its solution can, for example, consist in applying Euclid's algorithm for finding the greatest common divisor of given natural numbers X and Y; these numbers are relatively prime if and only if their greatest common divisor, found as the result of an application of Euclid's algorithm, is equal to one.

3. The situation will become considerably more complicated if we get interested in mass algorithmic problems of the following type. Any alphabet A and any normal algorithm $\mathfrak{A}$ in it are held fixed, while as F one takes the predicate

$$\ll \text{the algorithm } \mathfrak{A} \text{ is applicable to the word } X \gg .$$

In isolated cases, we may possibly be able, basing ourselves on concrete peculiarities of the alphabet A and the algorithm $\mathfrak{A}$, to develop a required method allowing us to recognize those words X to which the algorithm $\mathfrak{A}$ is applicable. However, *in the general case,* no other method comes to mind than attempting to determine whether the algorithm $\mathfrak{A}$ is applicable to the word X by means of a direct application of the algorithm to this word. But this "straightforward" method just will not achieve our goal if the algorithm $\mathfrak{A}$ is not applicable to the word X: no matter how far we might have progressed in the process of applying $\mathfrak{A}$ to X, if this process has not yet broken off, we are not in a position to say anything: perhaps the process will break off at the next step, but perhaps it will never break off. The following question arises: can we try—since the "direct" method fails to yield the desired result—to solve this problem by some other "indirect" method? One of the basic results of the general theory of algorithms states—and we shall establish this in §48—that for a suitable choice of the algorithm $\mathfrak{A}$, not only the "direct", but also every other method will fail to bring us the desired result: *such a method turns out to be simply impossible.*

4. Of course, in order to obtain this sort of result, the inexact, vague term ≪single general method≫, found in the formulation of a mass problem, must be made precise in a suitable manner. It is natural to make it precise as an ≪algorithm≫. This algorithm should give the correct answer, ≪yes≫ or ≪no≫, to each question making up the contents of any singular problem in the family under consideration. It is natural to treat this requirement as follows: the algorithm must be applicable to any value of the parameter individualizing a singular problem of the family under consideration, and should change this value into the word ≪yes≫, if the problem is solved in the positive sense, and into the word ≪no≫, if it is solved in the negative sense. In this connection,

the alphabet used for writing down values of the parameter is complemented by the Latin letters e, n, o, s, y (needed to make up the words ≪yes≫ and ≪no≫) and ≪algorithm≫ is understood as an algorithm over the alphabet A obtained in this way.

It is not hard, however, to see that in essence there is no need for the introduction of these Latin letters. In order to convince oneself of this, it is only necessary to bear in mind the presence of algorithms $\mathfrak{A}$ and $\mathfrak{B}$ in the alphabet A, performing in the following manner: $\mathfrak{A}$ changes ≪yes≫ into the empty word, and ≪no≫ into a non-empty word; $\mathfrak{B}$ changes the empty word into ≪yes≫, and every non-empty word in the alphabet A, into the word ≪no≫.

The presence of these algorithms makes it possible to replace the requirement formulated above for the desired algorithm by the following requirement: the algorithm must be applicable to any value of the parameter determining a singular problem of the family under consideration, and should change this value into the empty word if and only if the singular problem is decided in the positive sense.

The normalization principle for algorithms [§27.6] finally yields the possibility of making the formulation of a mass algorithmic problem even more precise. According to this principle, the desired algorithm, if it exists, is normalizable and, consequently, completely equivalent with respect to the parameter value alphabet under consideration to some normal algorithm over this alphabet. When applied to a parameter value, this normal algorithm performs just like the original one: it changes the parameter value into the empty word if and only if the singular problem individualized by this value is solved in the positive sense.

Thus, we have arrived at the following statement of the mass problem for a given family of singular problems. It is required to construct a normal algorithm over the parameter value alphabet, applicable to every parameter value and changing a parameter value into the empty word if and only if the singular problem individualized by this value is solved in the positive sense.

We shall call mass algorithmic problems stated in this manner *normal mass problems*. The normalization principle obviously yields the possibility of asserting that every mass algorithmic problem is reducible to a normal mass problem.

A normal mass problem is called *solvable* if a normal algorithm satisfying the condition in the above formulation can be constructed.

A number of normal mass problems naturally arise in the theory of algorithms, itself. These are, in the first place, problems related to the applicability of a normal algorithm to a word. In this chapter, we shall prove the unsolvability of certain problems of this sort—the non-existence of the desired normal algorithms.

§47 Self-Applicable and Non-Self-Applicable Algorithms

1. If an alphabet A contains more then one letter, then, as was already explained [§45], there is a simple way to write down the schemes of normal algorithms in A in the form of words in this very alphabet. To the record $\mathfrak{A}^{\mathrm{Rc}}$ of any normal algorithm $\mathfrak{A}$ in the alphabet A, one can then try to apply this very same algorithm $\mathfrak{A}$. In this section, we shall prove a simple and almost obvious non-existence theorem related to the application of normal algorithms to their own records. In what follows, it will serve us as the basis of many less trivial results.

2. We shall consider an alphabet A which is an extension of the alphabet A_0, i.e., one containing the Latin letters a and b. We define, as in [§45], the record of an arbitrary normal algorithm in A. It is a word in A_0 and, therefore, in A.

We shall say that a normal algorithm $\mathfrak{A}$ in A is *self-applicable* if it is applicable to its own record. We shall say that it is *non-self-applicable* if it is not applicable to its own record.

Self-applicable, as well as non-self-applicable, normal algorithms in A can be easily constructed. The identity algorithm in A [§28.4], for example, is self-applicable. It is applicable to every word in A, and, in particular to its own record. The empty algorithm in A [§28.5] is non-self-applicable. It is not applicable to any word in A, and in particular is inapplicable to its own record.

2.1. *There is no normal algorithm in* A, *applicable to the record of any normal algorithm* $\mathfrak{A}$ *in* A *if and only if* $\mathfrak{A}$ *is non-self-applicable.*

Indeed, let us assume that such an algorithm $\mathfrak{H}$ can be constructed. We shall show that it will then be self-applicable and non-self-applicable. But by the principle of reductio ad absurdum,[1] this will mean that it cannot exist.

Let us prove that $\mathfrak{H}$ is non-self-applicable. To this end, assume that $\mathfrak{H}$ is self-applicable. Then $\mathfrak{H}$ is applicable to $\mathfrak{H}^{\mathrm{Rc}}$, and so by assumption, $\mathfrak{H}^{\mathrm{Rc}}$ is the record of a non-self-applicable algorithm, i.e., $\mathfrak{H}$ is non-self-applicable. Thus, from the assumption of the self-applicability of $\mathfrak{H}$ it follows that $\mathfrak{H}$ is *non-self-applicable.* But the *self-applicability* of $\mathfrak{H}$ also follows, of course, from this assumption. These two consequences contradict one another; therefore, according to the principle of reductio ad absurdum, the assumption of the self-applicability of $\mathfrak{H}$ is false, i.e., the algorithm $\mathfrak{H}$ really is *non-self-applicable.*

But from this we immediately obtain that $\mathfrak{H}$ is *self-applicable.* In fact, it is by assumption applicable to the record of every non-self-applicable normal algorithm in A, and hence, to its own record, too.

By the same token, Theorem 2.1 is proven.

REMARK: The following question arises: what meaning should we read into the negation figuring in the assertion 2.1? An examination of the proof presented shows that it is natural to treat the given negation as *reductional* [see §15.4]. A reader experienced in mathematical logic has seen that at the final stage of the argument, we essentially come up with a derivation (by means of an institutionistic propositional calculus) of the contradiction

$$!\mathfrak{H}(\mathfrak{H}^{\mathrm{Rc}}) \;\&\; \rceil !\mathfrak{H}(\mathfrak{H}^{\mathrm{Rc}})$$

---

[1]See §15.4.

from the previously obtained assertion

$$!\mathfrak{H}(\mathfrak{H}^{\mathrm{Rc}}) \equiv \rceil ! \mathfrak{H}(\mathfrak{H}^{\mathrm{Rc}}).$$

The argument presented repeats the scheme of the so-called "Russell paradox"[2] and is in essence its algorithmic-theoretic version. The original—set-theoretic—version of the paradox is better known. It consists in the following. Let R denote the set of all sets which are not elements of themselves. This means that for each set X, the following assertion holds: $X \in R$ if and only if $X \notin X$. Consequently, $R \in R$ if and only if $R \notin R$. But from this, as in our case, one obtains a contradiction: $R \in R$ and $R \notin R$ simultaneously. (Note that in neither of the arguments did we make use of the law of the excluded middle, and hence it bears no responsibility for the final result.)

It should be emphasized that neither version considered here of the "paradox" contains anything contradicting common sense, so that the term "paradox", itself, should be understood with suitable stipulations. True, the arguments eventually lead to contradictions in both cases, but these contradictions are extracted from certain *assumptions:* In the first case—from the assumption of the existence of a normal algorithm, and in the second—from the assumption of the existence of a set satisfying a definite condition. In the cases considered, "overcoming" the paradox consists in stating the "illegality" of these assumptions, i.e., their falsity. The first statement yields our Theorem 2.1. The second shows that within the bounds of set-theoretic conceptions, too much freedom in defining sets, which are then regarded as indivisible objects, is inadmissible. It also shows that without giving way to contradictions, it is impossible to arbitrarily collect "objects" into "sets" which will themselves be regarded as "objects." Forbidding the use of the set R, we rid set theory of one specific difficulty, but do not create any reliable guarantees for the future. Indeed, the simple, although perhaps also somewhat bizarre, method of defining the set R apparently does not contradict Cantor's initial theses, and the "inadmissibility" of R is only revealed a posteriori.

3. The theorem just proven asserts that no normal algorithm *in the alphabet* A can possess a certain special property. Assuming now that the alphabet A contains not only the letters a and b, but also the letters c and d, we shall strengthen Theorem 2.1 by asserting that no normal algorithm *over the alphabet* A_0 can possess the same property. As before, we shall be dealing with records which are defined only for normal algorithms in A. The alphabet consisting of the four letters a, b, c, d will be denoted by A_1.[3]

3.1. *If* A *is an extension of* A_1, *then there is no normal algorithm* $\mathfrak{K}$ *over* A_0, *applicable to the record of a normal algorithm* $\mathfrak{A}$ *in* A *if and only if* $\mathfrak{A}$ *is non-self-applicable.*

Indeed, assume that such an algorithm $\mathfrak{K}$ has been constructed. Then Theorem §41.6.27 is applicable to the alphabet A_0, the letters c, d and the algorithm

[2]This paradox was also independently discovered by Zermelo.

[3]Of course, instead of the letters a, b, c, d, we could have taken any other four distinct letters.

$\mathfrak{K}$. According to this theorem, one can construct a normal algorithm $\mathfrak{H}_0$ in the alphabet A_1, applicable to those and only those words in A_0 to which the algorithm $\mathfrak{K}$ is applicable.

Bearing in mind that A is an extension of A_1, we construct the natural extension $\mathfrak{H}$ of the algorithm $\mathfrak{H}_0$ to the alphabet A [§35.3]. For any word P in A_1, the conditional equality

$$\mathfrak{H}_0(P) \simeq \mathfrak{H}(P)$$

holds, showing that the algorithm $\mathfrak{H}$ is applicable to those and only those words in A_1 to which the algorithm $\mathfrak{H}_0$ is applicable. Since A_1 is an extension of A_0, the algorithms $\mathfrak{H}$ and $\mathfrak{K}$ are applicable to one and the same words in A_0 and, therefore, to one and the same records of a normal algorithm in A. Consequently, $\mathfrak{H}$, as $\mathfrak{K}$, is applicable to the record of a normal algorithm $\mathfrak{A}$ in A if and only if the algorithm $\mathfrak{A}$ is non-self-applicable. Nevertheless, $\mathfrak{H}$ is a normal algorithm in A. Such an algorithm, however, cannot exist [2.1]. Consequently, no normal algorithm $\mathfrak{K}$ over A_0 can possess the property given in the statement of Theorem 3.1.

Q.E.D.

4. Along with the properties of self-applicability and non-self-applicability of normal algorithms, it is natural to consider the mass algorithmic problems of recognizing these properties. For each normal algorithm in the alphabet A, given by its record, one can pose the singular problem of determining whether or not this algorithm is self-applicable (non-self-applicable). To each given alphabet A which is an extension of the alphabet A_0, there corresponds the class of such singular problems. Each singular problem of this class is individualized by a certain word in the alphabet A_0—the record of the normal algorithm under consideration in the singular problem. The corresponding normal mass problem will consist in the construction of a normal algorithm over the alphabet A_0 of records, applicable to the record $\mathfrak{A}^{Rc}$ of any normal algorithm $\mathfrak{A}$ in A and changing $\mathfrak{A}^{Rc}$ into Λ if and only if $\mathfrak{A}$ is self-applicable (non-self-applicable). We now show that this mass problem is unsolvable if A is an extension of A_1.

4.1. *If* A *is an extension of* A_1, *then no normal algorithm* $\mathfrak{H}$ *over* A_0 *can satisfy the following condition: given any normal algorithm* $\mathfrak{A}$ *in* A, $\mathfrak{H}$ *changes* $\mathfrak{A}^{Rc}$ *into* Λ *if and only if* $\mathfrak{A}$ *is non-self-applicable.*

In fact, assume that a normal algorithm $\mathfrak{H}$ over A_0, satisfying this condition, has been constructed. Let Б be its alphabet. Construct the normal algorithm $\mathfrak{B}$ in Б with the scheme

$$\{\xi \rightarrow \xi,$$

where ξ runs through the alphabet Б. $\mathfrak{B}$ is obviously applicable to Λ and inapplicable to any non-empty word in Б.

Define the algorithm $\mathfrak{K}$ as the normal composition of the algorithms $\mathfrak{H}$ and $\mathfrak{B}$:

$$\mathfrak{K} \rightleftharpoons (\mathfrak{B} \circ \mathfrak{H}). \tag{1}$$

and, in particular,

$$\mathfrak{K}(\mathfrak{A}^{\mathrm{Rc}}) \simeq \mathfrak{B}(\mathfrak{H}(\mathfrak{A}^{\mathrm{Rc}})) \tag{2}$$

for any normal algorithm $\mathfrak{A}$ in A.

According to (2), for the applicability of $\mathfrak{K}$ to the record $\mathfrak{A}^{\mathrm{Rc}}$ of the normal algorithm $\mathfrak{A}$, it is necessary and sufficient that the algorithm $\mathfrak{H}$ be applicable to$\mathfrak{A}^{\mathrm{Rc}}$, and the algorithm $\mathfrak{B}$, to $\mathfrak{H}(\mathfrak{A}^{\mathrm{Rc}})$. But in case $\mathfrak{H}$ is applicable to $\mathfrak{A}^{\mathrm{Rc}}$, $\mathfrak{H}(\mathfrak{A}^{\mathrm{Rc}})$ is a word in Б, since $\mathfrak{H}$ is an algorithm in Б. Therefore, $\mathfrak{B}$ is applicable to $\mathfrak{H}(\mathfrak{A}^{\mathrm{Rc}})$ if and only if $\mathfrak{H}(\mathfrak{A}^{\mathrm{Rc}}) = \Lambda$. Thus, the algorithm $\mathfrak{K}$ is applicable to $\mathfrak{A}^{\mathrm{Rc}}$ if and only if $\mathfrak{H}(\mathfrak{A}^{\mathrm{Rc}}) = \Lambda$, i.e., if and only if $\mathfrak{A}$ is non-self-applicable. Such an algorithm $\mathfrak{K}$, however, cannot exist, since A is an extension of A_1 [3.1]. Consequently, there is no normal algorithm $\mathfrak{H}$ satisfying the condition formulated in Theorem 4.1. Q.E.D.

It was not even required in Theorem 4.1 that the algorithm $\mathfrak{H}$ be applicable to the record of every normal algorithm in A. It was merely required that, firstly, it change the record of every non-self-applicable algorithm into the empty word and, secondly, every normal algorithm in A, whose record is changed by $\mathfrak{H}$ into the empty word, be non-self-applicable. The theorem asserted the non-existence of an algorithm $\mathfrak{K}$ over A_0, satisfying these conditions. The theorem, of course, remains true if to the conditions imposed on $\mathfrak{H}$, we add the requirement of applicability to the record of every normal algorithm in A, in accordance with the formulation of the normal mass problem of recognizing non-self-applicability given above. Thus, we obtain the following modification of Theorem 4.1.

4.2. *If* A *is an extension of* A_1, *then there is no normal algorithm over* A_0, *applicable to the record* $\mathfrak{A}^{\mathrm{Rc}}$ *of any normal algorithm* $\mathfrak{A}$ *in* A *and changing* $\mathfrak{A}^{\mathrm{Rc}}$ *into* Λ *if and only if* $\mathfrak{A}$ *is non-self-applicable.*

Further, it is not difficult to see that it is possible to replace the word ≪non-self-applicable≫ in Theorem 4.2 by the word ≪self-applicable≫, i.e., the following theorem holds:

4.3. *If* A *is an extension of* A_1, *then there is no normal algorithm over* A_0, *applicable to the record* $\mathfrak{A}^{\mathrm{Rc}}$ *of any normal algorithm* $\mathfrak{A}$ *in* A *and changing* $\mathfrak{A}^{\mathrm{Rc}}$ *into* Λ *if and only if* $\mathfrak{A}$ *is self-applicable.*

Indeed, suppose that such an algorithm $\mathfrak{K}$ has been constructed. Let Б be its alphabet. Then Б is an extension of A_0, and so a is a letter of Б. Construct the branching normal algorithm $\mathfrak{A}_{Б,\Lambda,a,\Lambda}$ [§30.1] and define the algorithm $\mathfrak{H}$ as the normal composition of the algorithms $\mathfrak{K}$ and $\mathfrak{A}_{Б,\Lambda,a,\Lambda}$:

$$\mathfrak{H} \rightleftharpoons (\mathfrak{A}_{Б,\Lambda,a,\Lambda} \circ \mathfrak{K}). \tag{3}$$

$\mathfrak{H}$ is a normal algorithm over the alphabet Б (3) and, consequently, over A_0. The conditional equality

$$\mathfrak{H}(Q) \simeq \mathfrak{A}_{Б,\Lambda,a,\Lambda}(\mathfrak{K}(Q))$$

holds for any Q in Б and, in particular,

$$\mathfrak{H}(\mathfrak{A}^{\mathrm{Rc}}) \simeq \mathfrak{A}_{Б,\Lambda,a,\Lambda}(\mathfrak{K}(\mathfrak{A}^{\mathrm{Rc}})) \tag{4}$$

for any normal algorithm $\mathfrak{A}$ in the alphabet A. On the other hand,

(5)
$$\mathfrak{A}_{\text{Б},\Lambda,a,\Lambda}(\mathfrak{K}(\mathfrak{A}^{\text{Rc}})) \asymp a$$
$$\asymp \Lambda$$

if $\mathfrak{K}(\mathfrak{A}^{\text{Rc}}) \asymp \Lambda$ [§30.1(2)], and

(6)
$$\mathfrak{A}_{\text{Б},\Lambda,a,\Lambda}(\mathfrak{K}(\mathfrak{A}^{\text{Rc}})) \asymp \Lambda$$

if $\mathfrak{K}(\mathfrak{A}^{\text{Rc}}) \neq \Lambda$[§30.1(3)].

Since the algorithm $\mathfrak{K}$ is applicable to the record of any normal algorithm in A, the algorithm $\mathfrak{H}$ is also applicable to the record of any normal algorithm in A; moreover, $\mathfrak{H}(\mathfrak{A}^{\text{Rc}}) \asymp \Lambda$ if and only if $\mathfrak{K}(\mathfrak{A}^{\text{Rc}}) \neq \Lambda$ [(4)–(6)]. But $\mathfrak{K}(\mathfrak{A}^{\text{Rc}}) \asymp \Lambda$ if and only if the algorithm $\mathfrak{A}$ is self-applicable. Consequently, $\mathfrak{K}(\mathfrak{A}^{\text{Rc}}) \neq \Lambda$ if and only if $\mathfrak{A}$ is non-self-applicable. Hence, $\mathfrak{H}(\mathfrak{A}^{\text{Rc}}) \asymp \Lambda$ if and only if $\mathfrak{A}$ is non-self-applicable. The non-existence of a normal algorithm $\mathfrak{H}$ with this property was, however, just proven [4.2]. Consequently, no normal algorithm $\mathfrak{K}$ can satisfy the property stated in Theorem 4.3. Q.E.D.

§48 The Problem of Recognizing an Algorithm's Applicability to Initial Data

1. In relation to each normal algorithm $\mathfrak{G}$ over a given alphabet Γ, it is natural to pose the problem of recognizing the applicability of this algorithm to words in Γ, consisting of the following: it is required to construct a normal algorithm over Γ, applicable to every word in Γ and changing those and only those words in Γ, to which the algorithm is applicable, into the empty word. We shall show in this section that an alphabet Γ and an algorithm $\mathfrak{G}$ in it can be constructed in such a way that the problem of recognizing the applicability of $\mathfrak{G}$ to words in Γ proves unsolvable: the desired normal algorithm in this problem cannot exist. The construction will be based upon the modified Universal Algorithm Theorem, which we shall combine with the results of §47.

2. Let

$$A_2 \rightleftharpoons abcde.$$

We prove the following theorem:

2.1. *One can construct a normal algorithm $\mathfrak{W}$ over the alphabet A_2, satisfying the following condition: there is no normal algorithm $\mathfrak{K}$ over A_2, changing into Λ those and only those words in A_2, to which $\mathfrak{W}$ is not applicable.*

In fact, bearing in mind that the letter e does not belong to the alphabet A_1, i.e., to *abcd*, we construct a normal algorithm $\mathfrak{W}$ over A_1e, i.e., over A_2, by §45.5.1 in such a way that the condition

(1)
$$\mathfrak{W}(\mathfrak{A}^{\text{Rc}}eP) \simeq \mathfrak{A}(P)$$

is satisfied. Here A_1 plays the role of the alphabet A from Theorem §45.5.1, e, the role of δ; (1) holds for the words P in A_1 and the normal algorithms $\mathfrak{A}$ in A_1. We show that $\mathfrak{W}$ satisfies the condition stated in the theorem.

Suppose, in fact, that $\mathfrak{K}$ is a normal algorithm over A_2, changing into Λ those and only those words in A_2, to which $\mathfrak{W}$ is not applicable.

Construct the identity normal algorithm $\mathfrak{I}_{A_0}$ in the alphabet A_0 [§28.4]. Construct the normal algorithm $\mathfrak{B}$ over the alphabet A_0e by the Union Theorem [§38.1.1] in such a way that

$$(2) \qquad \mathfrak{B}(P) \simeq \mathfrak{I}_{A_0}(P)e\mathfrak{I}_{A0}(P)$$

for any word P in the alphabet A_0.

Construct, finally, the algorithm $\mathfrak{H}$ as the normal composition of the algorithms $\mathfrak{B}$ and $\mathfrak{K}$:

$$(3) \qquad \mathfrak{H} \rightleftharpoons (\mathfrak{K} \circ \mathfrak{B}).$$

$\mathfrak{H}$ is a normal algorithm over A, since $\mathfrak{K}$ is a normal algorithm over A_2 [(3), §37.4.2]. Consequently, $\mathfrak{H}$ is a normal algorithm over A_0. Since $\mathfrak{I}_{A_0}$ is the identity algorithm in A_0, we have, by (2),

$$(4) \qquad \mathfrak{B}(\mathfrak{A}^{\mathrm{Rc}}) \coloneqq \mathfrak{A}^{\mathrm{Rc}}e\mathfrak{A}^{\mathrm{Rc}}$$

for any normal algorithm $\mathfrak{A}$ in A_1. Therefore

$$\mathfrak{H}(\mathfrak{A}^{\mathrm{Rc}}) \simeq \mathfrak{K}(\mathfrak{A}^{\mathrm{Rc}}e\mathfrak{A}^{\mathrm{Rc}})$$

and, consequently, $\mathfrak{H}(\mathfrak{A}^{\mathrm{Rc}}) \coloneqq \Lambda$ if and only if $\mathfrak{K}(\mathfrak{A}^{\mathrm{Rc}}e\mathfrak{A}^{\mathrm{Rc}}) \coloneqq \Lambda$.

On the other hand, $\mathfrak{W}(\mathfrak{A}^{\mathrm{Rc}}e\mathfrak{A}^{\mathrm{Rc}}) \simeq \mathfrak{A}(\mathfrak{A}^{\mathrm{Rc}})$ for any normal algorithm $\mathfrak{A}$ in A_1 [(1)]. Hence the normal algorithm $\mathfrak{A}$ in A_1 is non-self-applicable if and only if the algorithm $\mathfrak{W}$ if not applicable to the word $\mathfrak{A}^{\mathrm{Rc}}e\mathfrak{A}^{\mathrm{Rc}}$.

But, according to our assumption concerning the algorithm $\mathfrak{K}$, $\mathfrak{W}$ is inapplicable to $\mathfrak{A}^{\mathrm{Rc}}e\mathfrak{A}^{\mathrm{Rc}}$ if and only if $\mathfrak{K}(\mathfrak{A}^{\mathrm{Rc}}e\mathfrak{A}^{\mathrm{Rc}}) \coloneqq \Lambda$ i.e., $\mathfrak{H}(\mathfrak{A}^{\mathrm{Rc}}) \coloneqq \Lambda$. Thus, $\mathfrak{H}(\mathfrak{A}^{\mathrm{Rc}}) \coloneqq \Lambda$ if and only if the algorithm $\mathfrak{A}$ is non-self-applicable. An algorithm $\mathfrak{H}$ with this property, however, cannot exist by Theorem §47.4.1, whose hypothesis is satisfied since the role of A is played by A_1.

By the same token, we have established that no normal algorithm $\mathfrak{K}$ over A_2 can change into Λ those and only those words in A_2, to which the algorithm $\mathfrak{W}$ is inapplicable. Q.E.D.

The algorithm just found, $\mathfrak{W}$, satisfying the condition formulated in Theorem 2.1, is a normal algorithm in an alphabet consisting of a large number of letters. We now show that a normal algorithm with the same property can even be constructed in the two-letter alphabet A_0.

2.2. *One can construct a normal algorithm* $\mathfrak{W}_0$ *in the alphabet* A_0, *satisfying the following condition: There is no normal algorithm* $\mathfrak{H}$ *over* A_0, *changing those and only those words in* A_0 *into* Λ, *to which* $\mathfrak{W}$ *is inapplicable.*

Let us construct, in fact, a normal algorithm $\mathfrak{W}$ over A_2 according to 2.1 and a normal algorithm $\mathfrak{W}_0$ in A_0 as the translation of the algorithm $\mathfrak{W}$ [§41.6]. Moreover, let the roles of A and Б be played by the empty alphabet and the alphabet of the algorithm $\mathfrak{W}$, respectively. It is obvious that Б is an extension of A_2.

Denoting the translation algorithm by $\mathfrak{T}$, we shall have

$$\mathfrak{W}_0(\mathfrak{T}(Q)) \simeq \mathfrak{T}(\mathfrak{W}(Q)) \quad [\S41.6.26]$$

for any Q in Б, where $\mathfrak{T}$ is a normal algorithm over Б. Since the algorithm $\mathfrak{T}$ is applicable to every word in Б, this conditional equality shows that $\mathfrak{W}_0$ is applicable to the translation of a word Q in Б if and only if $\mathfrak{W}$ is applicable to this same word.

$\mathfrak{W}_0$ is a normal algorithm in A_0. We show that it satisfies the condition formulated in 2.2.

Indeed, let us assume that $\mathfrak{H}$ is a normal algorithm over A_0, changing those and only those words in A_0, to which $\mathfrak{W}_0$ is inapplicable, into Λ.

We construct the algorithm $\mathfrak{K}$ as the normal composition of the algorithms $\mathfrak{T}$ and $\mathfrak{H}$:

$$(5) \qquad \mathfrak{K} \rightleftharpoons (\mathfrak{H} \circ \mathfrak{T}).$$

$\mathfrak{H}$ is, like $\mathfrak{T}$, a normal algorithm over Б [(5), §37.4.2] and, therefore, over A_2. For any Q in Б,

$$\mathfrak{K}(Q) \simeq \mathfrak{H}(\mathfrak{T}(Q)) \quad [(5), \S37.4.3]$$

and so $\mathfrak{K}(Q) \equiv \Lambda$ for those and only those words Q in A_2, for which

$$(6) \qquad \mathfrak{H}(\mathfrak{T}(Q)) \equiv \Lambda.$$

Here $\mathfrak{T}(Q)$ is always a word in A_0. Therefore, according to the assumption about the algorithm $\mathfrak{H}$, the equality (6) holds for those and only those words Q in A_2, for which the word $\mathfrak{T}(Q)$ is such that the algorithm $\mathfrak{W}_0$ is not applicable to it. But the applicability of $\mathfrak{W}_0$ to $\mathfrak{T}(Q)$, as was noted above, is equivalent to the applicability of the algorithm $\mathfrak{W}$ to Q. Thus, the normal algorithm $\mathfrak{K}$ over A_2 changes into Λ those and only those words in A_2, to which the algorithm $\mathfrak{W}$ is inapplicable. Such a normal algorithm, however, cannot exist [2.1]. Consequently, a normal algorithm $\mathfrak{H}$ over A_0, changing those and only those words in A_0 into Λ, to which the algorithm $\mathfrak{W}_0$ is inapplicable, cannot exist.

Q.E.D.

Analogous to the way in which we passed from Theorem §47.4.1 to Theorems §47.4.2 and §47.4.3, we can pass from Theorem 2.2 to the following results.

2.3. *One can construct a normal algorithm $\mathfrak{W}_0$ in the alphabet A_0, satisfying the following condition: there is no normal algorithm over A_0, applicable to every word in A_0 and changing those and only those words in A_0 into Λ, to which $\mathfrak{W}_0$ is inapplicable.*

2.4. *One can construct a normal algorithm $\mathfrak{W}_0$ in the alphabet A_0, satisfying the following condition: there is no normal algorithm over A_0, applicable to* *every word in A_0 and changing those and only those words in A_0 into Λ, to which $\mathfrak{W}_0$ is applicable.*

Theorem 2.3 is an obvious consequence of Theorem 2.2, while Theorem 2.4 is easily proven on the basis of 2.3 with the aid of the branching algorithm $\mathfrak{A}_{Б,\Lambda,a,\Lambda}$, where Б is the alphabet of the supposed algorithm $\mathfrak{H}$ [§30.1].

Theorem 2.4 shows that for some normal algorithm $\mathfrak{W}_0$ in A_0, the problem of recognizing applicability to words in A_0 is unsolvable: a normal algorithm required in this problem cannot exist.

§49 Set-Theoretic Commentary to §47 and §48

Theorem §48.2.4 is one of the most fundamental results of the general theory of algorithms. One can extract from it, in particular, negative solutions to such famous mass algorithmic problems of mathematics as the decision problem for the first order predicate calculus (Church [2] and Turing [1]; 1936), the identity problem for semigroups (Markov [5], [6] and Post [2]; 1947), the identity problem for groups (Novikov [1]; 1952), the homeomorphy problem for polyhedra (Markov [12], [13]; 1958), Hilbert's 10th problem (Matiyasevich [2]; 1970). It expresses a certain important fact of the theory of algorithms, discovered in 1936 by Church [1], [2], Kleene [1] and Turing [1], [2]. We presented this fact in the language of normal algorithms. Now, taking the interests of a set-theoretically inclined reader into account, we shall present it in the language of set theory. In this connection, we ask the reader who regards set-theoretic conceptions critically to bear in mind that we attribute only a purely heuristic significance to the present commentary, and that we shall not refer to the statements to be made in this section in any of our proofs. This commentary, just like the analogous commentaries in other sections (which are all explicitly mentioned as such), can be removed from the book *mechanically* without any consequences whatsoever for the rest of its contents.

1. A set M of words in the alphabet A is called *decidable in* A if there exists a normal algorithm $\mathfrak{A}$ *over* A, verifying whether or not words in A belong to M. This statement can be made precise in the following manner: it is required that the algorithm $\mathfrak{A}$ be applicable to *every* word in A and that it change *those and only those words in* A, which belong to M, into the empty word.

A set M of words in an alphabet A is called *enumerable in* A if there exists a normal algorithm $\mathfrak{A}$ *over* A, such that for any P in A, $\mathfrak{A}(P)$ is defined if and only if $P \in M$.

Accepting these definitions, the reader will notice without difficulty that if M is a set of words in A and the alphabet Б is an extension of A, then M is decidable (enumerable) in Б if and only if it is decidable (enumerable) in A. Therefore, the set-theoretic properties of being decidable and being enumerable are in essence independent of the alphabet in which a set is considered, so that mention of the alphabet can often be omitted.

It is not hard to show that

1.1. *Every decidable set of words is enumerable.*

Theorem §48.2.4 implies that

1.2. *There exists an enumerable, but undecidable set of words in the alphabet* A_0.

The fact that the alphabet figuring in 1.2 consists of two letters is dictated not by the essence of the matter, but merely by the technical details of the proof. An analogous theorem can also be proven for a one-letter alphabet, so that

1.3. *There exists an enumerable, but undecidable set on natural numbers.*

2. And now let us see what other results of §47 and §48 look like in set-theoretic terms.

Theorem §47.3.1 asserts that if the alphabet A is an extension of the alphabet A_1, then the set of records of the *non-self-applicable* normal algorithms in A is non-enumerable. It is, all the more so, undecidable (Theorem §47.4.2). The set of records of the *self-applicable* normal algorithms in A also is undecidable (Theorem §47.4.3). We suggest that the reader show that it is nevertheless enumerable.

Theorem §48.2.1 actually[4] asserts (the reader will convence her or himself of this without assistance) that the set of words in A_2, to which the algorithm $\mathfrak{W}$ is inapplicable, is non-enumerable. It is, all the more so, undecidable. The same is also asserted about the set of words in A_0, to which the algorithm $\mathfrak{W}_0$ is inapplicable (Theorems §48.2.2 and §48.2.3). We have already spoken of the set-theoretic interpretation of Theorem §48.2.4.

§50 Constructive Commentary to §47 and §48

1. Let F be any one-place verbal predicate in an alphabet A and let P be a word in this alphabet. Consider the disjunction

(1) ≪ P possesses the property F or it is not the case that P possesses this property.≫

Assume that this disjunction is false. Then both its terms are false, i.e., on one hand, it is not the case that P possesses the property F, while on the other hand, it is not the case that this is false. Consequently, our proposition leads to a contradiction for any P, and this means that

1.1. *Given any* P, *it is not the case that the disjunction* (1) *is false.*

This is one of the forms of one of the most important principles of so-called institutionistic logic—the ≪law of the double negation of the "law of the excluded middle"≫. This principle is usually associated with the name of the outstanding Dutch mathematician Brouwer (1881–1966), the first to subject the view on the "law of the excluded middle" as a logical principle of universal applicability to sharp and principled criticizm in his paper [1].[5]

[4] See Theorem §51.2.1 for comparison.

[5] The reader will find a detailed presentation of Brouwer's position in Heyting's book [1].

In fact, *in the general case* (i.e., when the predicate F is arbitrary), we are unable to single out the true term of the disjunction (1). But this is precisely what is required in order to recognize it as true [see §7.1]!

But if we also bear in mind the parametric nature of this disjunction (it depends on P), then it becomes clear that for establishing the truth of the statement

(2) $\ll$ given any P, the disjunction (1) is true $\gg$,

which is a universal statement [see §9], one must look for a general method (an algorithm) which starts with P and singles out the true term of the disjunction (1). If we adopt a constructive semantics of universal statements in such a way that it is required to present the desired algorithm in a precise form (for example, in the form of a normal algorithm), then it may happen that for a concrete predicate F, there is no such algorithm.[6] We shall then be obliged to recognize the negation of (2) to be true, and the statement (2), to be false.

We can get four examples of such predicates in §47 and §48.

These are the following predicates F_1–F_4 with the free verbal variable P for words in the alphabet A_0:

1°. F_1: $\ll P$ is the record of a non-self-applicable normal algorithm in the alphabet $A_1 \gg$ (see Theorem §47.4.2);

2°. F_2: $\ll P$ is the record of a self-applicable normal algorithm in the alphabet $A_1 \gg$ (see Theorem §47.4.3);

3°. F_3: $\ll$The normal algorithm $\mathfrak{W}_0$ is inapplicable to the word $P \gg$ (see Theorem §48.2.3);

4°. F_4: $\ll$The normal algorithm $\mathfrak{W}_0$ is applicable to the word $P \gg$ (see Theorem §48.2.4).

The reader will be able to find further examples in our subsequent presentation.

Constructively true negations of statements like (2) will probably seem "too exotic" (although there is actually nothing paradoxical in them) to a reader who is not experienced in constructive logic. We shall therefore make no use of them in this monograph, preferring to use their decipherings—assertions about the non-existence of the corresponding algorithms—instead of them. Nevertheless, the very fact of the presence of such statements is a matter of principle for constructive logic.

§51 The Problem of Recognizing Annihilation

1. We shall say that an algorithm $\mathfrak{K}$ *annihilates* a word P if it changes P into the empty word.

One can pose the *problem of recognizing the annihilation of words by a normal algorithm,* analogous to the problem of recognizing the applicability of a normal algorithm [§48.1] and consisting in the following. Let a normal algorithm $\mathfrak{G}$ be given over an alphabet Γ. It is required to construct a normal

[6] See, for example, Kleene [5, §82, Theorem 63^N].

algorithm over Γ, applicable to every word in Γ and annihilating those and only those words in Γ which are annihilated by $\mathfrak{G}$. We shall show in this section that it is possible to construct an alphabet Γ and an algorithm $\mathfrak{G}$ in it in such a way that the problem of recognizing the annihilation of words in Γ by the algorithm $\mathfrak{G}$ will be unsolvable.

2. Let us first of all prove the following lemma:

2.1. *Given any normal algorithm* $\mathfrak{A}$ *over the alphabet* A, *one can construct a normal algorithm* $\mathfrak{B}$ *over* A, *annihilating those and only those words in* A, *to which the algorithm* $\mathfrak{A}$ *is applicable.*

Let, in fact, $\mathfrak{A}$ be a normal algorithm over an alphabet A. Let Б be its alphabet. We construct a normal algorithm $\mathfrak{C}_Б$ in Б, annihilating every word in this alphabet [§29.2]. Define $\mathfrak{B}$ as the normal composition of the algorithms $\mathfrak{A}$ and $\mathfrak{C}_Б$:

$$\mathfrak{B} \rightleftharpoons (\mathfrak{C}_Б \circ \mathfrak{A}). \tag{1}$$

Let us show that it is the desired normal algorithm over A, annihilating those and only those words in A, to which the normal algorithm $\mathfrak{A}$ is applicable.

Indeed, $\mathfrak{B}$ is a normal algorithm over Б [(1), §37.4.2], and, since Б is an extension of A in an obvious way, $\mathfrak{B}$ is a normal algorithm over A. For any P in Б, we have, further,

$$\mathfrak{B}(P) \simeq \mathfrak{C}_Б(\mathfrak{A}(P)). \tag{2}$$

Here $\mathfrak{A}(P)$ is a word in Б whenever the algorithm $\mathfrak{A}$ is applicable to P. Consequently, $\mathfrak{C}_Б(\mathfrak{A}(P)) = \Lambda$ under the same condition. The conditional equality (2) shows, therefore, that the algorithm $\mathfrak{A}$ is applicable to a word P in Б if and only if the algorithm $\mathfrak{B}$ annihilates this word. Since Б is an extension of A, this is true a fortiori for words in A. Q.E.D.

3. Theorem §48.2.2 and Lemma 2.1 yield the following result:

3.1. *One can construct a normal algorithm* $\mathfrak{W}_1$ *over the alphabet* A_0, *satisfying the following condition: there is no normal algorithm over* A_0, *annihilating those and only those words in* A_0, *which the algorithm* $\mathfrak{W}_1$ *does not annihilate.*

Indeed, let us construct the normal algorithm $\mathfrak{W}_0$ in accordance with §48.2.2 and apply Lemma 2.1 to the alphabet A_0 and the algorithm $\mathfrak{W}_0$. In accordance with this lemma, we construct the normal algorithm $\mathfrak{W}_1$ annihilating those and only those words in A_0, to which the algorithm $\mathfrak{W}_0$ is applicable. By our construction, $\mathfrak{W}_1$ fails to annihilate an arbitrary word in A_0 if and only if the algorithm $\mathfrak{W}_0$ is inapplicable to this word, from which it follows that $\mathfrak{W}_1$ is the desired algorithm.

Theorem 3.1 admits the following strengthening:

3.2. *One can construct a normal algorithm* $\mathfrak{W}_2$ *in the alphabet* A_0, *satisfying the following condition: there is no normal algorithm over* A_0, *annihilating those and only those words in* A_0, *which the algorithm* $\mathfrak{W}_2$ *does not annihilate.*

By comparison with 3.1, the strengthening consists in the fact that $\mathfrak{W}_2$ is constructed as an algorithm *in* A_0, while $\mathfrak{W}_1$ was constructed as an algorithm *over* A_0. The transition from 3.1 to 3.2 is analogous to the transition from §48.2.1 to §48.2.2; it is accomplished by constructing the translation of the algorithm $\mathfrak{W}_1$ into the alphabet A_0, which yields precisely the desired algorithm $\mathfrak{W}_2$. We leave a detailed carrying out of the proof of Theorem 3.2 for the reader.

By analogy to the transition from Theorem §47.4.1 to Theorems §47.4.2 and §47.4.3, one can pass from 3.2 to the following results:

3.3. *One can construct a normal algorithm* $\mathfrak{W}_2$ *in the alphabet* A_0, *satisfying the following condition: there is no normal algorithm over* A_0, *applicable to every word in* A_0 *and annihilating those and only those words in* A_0, *which* $\mathfrak{W}_2$ *does not annihilate.*

3.4. *One can construct a normal algorithm* $\mathfrak{W}_2$ *in the alphabet* A_0, *satisfying the following condition: there is no normal algorithm over* A_0, *applicable to every word in* A_0 *and annihilating those and only those words in* A_0, *which* $\mathfrak{W}_2$ *annihilates.*

Theorem 3.3 is an obvious consequence of Theorem 3.2, while Theorem 3.4 is easily proven on the basis of 3.3 with the aid of the branching algorithm $\mathfrak{A}_{Б,\Lambda,a,\Lambda}$ [§30.1], where Б is the supposed algorithm's alphabet.

Theorem 3.4 shows that the problem of recognizing the annihilation of words in A_0 is unsolvable for some normal algorithm $\mathfrak{W}_2$ in A_0.

Theorem 3.2 will serve us as the basis for the proof of some some further theorems on the non-existence of algorithms.

4. Set-theoretic commentary.

Theorem 3.2 actually means that the set of words in the alphabet A_0, not annihilated by the algorithm $\mathfrak{W}_2$, is non-enumerable. All the more so is it undecidable (Theorem 3.3). The set of words in A_0, annihilated by the algorithm $\mathfrak{W}_2$, is also undecidable (Theorem 3.4). But it is enumerable.

§52 Complexity Approach to the Problem of Recognizing Applicability

1. Considering the importance of Theorem §48.2.4, we shall give another proof of it in this section, which in a certain sense will reveal the *cause* of the unsolvability of the problem considered in this theorem. The proof will be based on the special approach developed by Markov in his research[7] on the complexity of algorithms. This approach consists in the following.

Suppose we want to prove the unsolvability of a certain normal mass problem $\mathcal{P}$ defined by a verbal predicate F in an alphabet A. We take an arbitrary number N and consider the words in A of length not exceeding N. There are a finite number of such words, and therefore, as is not difficult to understand, the assumption that there does not exist a normal algorithm over A, recognizing which of the words among these satisfy the predicate F, leads to a contradiction. It is quite possible that we will not know how to construct such an algorithm (the point is that in the general case, nothing more is being refuted than the

[7] The main results are presented in the papers [16] and [17].

assumption of its non-existence and this, generally speaking, does not yield any method for constructing it). But perhaps we will succeed in proving that every such algorithm has complexity, i.e., length of representation[8] at least $f(N)$, where f is a certain unbounded, i.e., assuming arbitrarily large values, computable arithmetical function. We shall then be able to assert that the problem under consideration is unsolvable.

In fact, if a normal algorithm solving this problem existed, then it would, in particular, solve it for the words of length at most N, where N is arbitrary. But then the complexity of this algorithm would be greater than $f(N)$ for any N, which is impossible in view of the fact that f is unbounded.

Thus, the idea behind this approach consists in attempting to approximate the original normal mass problem $\mathcal{P}$ by a sequence of "curtailed" problems $\mathcal{P}_N$ (each of which cannot be unsolvable in view of its ≪finiteness≫) in such a way that the minimal complexity of a solution to particular problems $\mathcal{P}_N$ is bounded below by some unbounded computable function of the variable N. If this plan of action is successfully realized, then by the same token, the unsolvability of the problem $\mathcal{P}$ will have been established.

Let us now illustrate the approach we have projected by means of a simple example of the problem considered in §48 of recognizing the applicability of an algorithm to initial data. An example of an algorithm with a linear lower bound for its complexity was first published in Barzdin's paper [1]. The plan of our presentation goes back to Ilyushkin's senior thesis, written under the guidance of Markov at the Moscow Institute of Physics and Engineering and reported on in the fall of 1967 at the joint seminar on constructive mathematics of the University of Moscow and the Computing Center of the USSR Academy of Sciences, which had been meeting at that time under the authors' guidance.

2. As in previous sections, A_0 and A_1 will denote the alphabets *ab* and *abcd*, respectively. For normal algorithms $\mathfrak{A}$ in A_1, we define their representations $\mathfrak{A}^{Rp}$ [§42.1] and records $\mathfrak{A}^{Rc}$ [§45.2]. By the *complexity* of a normal algorithm $\mathfrak{A}$ in A_1, to be denoted in the sequel by the symbol compl($\mathfrak{A}$), we shall mean the natural number $|\mathfrak{A}^{Rp\delta}$ [cf. §31.3].

Like the assertion §47.2.1, it is easy to prove the following assertion:

2.1. *Let N be a natural number and let $\mathfrak{B}$ be a normal algorithm in* A_1, *such that for every normal algorithm $\mathfrak{A}$ in* A_1 *with* compl($\mathfrak{A}$) $\leqslant N$, *the following condition is satisfied: $\mathfrak{B}$ is applicable to $\mathfrak{A}^{Rc}$ if and only if $\mathfrak{A}$ is non-self-applicable. Then* compl($\mathfrak{B}$) $> N$.

Indeed, if compl($\mathfrak{B}$) $\leqslant N$, then $\mathfrak{B}$ is self-applicable if and only if it is non-self-applicable. But from this, as we had already seen in the proof of Lemma §47.2.1, it is easy to derive a contradiction, which is what yields compl($\mathfrak{B}$) $>$ N.

The reader can see that ≪Russell's paradox≫ is again being simulated in the given case. We think that an even stronger association should be evoked by the well-known ≪paradox of the village barber≫: The barber shaving those and only those men who do not shave themselves does not live in this village.

---

[8]See §31.3.

3. Now let $\mathfrak{C}$ be any normal algorithm in A_1 and let ε be a letter which is not a letter of the alphabet A_1. Consider the normal algorithm $\mathfrak{D}$ in the alphabet $A_1\varepsilon$, defined by the scheme which can be written down in the following abbreviated form:

$$(1) \qquad \begin{cases} \varepsilon\xi \rightarrow\cdot \\ \xi\varepsilon \rightarrow\cdot \\ \quad \varepsilon \rightarrow \varepsilon \\ \mathfrak{C}^{\varepsilon}, \end{cases}$$

where the literal variable ξ runs through the alphabet A_1 and the construction $\mathfrak{C}^{\varepsilon}$ has the same meaning as in the proof of the Composition Theorem.[9] The following assertion can be easily proven (we leave this to the reader):

3.1. *If P is a word in A_1, then $\mathfrak{D}(P)$ is defined if and only if $\mathfrak{C}(P) \neq \Lambda$.*[10]

Now let $\mathfrak{B}$ denote the translation of the algorithm $\mathfrak{D}$ into the alphabet A_1 [see §41.6], defined in such a way that the roles of the alphabets A and Б are played by A_0 and *cde*, respectively, and the roles of the A_0-literoids assigned to the letters c, d and ε are played by the words *cdc*, *cddc* and *cdddc*. We shall find an upper bound for compl($\mathfrak{D}$) in terms of compl($\mathfrak{C}$).

Let the scheme of the algorithm $\mathfrak{C}$ consist of k substitution formulas. Then the number of letters of the alphabet A_1 in the scheme $\mathfrak{C}$ equals

$$\mathrm{compl}(\mathfrak{C}) - (2k+1).$$

The scheme of the algorithm $\mathfrak{C}^{\varepsilon}$ has just as many letters of the alphabet A_1. But the number of its substitution formulas has been increased by one in comparison with $\mathfrak{C}$'s scheme. Besides, occurrences of the letter ε, whose number can be as great as $k+1$, have appeared in the scheme of $\mathfrak{C}^{\varepsilon}$.

In passing to the translation of the algorithm $\mathfrak{C}^{\varepsilon}$, the letters of the alphabet A_1 will be replaced by words of length at most 4. The letters ε are replaced by words of length 5. Thus, the complexity of $\mathfrak{C}^{\varepsilon}$'s translation is bounded above by the number

$$4[\mathrm{compl}(\mathfrak{C}) - (2k+1)] + 5(k+1) + 2(k+1) + 1,$$

i.e., by the number

$$4\,\mathrm{compl}(\mathfrak{C}) - k + 4.$$

Furthermore, it is not hard to verify that the first three lines of the scheme (1) contribute to a total of 86 to the complexity of $\mathfrak{B}$. Therefore,

$$\mathrm{compl}(\mathfrak{B}) \leqslant 4\,\mathrm{compl}(\mathfrak{C}) - k + 90,$$

and so

$$(2) \qquad \mathrm{compl}(\mathfrak{B}) \leqslant 4\,\mathrm{compl}(\mathfrak{C}) + 89.$$

4. We now prove that

---

[9] See p. 159.

[10] As usual, $\mathfrak{C}(P) \neq \Lambda$ means that $\mathfrak{C}(P)$ is defined and different from the empty word.

4.1. *If the natural number N and the normal algorithm $\mathfrak{C}$ in A_1 are such that for every normal algorithm $\mathfrak{A}$ in A_1 with* $\mathrm{compl}(\mathfrak{A}) \leqslant N$*: 1)* $\mathfrak{C}(\mathfrak{A}^{\mathrm{Rc}})$ *is defined and 2)* $\mathfrak{C}(\mathfrak{A}^{\mathrm{Rc}}) = \Lambda$ *if and only if $\mathfrak{A}$ is self-applicable, then* $\mathrm{compl}(\mathfrak{C}) > \frac{1}{4}N - \frac{89}{4}$.

Indeed, let N and $\mathfrak{C}$ be as indicated in the formulation. Construct $\mathfrak{D}$ and $\mathfrak{B}$ by the method given in Part 3. Then for a normal algorithm $\mathfrak{A}$ in A_1 with $\mathrm{compl}(\mathfrak{A}) \leqslant N$, $\mathfrak{D}(\mathfrak{A}^{\mathrm{Rc}})$—and hence also $\mathfrak{B}(\mathfrak{A}^{\mathrm{Rc}})$ by virtue of §41.6.27—is defined if and only if $\mathfrak{A}$ is non-self-applicable [3.1]. Since $\mathfrak{B}$ is a normal algorithm in A_1,

$$(1) \qquad \mathrm{compl}(\mathfrak{B}) > N \quad [2.1].$$

But then

$$N < 4\,\mathrm{compl}(\mathfrak{C}) + 89 \quad [(1), 3(2)]$$

and

$$\mathrm{compl}(\mathfrak{C}) > \frac{1}{4} - \frac{89}{4}.$$

Q.E.D.

5. Let $\mathfrak{G}$ be a normal algorithm over an alphabet A and let N be a natural number. We shall say that the *normal algorithm $\mathfrak{C}$ over* A *solves the N-restricted problem of recognizing the applicability of the algorithm $\mathfrak{G}$ to words in the alphabet* A if for any word P in A such that $[P^{\delta} \leqslant N$, the following two conditions are fulfilled: 1) the algorithm $\mathfrak{C}$ is applicable to P and 2) $\mathfrak{C}(P) = \Lambda$ if and only if $\mathfrak{G}$ is applicable to P.

Let us now take the alphabet A_1 and construct, according to the modified Universal Algorithm Theorem [§45.5.1], a normal algorithm $\mathfrak{W}$ over $\mathrm{A}_1 e$, such that for any normal algorithm $\mathfrak{A}$ in A_1 and any word P in this alphabet, the conditional equality

$$(1) \qquad \mathfrak{W}(\mathfrak{A}^{\mathrm{Rc}} e P) \simeq \mathfrak{A}(P)$$

holds. By the Union Theorem, we construct a normal algorithm $\mathfrak{K}$ over the alphabet $\mathrm{A}_0 e$, such that for any word P in A_0, the graphic equality

$$(2) \qquad \mathfrak{K}(P) = PeP$$

is fulfilled. We now define the algorithm $\mathfrak{F}$ as the normal composition of the algorithms $\mathfrak{K}$ and $\mathfrak{W}$:

$$(3) \qquad \mathfrak{F} \rightleftharpoons (\mathfrak{W} \circ \mathfrak{K}).$$

Then $\mathfrak{F}$ is a normal algorithm over $\mathrm{A}_0 e$ and for any word P in A_0,

$$(4) \qquad \begin{aligned} \mathfrak{F}(P) &\simeq \mathfrak{W}(\mathfrak{K}(P)) \quad [(3)] \\ &\simeq \mathfrak{W}(PeP) \quad [(2)]. \end{aligned}$$

In particular, for any normal algorithm $\mathfrak{A}$ in A_1, we have

$$(5) \qquad \begin{aligned} \mathfrak{F}(\mathfrak{A}^{\mathrm{Rc}}) &\simeq \mathfrak{W}(\mathfrak{A}^{\mathrm{Rc}} e \mathfrak{A}^{\mathrm{Rc}}) && [(4)] \\ &\simeq \mathfrak{A}(\mathfrak{A}^{\mathrm{Rc}}) && [(1)]. \end{aligned}$$

According to Theorem §41.6.27, one can construct a normal algorithm $\mathfrak{G}$ in a two-letter extension of the alphabet A_0 (i.e., for example, in the alphabet A_1), such that for any word P in A_0, $\mathfrak{F}(P)$ is defined if and only if $\mathfrak{G}(P)$ is defined, so that, in particular, the algorithm $\mathfrak{G}$ is applicable to the word $\mathfrak{A}^{\mathrm{Rc}}$, where $\mathfrak{A}$ is a normal algorithm in A_1, if and only if $\mathfrak{A}(\mathfrak{A}^{\mathrm{Rc}})$ is defined, i.e., if $\mathfrak{A}$ is self-applicable.

We now show that the following assertion is valid:

5.1. *Let N be an arbitrary number. Whatever be the normal algorithm $\mathfrak{C}$ in the alphabet A_1, solving the N-restricted problem of recognizing the applicability of the algorithm $\mathfrak{G}$ to words in the alphabet A_0,*

$$\mathrm{compl}(\mathfrak{C}) > \frac{1}{36}N - 22\frac{1}{2}.$$

In fact, if the normal algorithm $\mathfrak{C}$ in A_1 solves the problem of recognizing the applicability of the algorithm $\mathfrak{G}$ to words in A_0, then, in particular, for any normal algorithm $\mathfrak{A}$ in A_1, such that $[\mathfrak{A}^{\mathrm{Rc}\delta} \leqslant N$, we have 1) $\mathfrak{C}(\mathfrak{A}^{\mathrm{Rc}}$ is defined and 2) $\mathfrak{C}(\mathfrak{A}^{\mathrm{Rc}}) \rightleftharpoons \Lambda$ if and only if $\mathfrak{A}$ is self-applicable.

Representations of normal algorithms in the alphabet A_1 are words in a seven-letter alphabet. We therefore have $[\mathfrak{A}^{\mathrm{Rc}\delta} \leqslant 9[\mathfrak{A}^{\mathrm{Rp}\delta}$, that is, $[\mathfrak{A}^{\mathrm{Rc}\delta} \leqslant 9\,\mathrm{compl}(\mathfrak{A})$ for any such algorithm $\mathfrak{A}$.

Now let $\mathfrak{A}$ be a normal algorithm in A_1, such that[11]

$$\mathrm{compl}(\mathfrak{A}) \leqslant \left[\frac{N}{9}\right].$$

Then

$$9\,\mathrm{compl}(\mathfrak{A}) \leqslant N,$$

and so

$$[\mathfrak{A}^{\mathrm{Rc}\delta} \leqslant N.$$

Then: 1) $\mathfrak{C}(\mathfrak{A}^{\mathrm{Rc}})$ is defined and 2) $\mathfrak{C}(\mathfrak{A}^{\mathrm{Rc}}) \rightleftharpoons \Lambda$ if and only if $\mathfrak{A}$ is self-applicable. Consequently,

$$\mathrm{compl}(\mathfrak{C}) > \frac{1}{4}\left[\frac{N}{9}\right] - \frac{89}{4} \quad [4.1],$$

and so

$$\begin{aligned} \mathrm{compl}(\mathfrak{C}) &> \frac{1}{4}\left(\frac{N}{9} - 1\right) - \frac{89}{4} \\ &= \frac{1}{36} \cdot N - 22\frac{1}{2}. \end{aligned}$$

[11]Here the square brackets denote the integral part of the number enclosed in them.

Q.E.D.

6. Thus, if the problem of recognizing the applicability of the algorithm $\mathfrak{G}$ to words in the alphabet A_0 were in fact solvable, it would be solved with the aid of a normal algorithm $\mathfrak{C}$ over the alphabet A_0. But then, according to the Reduction Theorem [§41.7.1], one could find a normal algorithm $\mathfrak{C}$ in A_1, which would solve it. This algorithm's complexity, according to 5.1, would be greater than any of the numbers

$$\frac{1}{36} \cdot N - 22\frac{1}{2},$$

which is obviously impossible.

REMARK: The function $f(N) = \frac{1}{36} \cdot N - 22\frac{1}{2}$ is, therefore, a "lower bound" for the complexity of the problem of recognizing the applicability of the algorithm $\mathfrak{G}$. The question of an "upper bound" for this problem's complexity for an arbitrary normal algorithm is of interest. The nature of this problem can be perceived from the following proposition:

6.1. *Given any normal algorithm $\mathfrak{G}$ over the alphabet A_0, one can find a natural number C, such that for every natural number N, one can derive a contradiction from the assumption that there exists no normal algorithm $\mathfrak{C}$ in the alphabet A_1, solving the N-restricted problem of recognizing the applicability of $\mathfrak{G}$ to words in the alphabet A_0 and such that*

$$\mathrm{compl}(\mathfrak{C}) \leqslant N + C.$$

As regards the proof of this assertion, see Barzdin [1] and Petri [1].

§53 Non-Completable Algorithm

1. Let A be any alphabet. We shall call a normal algorithm $\mathfrak{A}$ over the alphabet A *complete with respect to* A if it is applicable to every word in A.

Let $\mathfrak{A}$ and $\mathfrak{B}$ be two algorithms over A. We shall call $\mathfrak{B}$ a *completion of* $\mathfrak{A}$ *with respect to* A if: 1) $\mathfrak{B}$ is complete with respect to A and 2) whenever $\mathfrak{A}$ is applicable to any word P in A, the graphic equality $\mathfrak{B}(P) \coloneqq \mathfrak{A}(P)$ holds.

We shall call a normal algorithm $\mathfrak{A}$ over A *completable with respect to* A if its completion with respect to A exists.

Consider the following two examples:

1°. Let $\mathfrak{A}$ be a normal algorithm over A which is not applicable to any word in A (in other words, an algorithm assuming *no value* for words in A). It is easy to see that $\mathfrak{A}$ is completable with respect to A: any normal algorithm over A, complete with respect to A, is its completion.

2°. Let $\mathfrak{B}$ be a normal algorithm over A, assuming *exactly one value* for words in A, i.e., such that whenever $\mathfrak{B}$ is applicable to any word P in A, $\mathfrak{B}(P) \coloneqq Q$, where Q is some fixed word in A. It is easy to see that $\mathfrak{B}$ is also completable with respect to A: every normal algorithm over A, changing each word P in A into the word Q, is its completion (it is not hard to construct such a constant algorithm; see §29.3).

The following question naturally arises: how long will the situation described in these examples continue? The answer to it is given by Theorem 1.1, presented below: it turns out that even among algorithms assuming only *two distinct values,* there are non-completable ones. Thus,

1.1. [12]*One can construct a normal algorithm* $\mathfrak{G}$ *over the alphabet* A_0, *such that:* 1) *whenever* $\mathfrak{G}$ *is applicable to any word* P *in* A_0, $\mathfrak{G}(P) \leftrightharpoons a$ *or* $\mathfrak{G}(P) \leftrightharpoons b$ *and* 2) $\mathfrak{G}$ *is non-completable with respect to* A_0.

PROOF: We construct, as was done already in §52.5, a normal algorithm $\mathfrak{F}$ over the alphabet A_1, such that for any normal algorithm $\mathfrak{A}$ in A_1, we have the conditional equality

$$(1) \qquad \mathfrak{F}(\mathfrak{A}^{\mathrm{Rc}}) \simeq \mathfrak{A}(\mathfrak{A}^{\mathrm{Rc}}).$$

Let Φ be the alphabet of the algorithm $\mathfrak{F}$. It is easy to see that a and b are letters of Φ. Denote the branching normal algorithm $\mathfrak{A}_{\Phi,a,a,b}$ [§30.1] by the symbol $\mathfrak{C}_1$. By construction,

$$(2) \qquad \mathfrak{C}_1(a) \leftrightharpoons a$$

and

$$(3) \qquad \mathfrak{C}_1(P) \leftrightharpoons b$$

for any word P in Φ, distinct from a.

We define the desired algorithm $\mathfrak{G}$ as the normal composition of the algorithms $\mathfrak{F}$ and $\mathfrak{C}_1$:

$$(4) \qquad \mathfrak{G} \rightleftharpoons (\mathfrak{C}_1 \circ \mathfrak{F}).$$

$\mathfrak{G}$ is a normal algorithm over Φ and, by the same token, over A_0, and for any P in A_0, the conditional equality

$$(5) \qquad \mathfrak{G}(P) \simeq \mathfrak{C}_1(\mathfrak{F}(P)) \quad [(4)]$$

holds, so that indeed, if $\mathfrak{G}(P)$ is defined, then $\mathfrak{G}(P) \leftrightharpoons a$ or $\mathfrak{G}(P) \leftrightharpoons b$ [(5), (2), (3)].

Let us now show that the algorithm $\mathfrak{G}$ is non-completable with respect to A_0. Indeed, assume that $\mathfrak{G}$ is completable with respect to A_0, and let $\mathfrak{L}$ be a completion of $\mathfrak{G}$ with respect to A_0. Then $\mathfrak{L}$ is a normal algorithm over A_0, complete with respect to A_0 and such that if $\mathfrak{G}(P)$ is defined for some P in A_0, then $\mathfrak{L}(P) \leftrightharpoons \mathfrak{G}(P)$.

Let Л be the alphabet of the algorithm $\mathfrak{L}$. The letters a and b occur in Л. Denote the branching algorithm $\mathfrak{A}_{Л,a,b,a}$ [§30.1] by $\mathfrak{C}_2$. By construction,

$$(6) \qquad \mathfrak{C}_2(a) \leftrightharpoons b$$

---

[12]See commentaries in Part 3.

and

$$(7) \qquad \mathfrak{C}_2(P) \mathrel{\overline{\overline{\cdot}}} a$$

for any word P in Π, distinct from a.

Construct a normal algorithm $\mathfrak{N}$, defining it as the normal composition of the algorithms $\mathfrak{L}$ and $\mathfrak{C}_2$:

$$(8) \qquad \mathfrak{N} \rightleftharpoons (\mathfrak{C}_2 \circ \mathfrak{L}).$$

$\mathfrak{N}$ is a normal algorithm over Π and, therefore, over A_0, and so for any word P in A_0, we have the conditional equality

$$(9) \qquad \mathfrak{N}(P) \mathrel{\overline{\overline{\cdot}}} \mathfrak{C}_2(\mathfrak{L}(P))[(8)].$$

Since $\mathfrak{L}$ is complete with respect to A_0 and $\mathfrak{C}_2$ is applicable to any word in Π, the conditional equality (9) may be replaced by the graphic equality

$$(10) \qquad \mathfrak{N}(P) \mathrel{\overline{\overline{\cdot}}} \mathfrak{C}_2(\mathfrak{L}(P)),$$

so that the algorithm $\mathfrak{N}$ also is complete with respect to A_0. By virtue of (10), (6) and (7), its values are the words a and b, so that one can construct a normal algorithm $\mathfrak{N}_1$ in the alphabet A_1, which is completely equivalent to it with respect to A_0. Let us consider the word $\mathfrak{N}_1^{Rc}$ and show that[13]

$$(11) \qquad \mathfrak{G}(\mathfrak{N}_1^{Rc}) \not\mathrel{\overline{\overline{\cdot}}} \mathfrak{L}(\mathfrak{N}_1^{Rc}).$$

By the same token, the assumption that $\mathfrak{L}$ is a completion of $\mathfrak{G}$ will be refuted.

In fact, $\mathfrak{L}(\mathfrak{N}_1^{Rc})$ is defined, since $\mathfrak{N}_1^{Rc}$ is a word in the alphabet A_0 and $\mathfrak{L}$ is complete with respect to A_0 by assumption. Moreover,

$$(12) \qquad \begin{aligned} \mathfrak{G}(\mathfrak{N}_1^{Rc}) &\simeq \mathfrak{C}_1(\mathfrak{F}(\mathfrak{N}_1^{Rc})) \quad [(5)] \\ &\simeq \mathfrak{C}_1(\mathfrak{N}_1(\mathfrak{N}_1^{Rc})) \quad [(1)] \end{aligned}$$

and, by the construction of $\mathfrak{N}_1$,

$$(13) \qquad \mathfrak{N}_1(\mathfrak{N}_1^{Rc}) \simeq \mathfrak{N}(\mathfrak{N}_1^{Rc}),$$

so that

$$(14) \qquad \mathfrak{G}(\mathfrak{N}_1^{Rc}) \simeq \mathfrak{C}_1(\mathfrak{N}(\mathfrak{N}_1^{Rc})) \quad [(12, (13)].$$

As was already noted, $\mathfrak{N}(\mathfrak{N}_1^{Rc})$ is defined; moreover, $\mathfrak{N}(\mathfrak{N}_1^{Rc}) \mathrel{\overline{\overline{\cdot}}} a$, or $\mathfrak{N}(\mathfrak{N}_1^{Rc}) \mathrel{\overline{\overline{\cdot}}} b$. Therefore, the symbol $\ll\simeq\gg$ in (14) may be replaced by the symbol $\ll\mathrel{\overline{\overline{\cdot}}}\gg$ and, furthermore, by virtue of (2) and (3),

$$(15) \qquad \mathfrak{C}_1(\mathfrak{N}(\mathfrak{N}_1^{Rc})) \mathrel{\overline{\overline{\cdot}}} \mathfrak{N}(\mathfrak{N}_1^{Rc}),$$

[13] We recall that the graphic difference between two expressions presupposes their meaningfulness.

so that we have

$$\mathfrak{G}(\mathfrak{N}_1^{\mathrm{Rc}}) \eqcirc \mathfrak{N}(\mathfrak{N}_1^{\mathrm{Rc}}) \quad [(14),(15)] \tag{16}$$

and, therefore,

$$\mathfrak{G}(\mathfrak{N}_1^{\mathrm{Rc}}) \eqcirc a \tag{17}$$

or

$$\mathfrak{G}(\mathfrak{N}_1^{\mathrm{Rc}}) \eqcirc b. \tag{18}$$

But

$$\mathfrak{N}(\mathfrak{N}_1^{\mathrm{Rc}}) \eqcirc \mathfrak{C}_2(\mathfrak{L}(\mathfrak{N}_1^{\mathrm{Rc}})) \quad [(10)], \tag{19}$$

and hence

$$\mathfrak{G}(\mathfrak{N}_1^{\mathrm{Rc}}) \eqcirc \mathfrak{C}_2(\mathfrak{L}(\mathfrak{N}_1^{\mathrm{Rc}})) \quad [(16),\ (19)]. \tag{20}$$

Now let

$$\mathfrak{G}(\mathfrak{N}_1^{\mathrm{Rc}}) \eqcirc \mathfrak{L}(\mathfrak{N}_1^{\mathrm{Rc}}). \tag{21}$$

Then

$$\mathfrak{G}(\mathfrak{N}_1^{\mathrm{Rc}}) \eqcirc \mathfrak{C}_2(\mathfrak{G}(\mathfrak{N}_1^{\mathrm{Rc}})).$$

But this equality is possible neither for $\mathfrak{G}(\mathfrak{N}_1^{\mathrm{Rc}}) \eqcirc a$ [(6)] nor for $\mathfrak{G}(\mathfrak{N}_1^{\mathrm{Rc}}) \eqcirc b$ [(7)]. Consequently, we have (11) [(17), (18)], which is just what remained to be proven.

2. Theorem 1.1 expresses an essentially stronger fact of the theory of algorithms than Theorem §48.2.4. Namely, the following assertion holds:

2.1. *Let $\mathfrak{A}$ be a normal algorithm over the alphabet* A, *such that the problem of recognizing the applicability of $\mathfrak{A}$ to words in* A *is solvable. Then $\mathfrak{A}$ is completable with respect to* A.

Indeed, let the hypothesis of our assertion be fulfilled. Then one can find a normal algorithm $\mathfrak{C}$ over the alphabet A, such that for any word P in A, we will have the conditions

$$!\,\mathfrak{C}(P) \tag{1}$$

and

$$\mathfrak{C}(P) \eqcirc \Lambda \text{ if and only if } !\,\mathfrak{A}(P). \tag{2}$$

Let us take any normal algorithm $\mathfrak{B}$ over A, such that for all P in A, we will have

$$!\,\mathfrak{B}(P), \tag{3}$$

and in accordance with the Branching Theorem, construct the normal algorithm

$$(\mathfrak{A} \curlyvee \mathfrak{B} | \mathfrak{C}),$$

which we shall denote by the letter $\mathfrak{D}$. $\mathfrak{D}$ is a normal algorithm over A and for any P in A,

$$\mathfrak{D}(P) \simeq \mathfrak{A}(P) \tag{4}$$

if $\mathfrak{C}(P) = \Lambda$, and

$$\mathfrak{D}(P) \simeq \mathfrak{B}(P) \tag{5}$$

if $\mathfrak{C}(P) \neq \Lambda$.

Let us note that if P is a word in A and $\mathfrak{C}(P) = \Lambda$, then

$$\mathfrak{D}(P) = \mathfrak{A}(P). \tag{6}$$

In fact, $!\,\mathfrak{A}(P)$ in this case [(2)], and so (6) holds by virtue of (4).

Let us show that $\mathfrak{D}$ is complete with respect to A. Indeed, let P be an arbitrary word in A. Then $!\,\mathfrak{C}(P)$ by virtue of (1). If $\mathfrak{C}(P) = \Lambda$, then $!\,\mathfrak{D}(P)$ by virtue of (6). But if $\mathfrak{C}(P) \neq \Lambda$, then $!\,\mathfrak{D}(P)$ by virtue of (5) and (3).

We now show that if P is in A and $!\,\mathfrak{A}(P)$, then $\mathfrak{D}(P) = \mathfrak{A}(P)$. Indeed, let P be in A and $!\,\mathfrak{A}(P)$. Thus, according to (2), $\mathfrak{C}(P) = \Lambda$, but then (6) holds.

Therefore, the algorithm $\mathfrak{D}$ is a completion of $\mathfrak{A}$ with respect to A, which proves 2.1.

From this follows

2.2. *If a normal algorithm $\mathfrak{A}$ over an alphabet A is non-completable with respect to A, then the problem of recognizing the applicability of $\mathfrak{A}$ to words in A is unsolvable.*

The fact that the strengthening is essential can be perceived from

2.3. *One can find an alphabet A and a normal algorithm $\mathfrak{A}$ over A, such that the problem of recognizing the applicability of $\mathfrak{A}$ to words in A is unsolvable, but $\mathfrak{A}$ is completable with respect to A.*

For the proof, it is sufficient to take any normal algorithm $\mathfrak{B}$ in an alphabet Б with an unsolvable problem of recognizing the applicability of $\mathfrak{B}$ to words in some subalphabet A of Б and construct the normal composition $\mathfrak{A}$ of this algorithm with the ≪annihilating≫ normal algorithm in Б. It is easy to see that for any P in A, the following condition is fulfilled: $!\,\mathfrak{A}(P)$ if and only if $!\,\mathfrak{B}(P)$. Furthermore, if P is in A and $!\,\mathfrak{A}(P)$, then $\mathfrak{A}(P) = \Lambda$. It follows from this that $\mathfrak{A}$ is completable with respect to A [1.2°], which is just what remained to be proven.

3. Set-theoretic commentary to Theorem 1.1. Let A be an alphabet and let M, N be any disjoint sets of words in it. We shall say that M and N are *separable by decidable sets* if there exist disjoint decidable sets M_1 and N_1 of words in A, such that $M \subseteq M_1$ and $N \subseteq N_1$.

It is not difficult to show that for the normal algorithm $\mathfrak{G}$ constructed in the proof of Theorem 1.1, the sets M_a and M_b of all the words in the alphabet A_0, such that $\mathfrak{G}(P) \eqcirc a$ and $\mathfrak{G}(P) \eqcirc b$, respectively, are *enumerable*. Nevertheless, they are *not separable by decidable sets*. We shall show this.

Indeed, let N_a and N_b be disjoint decidable sets of words in the alphabet A_0, such that $M_a \subseteq N_a$ and $M_b \subseteq N_b$. Let the normal algorithm $\mathfrak{C}$ over A_0 verify the belonging of words in A_0 to the set N_a. This means that for any P in A_0: 1) $\mathfrak{C}(P)$ is defined and 2) $\mathfrak{C}(P) \eqcirc \Lambda$ if and only if $P \in N_a$. Let $\mathfrak{J}_a$ and $\mathfrak{J}_b$ be normal algorithms in A_0 satisfying the conditions $\mathfrak{J}_a(P) \eqcirc a$ and $\mathfrak{J}_b(P) \eqcirc b$ for any P in A_0. We construct the normal algorithm $\mathfrak{L}$ as the normal branching of the algorithms $\mathfrak{J}_a$ and $\mathfrak{J}_b$, directed by the algorithm $\mathfrak{C}$:

$$\mathfrak{L} \rightleftharpoons (\mathfrak{J}_a \curlyvee \mathfrak{J}_b | \mathfrak{C}).$$

We shall show that $\mathfrak{L}$ is a completion of $\mathfrak{G}$ with respect to A_0. In fact, $\mathfrak{L}$ is complete with respect to A_0, since the algorithms $\mathfrak{C}$, $\mathfrak{J}_a$ and $\mathfrak{J}_b$ are applicable to any word in A_0. Now let P be a word in A_0 and let $\mathfrak{G}(P)$ be defined. Then $\mathfrak{G}(P) \eqcirc a$ or $\mathfrak{G}(P) \eqcirc b$. In the first case, $P \in M_a$; therefore, $P \in N_a$ and so $\mathfrak{C}(P) \eqcirc \Lambda$. Consequently, $\mathfrak{L}(P) \eqcirc \mathfrak{J}_a(P)$ in this case, i.e., $\mathfrak{L}(P) \eqcirc a$. In the second case, $P \in M_b$ and so $P \in N_b$. But N_a and N_b do not intersect, and hence $P \notin N_a$. Consequently, we get $\mathfrak{C}(P) \not\eqcirc \Lambda$ and then $\mathfrak{L}(P) \eqcirc \mathfrak{J}_b(P)$, i.e., $\mathfrak{L}(P) \eqcirc b$. Thus $\mathfrak{L}$ really is a completion of $\mathfrak{G}$, which contradicts Theorem 1.1. Hence, M_a and M_b are not separable by decidable sets. Q.E.D.

Theorem 1.1 therefore yields an example of two disjoint enumerable sets of words in the alphabet A_0, which are not separable by decidable sets. The circumstance that these sets turned out to be sets of words in a two-letter alphabet depends not on the essence of the matter, but on the technical details of the proof: analogous sets can also be constructed in a one-letter alphabet, and then may be regarded as sets of natural numbers possessing the corresponding properties. It was precisely in such a form that this result was first published by Kleene in 1950 [3].

The above result is an analogue of Novikov's well-known theorem on the existence of two B-inseparable CA-sets. An example of two enumerable sets of natural numbers which are not separable by decidable sets was presented by Novikov while conducting a seminar devoted to a clarification of the analogy between descriptive set theory and the theory of enumerable sets during the academic year of 1946–47. Unfortunately, Novikov's results related to this range of questions remained unpublished (for more about this, see the 277th page of Uspenskii's book [2], as well as the footnote appended to the very same page of the Russian translation of Kleene's book [5] by its editor, Uspenskii).

CHAPTER VII

Computable Verbal Functions

In this chapter, we present certain facts of the general theory of algorithms whose most natural formulation is in terms of "algorithms of several variables". The concept of a γ-system of words in an alphabet A, introduced in §24, can serve here as a sufficiently convenient technical tool. The letter γ, however, plays a double role in γ-systems: on one hand, it separates a system's terms from each other, while on the other, it replaces "outer parentheses"; in combination with the traditional functional notation, where the collection of variables of a function is enclosed in parentheses, the use of such a concept of a system would lead to notation to which the eye is unaccustomed. We therefore agree to discard the bounding letters γ from non-empty γ-systems. Thus, we shall employ the word P instead of the one-term γ-system $\gamma P\gamma$, the word $P\gamma Q$ instead of the two-term γ-system $\gamma P\gamma Q\gamma$, etc.

§54 Computable Verbal Functions

1. We shall regard as fixed an alphabet A containing the letters a and b. We shall consider systems of words in this alphabet to be words in the alphabet Aγ, taking into account the remark just made. Besides, we shall add any two new letters, say α and β, to Aγ in order to have the possibility of treating arbitrary normal algorithms over Aγ as normal algorithms in the alphabet A$\alpha\beta\gamma$, which we shall denote by the symbol A^+. Records of normal algorithms will be defined precisely for this alphabet (in the usual manner), so that, in particular, normal algorithms in A^+ can also appear as values assumed by variables of the functions introduced below.

A *computable verbal function of* N *variables* $(N \geqslant 1)$ *in the alphabet* A will mean any normal algorithm in the alphabet A^+, changing each N-term system of words in A to which it is applicable into some word in A, which will be regarded as the *value of the function* for the given system.

Since the alphabet A has been fixed, mention of it will often be omitted in what follows.

Let us illustrate this definition by means of several simple examples.

1°. The normal algorithm in A^+, whose scheme can be written down in the following abbreviated form:

$$\{\xi \to$$

(ξ runs through the alphabet $A\gamma$) for any $N \geqslant 1$ is a computable verbal function of N variables. Its value for any N-term system of words is equal to the empty word.

2°. The normal algorithm in A^+ with the scheme

$$\{\to$$

for any $N \geqslant 1$ is a computable verbal function of N variables. It is the "nowhere defined" function.

3°. The normal algorithm in A^+ with the scheme

$$\{\to\cdot$$

is a computable verbal function of one variable (the "identity function"), but this algorithm is not a computable verbal function of N variables for $N > 1$ (because when $N > 1$, the letter γ will always be present in a word which is the result of applying this algorithm to an N-term system).

When speaking of computable verbal functions (and in the sequel, also about computable operators over them), we shall sometimes, in cases of unwieldy phrases, make use of elliptical terminology, abbreviating part of a term being employed. Thus, instead of ≪computable verbal functions≫, we shall sometimes simply speak of ≪functions≫. In connection with this, we should like to particularly emphasize the fact that a computable verbal function is *not a function* satisfying such and such conditions. A computable verbal function is a *normal algorithm* subject to definite restrictions. We introduce this concept without reference to any general concept of a function, and the term denoting it should be perceived as a unified and indivisible phrase, not ≪through the nearest genus and specific difference≫.

We are particularly noting this circumstance in view of the fact that in the literature one can also come across a different approach, according with which a computable verbal function of one variable will be defined as a ≪mapping of a set of words into itself, such that its values can be computed with the aid of an algorithm≫. We should like our approach to be clearly distinguishable from the one just mentioned in view of the close connection between the latter and set-theoretical conceptions.

There is one more argument in favor of our definition. We would like it to be possible (at least in principle) to compute a computable function, to be known how to do so. Our definition satisfies this requirement: computing the value of a computable function means applying the corresponding normal algorithm to the appropriate initial data. But as regards the definition just formulated, here the situation is not so simple. Indeed, let Φ be a statement whose truth is an unsolved mathematical problem (we take Fermat's Problem for concreteness). We define the arithmetical function f by setting

$$f(N) = \begin{cases} 1 & \text{if } \Phi \text{ is true,} \\ 0 & \text{if } \Phi \text{ is false,} \end{cases}$$

for an arbitrary natural number N. A mathematician accepting the ≪law of the excluded middle≫ must regard this function as computable. However, it is clear that even finding $f(0)$ is equivalent to solving Fermat's Problem. One can also define a similar function for any other "hard" problem (any unsolvable normal mass problem gives one an inexhaustible source of such problems). It seems paradoxical to us to regard as computable an everywhere defined constant function (its value is equal to 0 or 1) when there is not even any visible approach, for example, to computing $f(0)$. Of course, there is no formal contradiction here, but there is a definite divergence from ordinary word usage, according to which the suffix ≪-able≫ carries with it a certain meaning. Taking such a position, one must ignore the natural, "everyday" meaning of this linguistic construction.

The situation we described is of a general nature. We shall come across analogous definitions in the future when we will consider constructive real numbers, constructive functions of a real variable, etc. All these concepts will be introduced independently, without any references to the corresponding traditional concepts.

2. Let us consider a standard way of representing computable verbal functions.

Let Б be any alphabet which is an extension of the alphabet $\text{A}\gamma$. Consider the "filtrating" normal algorithm $\mathfrak{F}_{\text{Б},\text{A}}$ in the alphabet Б, whose scheme can be written down in the following abbreviated form:

$$\{\xi \rightarrow \xi,$$

where ξ runs through the alphabet $\text{Б}\setminus\text{A}$. It is easy to see that $\mathfrak{F}_{\text{Б},\text{A}}$ is applicable to a word Q in Б if and only if it is a word in A; moreover, in this case we have the graphic equality

$$\mathfrak{F}_{\text{Б},\text{A}}(Q) \rightleftharpoons Q.$$

Now let $\mathfrak{B}$ be any normal algorithm in Б. Consider the normal composition $(\mathfrak{F}_{\text{Б},\text{A}} \circ \mathfrak{B})$ of the algorithms $\mathfrak{B}$ and $\mathfrak{F}_{\text{Б},\text{A}}$. For any N-term system S of words in A, we have

$$(\mathfrak{F}_{\text{Б},\text{A}} \circ \mathfrak{B})(S) \simeq \mathfrak{F}_{\text{Б},A}(\mathfrak{B}(S)),$$

so that in case the expression $(\mathfrak{F}_{\text{Б},\text{A}} \circ \mathfrak{B})(S)$ is defined, its value is a word in A.

On the other hand, if the algorithm $\mathfrak{B}$ changes each N-term system of words to which it is applicable into a word in A, then for any such system S, we have the conditional equality

$$(\mathfrak{F}_{\text{Б},\text{A}} \circ \mathfrak{B})(S) \simeq \mathfrak{B}(S).$$

Let us now translate the algorithm $(\mathfrak{F}_{\text{Б},\text{A}} \circ \mathfrak{B})$ into the alphabet A^+, taking as the "base" alphabet i.e., the alphabet which is unaffected by the translation, the alphabet $\text{A}\gamma$. We denote the resulting "reduced" algorithm by the symbol $\mathfrak{B}'$

One can, in particular, apply the procedure just described to the normal algorithm $\mathfrak{B}$ in the alphabet A^+. Moreover, every algorithm $\mathfrak{B}'$ will be a computable verbal function of N variables for $N \geqslant 1$ and, if the algorithm $\mathfrak{B}$ already was such a function, itself, then for any N-term system S of words in A, the conditional equality

$$\mathfrak{B}'(S) \simeq \mathfrak{B}(S)$$

will be fulfilled, so that among the "reduced" functions like $\mathfrak{B}'$ there will be (up to an equivalence) all the computable verbal functions of any number of variables.

3. The following method of fixing the values of the first variable of functions of the type we are considering will be convenient in the sequel.

Let P be any word in A. Consider the normal algorithm $\mathfrak{F}_P$ in $A\gamma$ with the scheme

$$\{\rightarrow\cdot P\gamma.$$

This algorithm changes each word Q in the alphabet $A\gamma$ into the word $P\gamma Q$ and, therefore, each N-term system S of words, into the (N+1)-term system $P\gamma S$.

Now let $\mathfrak{A}$ be any normal algorithm over the alphabet $A\gamma$. Consider the normal composition $(\mathfrak{A} \circ \mathfrak{F}_P)$ of the normal algorithms $\mathfrak{F}_P$ and $\mathfrak{A}$. It is a normal algorithm over the alphabet $A\gamma$, such that for any word Q in $A\gamma$, we have

$$(\mathfrak{A} \circ \mathfrak{F}_P)(Q) \simeq \mathfrak{A}(P\gamma Q).$$

Let us translate the algorithm $(\mathfrak{A} \circ \mathfrak{F}_P)$ into the alphabet A^+ in such a way that the role of the "base" alphabet is played by the alphabet $A\gamma$. Following Tseitin [3], we shall denote the normal algorithm obtained in this manner by the symbol $\widetilde{\mathfrak{A}}_P$.

If $\mathfrak{A}$ is a computable verbal function of $N+1$ variables, then for any N-term system S, we have

$$\text{(1)} \qquad \widetilde{\mathfrak{A}}_P(S) \simeq \mathfrak{A}(P\gamma S).$$

Therefore, the algorithm $\widetilde{\mathfrak{A}}_P$ changes each N-term system of words to which it is applicable into a word in the alphabet A, and so $\widetilde{\mathfrak{A}}_P$ is a computable verbal function of N variables. In the natural sense of the word, this function is obtained from the function $\mathfrak{A}$ by fixing the first variable.

4. Analyzing the method of constructing the scheme of the algorithm $(\mathfrak{A} \circ \mathfrak{F}_P)$ [§37.1] and then of $\widetilde{\mathfrak{A}}_P$ [§41.6], the reader can convince her or himself without difficulty that one can construct a normal algorithm $\mathfrak{K}$ over the alphabet $A\gamma$, such that for any normal algorithm $\mathfrak{A}$ in A^+ and for any word P in A, the following graphic equality will hold:

$$\text{(1)} \qquad \mathfrak{K}(\mathfrak{A}^{\mathrm{Rc}}\gamma P) = \widetilde{\mathfrak{A}}_P^{\mathrm{Rc}}.$$

It is easy to comprehend that by making use of the procedure indicated in Part 2, the algorithm $\mathfrak{K}$ can even be given in the form of a computable verbal function of two variables in the alphabet A.

5. REMARK: The existence of a normal algorithm $\mathfrak{K}$ carrying out the construction of $\widetilde{\mathfrak{A}}_P$ is not, of course, any kind of exception. Analogous (normal)

algorithms can also be constructed for the other combinations of normal algorithms considered by us. Thus, for example, when holding the alphabets of the component algorithms fixed, one can construct a normal algorithm changing the pair of records of any two normal algorithms $\mathfrak{A}$ and $\mathfrak{B}$ into the record of their normal composition $(\mathfrak{B} \circ \mathfrak{A})$, when the alphabets of the united algorithms are held fixed, one can construct a normal algorithm changing the pair of records of normal algorithms $\mathfrak{A}$ and $\mathfrak{B}$ into the record of their normal union $(\mathfrak{A} \frown \mathfrak{B})$, etc. The existence of such algorithms is suggested by the normalization principle, and their schemes can actually be constructed. When the concept of a ≪letter≫ is suitably regulated (it is possible, for example, to agree to regard words of the form ab^ia, $i > 0$, as ≪letters≫), one can also construct a normal algorithm which will change each natural number N into the record of a universal algorithm for an N-letter alphabet. The actual construction of such algorithms would be a definite strengthening of the theorems proven by us in Chapters IV and V.

§55 The Fixed Point Theorem

In this section, we shall present one of the most remarkable results of the general theory of algorithms—Kleene's Fixed Point Theorem. In spite of the proof's relative simplicity, this theorem uncovers a very unexpected fact and has numerous and important applications. Unfortunately, we are unable to touch on enough of them in this monograph because of insufficient space (nevertheless, see Theorem §56.1.1).

1. Let $\mathfrak{A}$ and $\mathfrak{B}$ be computable verbal functions of one and two variables, respectively. We shall say that $\mathfrak{B}$ *computes* $\mathfrak{A}$ if for any Q in A, the conditional equality

$$\mathfrak{B}(\mathfrak{A}^{\mathrm{Rc}}\gamma Q) \simeq \mathfrak{A}(Q)$$

holds. Considering the formulation of the Universal Algorithm Theorem [§45.5.1] from the point of view enunciated in the preceding section, we perceive without difficulty that

1.1. *One can construct a computable verbal function $\mathfrak{W}$ of two variables, computing every computable verbal function of one variable.*

The construction of such a "universal" function, as we at one time convinced ourselves, requires a definite effort. Another remarkable result is (cf. Kleene [5], p. 352, Theorem XXVII) that

1.2. *Each computable verbal function of two variables computes some computable verbal function of one variable.*

Indeed, let $\mathfrak{B}$ be an arbitrary computable verbal function of two variables. Using the computable verbal function $\mathfrak{K}$ from §54.4, as well as the severing normal algorithms $\mathfrak{J}_{A\gamma,\gamma}$ and $\mathfrak{G}_{A\gamma,\gamma}$ [§29.4], and applying the union and the composition theorems, we construct a computable verbal function $\mathfrak{C}$ of two variables, such that for any words P and Q in the alphabet A, the conditional equality

$$\mathfrak{C}(P\gamma Q) \simeq \mathfrak{B}(\mathfrak{K}(P\gamma P)\gamma Q)$$

holds. Then, in particular, for any Q in A, we have

$$\begin{aligned}\mathfrak{C}(\mathfrak{C}^{\mathrm{Rc}}\gamma Q) &\simeq \mathfrak{B}(\mathfrak{K}(\mathfrak{C}^{\mathrm{Rc}}\gamma\mathfrak{C}^{\mathrm{Rc}})\gamma Q)\\ &\simeq \mathfrak{B}(\widetilde{\mathfrak{C}}^{\mathrm{Rc}}_{\mathfrak{C}^{\mathrm{Rc}}}\gamma Q) \qquad [\S 54.4(1)].\end{aligned}$$

But

$$\mathfrak{C}(\mathfrak{C}^{\mathrm{Rc}}\gamma Q) \simeq \widetilde{\mathfrak{C}}_{\mathfrak{C}^{\mathrm{Rc}}}(Q) \quad [\S 54.3(1)]$$

and, therefore,

$$\widetilde{\mathfrak{C}}_{\mathfrak{C}^{\mathrm{Rc}}}(Q) \simeq \mathfrak{B}(\widetilde{\mathfrak{C}}^{\mathrm{Rc}}_{\mathfrak{C}^{\mathrm{Rc}}}\gamma Q).$$

Thus, $\mathfrak{B}$ really does compute the computable verbal function $\widetilde{\mathfrak{C}}_{\mathfrak{C}^{\mathrm{Rc}}}$ of one variable. The theorem is proven.

Note that starting with $\mathfrak{B}$, one can effectively construct the function $\widetilde{\mathfrak{C}}_{\mathfrak{C}^{\mathrm{Rc}}}$, and we could have even constructed a normal algorithm changing the record of any such function $\mathfrak{B}$ into the record of the corresponding function $\widetilde{\mathfrak{C}}_{\mathfrak{C}^{\mathrm{Rc}}}$, so that the proof of Theorem 1.2 is entirely constructive.

2. We shall say that a normal algorithm $\mathfrak{G}$ in the alphabet A^+ is a *computable operator over the computable verbal functions of one variable* if: 1) $\mathfrak{G}$ is a computable verbal function of one variable and 2) $\mathfrak{G}$ changes the record of each computable verbal function of one variable into the record of some other function of the same type.

If a word P in the alphabet A_0 is the record of a normal algorithm in A^+, then we shall denote this same normal algorithm in A^+ whose record is P, by the symbol $\langle P\rangle$.

The following "Fixed Point" Theorem holds:

2.1. *Given any computable operator $\mathfrak{G}$ over the computable verbal functions of one variable, one can find a computable verbal function $\mathfrak{C}$ of one variable, such that for any word P in the alphabet* A, *we have the conditional equality*

$$\mathfrak{C}(P) \simeq \langle\mathfrak{G}(\mathfrak{C}^{\mathrm{Rc}})\rangle(P).$$

Indeed, let $\mathfrak{G}$ be an arbitrary computable operator over the computable verbal functions of one variable. Let $\mathfrak{W}$ be a "universal function" whose existence is asserted in 1.1. Analogously to the way in which this was done in the proof of Theorem 1.2, we construct a computable verbal function $\mathfrak{B}$ of two variables, such that for any P and Q in A, the conditional equality

$$\mathfrak{B}(P\gamma Q) \simeq \mathfrak{W}(\mathfrak{G}(P)\gamma Q)$$

will be fulfilled. According to Theorem 1.2, one can find a computable verbal function $\mathfrak{C}$ of one variable, such that for any P in A, we have

$$\mathfrak{C}(P) \simeq \mathfrak{B}(\mathfrak{C}^{\mathrm{Rc}}\gamma P).$$

But

$$\begin{aligned}\mathfrak{B}(\mathfrak{C}^{\mathrm{Rc}}\gamma P) &\simeq \mathfrak{W}(\mathfrak{G}(\mathfrak{C}^{\mathrm{Rc}})\gamma P)\\ &\simeq \langle\mathfrak{G}(\mathfrak{C}^{\mathrm{Rc}})\rangle(P),\end{aligned}$$

and hence

$$\mathfrak{C}(P) \simeq \langle\mathfrak{G}(\mathfrak{C}^{\mathrm{Rc}})\rangle(P).$$

Q.E.D.

Therefore, every computable operator over the computable verbal functions of one variable has a fixed point.

3. REMARKS: 1°. The reader can state and prove without difficulty the corresponding assertion for the case of functions of N variables.

2°. A fixed point of the operator $\mathfrak{G}$ can be effectively constructed. In fact, one can even construct a normal algorithm changing the record of an arbitrary operator of the type considered into its fixed point.

3°. It is interesting to note that in the definition of an operator given by us, there is no "compatibility" requirement whatsoever: it is not required that an operator take "equal" (in the sense of coinciding values) functions into those which again are "equal".

4°. Let us illustrate Theorem 2.1 with the following example.

The method considered in §54.2 of representing computable verbal functions allows us to "arrange" the records of the reduced functions of one variable in an effective sequence. For example, we can order the records according to length, and for equal length, lexicographically. It is also possible, of course, to take any other "more intricate" method of writing down the terms of this sequence, as long as it admitted a realization in the form of a normal algorithm. It is asserted that in any such sequence one will be able to find neighboring records of two equal (i.e., assuming equal values when the variables are equal) functions.

In fact, without difficulty one can construct a computable operator $\mathfrak{G}$ converting each record of this sequence into its neighbor and extended to the records of the remaining verbal functions with due regard for their being equal to reduced functions. According to Theorem 2.1, this operator has a fixed point, but then there will also be one among the functions of the sequence constructed by us. Q.E.D.

It is easy to see that the records of equal functions will be found not only next to each other, but also at any given distance from each other.

The reader will undoubtedly find other interesting examples.

§56 Recognizing Invariant Properties of Computable Verbal Functions

In this section, we shall present the Uspenskii–Rice Theorem on the unsolvability of the problem of recognizing non-trivial invariant properties of computable verbal functions. For definiteness, we shall again consider functions of only one variable, sometimes without saying so explicitly. A transfer to the general case involves no difficulty.

1. When speaking about properties of computable verbal functions, we shall have in mind properties of their records, i.e., actually one-place verbal predicates for words in the alphabet A_0. The expressions ≪the function $\mathfrak{A}$ possesses the property F≫ and ≪the record of the function $\mathfrak{A}$ possesses the property F≫ will be regarded as synonymous.

Let $\mathfrak{A}$ and $\mathfrak{B}$ be computable verbal functions (of one variable). We shall say that $\mathfrak{A}$ and $\mathfrak{B}$ are *equivalent* and shall write

$$\mathfrak{A} \approx \mathfrak{B}$$

if for every P in A, we have the conditional equality

$$\mathfrak{A}(P) \simeq \mathfrak{B}(P).$$

Let F be a property of computable verbal functions. We shall call it *invariant* if from $\mathfrak{A}$'s possession of it and $\mathfrak{B} \approx \mathfrak{A}$ follows $\mathfrak{B}$'s possession of F. We shall call F *non-trivial* if one can find two computable verbal functions, such that one of them possesses it, but the other does not.

The following assertion holds (Uspenskii [1][1]–Rice [1]).

1.1. *Every non-trivial invariant property F of computable verbal functions of one variable is undecidable in the sense that there does not exist a normal algorithm $\mathfrak{C}$ over the alphabet A_0, applicable to the record of every computable verbal function $\mathfrak{A}$ of one variable and changing this record into the empty word if and only if $\mathfrak{A}$ possesses the property F.*

Indeed, let F be a non-trivial invariant property of the kind stated in the theorem's formulation. We shall show that a contradiction follows from the assumption of the existence of a normal algorithm $\mathfrak{C}$ with the properties stated in the theorem's formulation.

In fact, let $\mathfrak{C}$ be a normal algorithm over the alphabet A_0, verifying (in the above indicated sense of the word) the possession of the property F by an arbitrary function $\mathfrak{A}$. Assume that the function $\mathfrak{A}_1$ possesses the property F, but the function $\mathfrak{A}_2$ does not possess it, so that

(1) $$\mathfrak{C}(\mathfrak{A}_1^{\mathrm{Rc}}) = \Lambda,$$

but

(2) $$\mathfrak{C}(\mathfrak{A}_2^{\mathrm{Rc}}) \neq \Lambda.$$

We construct two normal algorithms, $\mathfrak{I}_1$ and $\mathfrak{I}_2$, in the alphabet A_0, in such a way that for any P in A_0, the following graphic equalities are fulfilled:

(3) $$\mathfrak{I}_1(P) = \mathfrak{A}_1^{\mathrm{Rc}}$$

and

(4) $$\mathfrak{I}_2(P) = \mathfrak{A}_2^{\mathrm{Rc}}.$$

Then by the Branching Theorem, we construct the normal algorithm

(5) $$\mathfrak{G} \rightleftharpoons (\mathfrak{I}_2 \curlyvee \mathfrak{I}_1 \mid \mathfrak{C}).$$

[1]See also Uspenskii [2], p.310, Theorem 9.

It is immediately obvious that for any word P in A_0, we have

$$(6) \qquad \mathfrak{G}(P) \simeq \begin{cases} \mathfrak{A}_2^{\mathrm{Rc}} & \text{if } \mathfrak{C}(P) = \Lambda \quad [(5), (4)], \\ \mathfrak{A}_1^{\mathrm{Rc}} & \text{if } \mathfrak{C}(P) \neq \Lambda \quad [(5), (3)], \end{cases}$$

and if $\mathfrak{G}$ is applicable to P, then $\mathfrak{C}$ is also applicable to this word. Therefore, in case of applicability, the algorithm $\mathfrak{G}$ changes words in the alphabet A_0 into words in A_0 once more.

Taking into account the procedure described in §54.2, we may assume that the algorithm $\mathfrak{G}$ is a computable verbal function of one variable. Since the algorithm $\mathfrak{C}$ is applicable to the record of any computable verbal function $\mathfrak{A}$ of one variable, this very same property is also possessed by $\mathfrak{G}$ [(6)]. Furthermore, $\mathfrak{G}(\mathfrak{A}^{\mathrm{Rc}})$ is always the record of some computable verbal function ($\mathfrak{A}_1$ or $\mathfrak{A}_2$), so that $\mathfrak{G}$ is a computable operator over the computable verbal functions of one variable. According to Theorem §55.2.1, this operator has a certain fixed point $\mathfrak{B}$. In the obvious manner, we have

$$(7) \qquad \mathfrak{B} \approx \langle \mathfrak{G}(\mathfrak{B}^{\mathrm{Rc}}) \rangle.$$

Therefore, by virtue of F's invariance, $\mathfrak{B}$ possesses this property if and only if $\langle \mathfrak{G}(\mathfrak{B}^{\mathrm{Rc}}) \rangle$ possesses it.

Let us now assume that $\mathfrak{B}$ possesses the property F. Then $\mathfrak{C}(\mathfrak{B}^{\mathrm{Rc}}) = \Lambda$. But then $\mathfrak{G}(\mathfrak{B}^{\mathrm{Rc}}) = \mathfrak{A}_2^{\mathrm{Rc}}$ [(6)] and, therefore, $\langle \mathfrak{G}(\mathfrak{B}^{\mathrm{Rc}}) \rangle$ is $\mathfrak{A}_2$, but $\mathfrak{A}_2$ does *not possess* F. Hence $\mathfrak{B}$ does not possess F.

On the other hand, assume that $\mathfrak{B}$ does not possess F. Then $\mathfrak{C}(\mathfrak{B}^{\mathrm{Rc}}) \neq \Lambda$. But then $\mathfrak{G}(\mathfrak{B}^{\mathrm{Rc}}) = \mathfrak{A}_1^{\mathrm{Rc}}$ [(6)] and, therefore, $\langle \mathfrak{G}(\mathfrak{B}^{\mathrm{Rc}}) \rangle$ is $\mathfrak{A}_1$, but $\mathfrak{A}_1$ possesses F. It is therefore *not the case that* $\mathfrak{B}$ *does not possess* F.

The contradiction obtained here—≪ $\mathfrak{B}$ *does not possess* F *and it is not the case that* $\mathfrak{B}$ *does not possess* F ≫—is what proves Theorem 1.1.

Note that by slightly changing our argument, we could have obtained the "stronger" and, in a certain respect, more natural contradiction ≪ $\mathfrak{B}$ *possesses* F *and it is not the case that* $\mathfrak{B}$ *possessesF* ≫. But this is immaterial for our purpose, so we shall not indulge in it.

2. Let us consider the following examples.

1°. The property F_P of a computable verbal function of ≪being applicable to a given fixed word P ≫ is invariant and non-trivial for any P. Consequently, it is undecidable.[2] (This example is an interesting supplement to Theorem §48.2.4.)

2°. The property of a function of ≪having a solvable problem of recognizing its applicability to words in the alphabet A≫ is invariant and, as can be seen without difficulty [§48.2.4], non-trivial. Consequently, it is undecidable.

3°. A function's property of ≪being complete with respect to A≫ is invariant and obviously non-trivial. Consequently, it is undecidable.

4°. A function's property of ≪being completable with respect to the alphabet A_0 ≫ is obviously invariant and non-trivial [§53.1.1]. Consequently, it is also undecidable.

[2]In examples 1°–5°, undecidability is understood in the sense defined in the statement of Theorem 1.1

In conclusion, we shall consider one more example, requiring, in our opinion, definite carefulness in our reasoning.

5°. Let $\mathfrak{B}$ be an arbitrary computable verbal function of one variable. We shall consider the property $G_{\mathfrak{B}}$ of ≪being a function equivalent to $\mathfrak{B}$ ≫ and show that it is undecidable.

This property is obviously invariant. Hence there arises a natural desire to apply Theorem 1.1 to it, all the more so since there is a function possessing the property $G_{\mathfrak{B}}$: the function $\mathfrak{B}$.

The situation is more complicated, however, with a function not possessing this property, i.e., with a function non-equivalent to $\mathfrak{B}$. In fact, because of Theorem §55.2.1, we can not expect to develop an effective method for finding such a function *in the general case*: such a method would generate a computable operator over the computable verbal functions without a fixed point.

And nevertheless, the following argument can be applied here:

If $\mathfrak{B}(\Lambda)$ is defined, then one can find a function not possessing the property $G_{\mathfrak{B}}$: as such a function, one may take, for example, the nowhere defined function from example §54.1.2°. And in the case when $\mathfrak{B}(\Lambda)$ is not defined, such a function can also be found: for example, one may take the function from example §54.1.1°.

Therefore, if the property $G_{\mathfrak{B}}$ were verifiable, then by virtue of Theorem 1.1, there would not exist a computable verbal function not possessing the property $G_{\mathfrak{B}}$. But by virtue of the argument just given, this would mean that: 1) $\mathfrak{B}(\Lambda)$ is not defined and 2) it is not the case that $\mathfrak{B}(\Lambda)$ is not defined. The contradiction we obtained shows that $G_{\mathfrak{B}}$ is undecidable. Q.E.D.

A mathematician sharing the belief in the firm nature of the "law of the excluded middle", having analyzed the cases ≪ $\mathfrak{B}(\Lambda)$ is defined≫ and ≪ $\mathfrak{B}(\Lambda)$ is not defined≫, would probably finish the proof on the undecidability of the property $G_{\mathfrak{B}}$ in the following manner: ≪Thus, if $\mathfrak{B}(\Lambda)$ is defined or $\mathfrak{B}(\Lambda)$ is not defined, then there exists a function not possessing the property $G_{\mathfrak{B}}$. But this implication's premise is true *by virtue of the law of the excluded middle.* Consequently, its conclusion is also true. Therefore, there exists a function not possessing the property $G_{\mathfrak{B}}$. Hence, this property is non-trivial, and so by virtue of Theorem 1.1, it is undecidable≫.

But why on earth must we believe in this "law's" firmness? It would seem, on the contrary, that example 1° shows that there exists no method allowing one to determine, starting with a function $\mathfrak{B}$, just which term of the disjunction

(1) ≪ $\mathfrak{B}(\Lambda)$ is defined or $\mathfrak{B}(\Lambda)$ is not defined ≫

holds,[3] but according to our agreement (see §7.1), this is exactly what is required in order to recognize the disjunction (1) as true. Yes, the impossibility of a normal algorithm recognizing the property F_{Λ} is asserted in example 1°. But the normalization principle [§27.6] leaves us no hope for the existence of any other effective method whatsoever of solving this problem. In the light of what has been said, the argument developed by us seems to be more careful, and hence also more convincing, than the argument of our imaginary colleague.

[3]Of course, they can not both be false simultaneously, since then a conjunction whose second term is a negation of its first term would be true.

CHAPTER VIII

THE IDENTITY PROBLEM FOR SEMIGROUPS (THUE'S PROBLEM)

The unsolvability of certain normal mass problems arising naturally within the theory of algorithms was established in Chapter VI. These results open up the possibility of turning to the establishment of the unsolvability of an entire series of mass algorithmic problems concerning such traditional branches of mathematics as algebra, topology and mathematical analysis. The problems considered in this connection should, naturally, be understood as normal mass problems, i.e., as problems of finding normal algorithms which would enable us to recognize one or another property of objects being studied. In this chapter, we shall consider one of the best known and most natural problems of this kind in algebra—the identity problem for semigroups. The central result of the chapter will actually consist in the construction of an example of a finitely presented (i.e., given in terms of a finite number of generating elements and defining relations) semigroup with an unsolvable identity problem.

The concept of a finitely presented semigroup admits a simple and natural constructivization in the form of the concept of an associative calculus. This transforms the identity problem under consideration into the equivalence problem for associative calculi. This problem had been formulated back in 1914 by the Norwegian mathematical Axel Thue [1]. Thue, himself, succeeded in solving it only for very special cases. Subsequent attempts to find a general solution to this problem, in spite of the efforts of well-known mathematicians, were unsuccessful, and the reason for this state of affairs later became clear: it turned out that concrete examples of associative calculi for which this problem in unsolvable can be constructed. The first examples of such calculi were simultaneously and independently published in 1947 by Markov [5], [6] and Post [2]. By the same token, the first example of an "intramathematical" unsolvable mass algorithmic problem was given. In the course of time, the unsolvability of other well-known problems of this kind was also established. In the first place, let us mention the identity problem for groups (Novikov [1], [2]), the homeomorphy problem for polyhedra (Markov [12], [13]) and Hilbert's 10th problem (Matiyasevich [2]).

The above-mentioned identity problem is settled in this chapter on the basis of the theory of normal algorithms presented in this book. Historically, however, the situation was somewhat different. Namely, the concept of a normal algorithm arose precisely in the process of a search for the simplest possible account of the unsolvability of this problem.

§57 Associative Calculi

1. Assume that the symbol ≪↔≫ is not a letter of the alphabet A. We shall then call every word in the alphabet A↔ of the form

$$T \leftrightarrow U, \tag{1}$$

where T and U are words in A, a *relation* in the alphabet A.

In particular,

$$\leftrightarrow \tag{2}$$

is a relation in A.

We shall call the word T the *left side of the relation* (1), and the word U—its *right side.*

2. Let us now define the concept of an ≪associative calculus≫. This concept will have a great deal in common with the concept of a normal algorithm. The essential difference will, however, consist in the fact that while every normal algorithm is a certain prescription to act in exactly this and that way and in such an order, an associative calculus will be not a prescription, but merely *permission* to perform such operations. In addition, there will ordinarily be several such permissible operations, and no restrictions will be imposed on the choice of one or another of them.

3. In analogy with a normal algorithm, every associative calculus will be connected to some alphabet. Similarly to the way in which the essential component part of a normal algorithm is its scheme, i.e., the system of its substitution formulas, the essential component part of each associative calculus will be its ≪defining system≫—a certain list of relations in the alphabet of the calculus. Unlike the situation for formulas in the scheme of a normal algorithm, the order in which relations occur in the defining system of an associative calculus will have no essential significance.

4. Suppose we have a (possibly empty) list of relations $\mathfrak{S}$ in an alphabet A. The following operations on words in A will be called *admissible operations with respect to* $\mathfrak{S}$:

1) substituting the right side of one of the relations belonging to $\mathfrak{S}$ in place of any occurrence of the left side of the same relation;

2) substituting the left side of one of the relations belonging to $\mathfrak{S}$ in place of any occurrence of the right side of the same relation.

By the associative calculus in the alphabet A, *defined by the system* $\mathfrak{S}$, we shall mean permission to perform in turn, starting with an arbitrary word P in A, any admissible operations with respect to $\mathfrak{S}$. All words which will be obtained in this manner, including the initial word P, will be called *equivalent to P* in the associative calculus under consideration.

Thus, every associative calculus is determined by some alphabet—the *alphabet of this calculus*—and a certain (possibly empty) system of relations in this alphabet—the *defining system of this calculus.* If the alphabet of an associate calculus is an extension of A, we shall say that it is an *associate calculus over the alphabet* A.[1]

5. Let $\mathfrak{A}$ be the associative calculus in the alphabet A, defined by the system of relations $\mathfrak{S}$. We shall say that the word Q in A is *adjacent* to the word P in the calculus $\mathfrak{S}$ if Q can be obtained from P as a result of a single admissible operation with respect to $\mathfrak{S}$. Assuming that the symbol $\ll \perp \gg$ is not a letter of the alphabet A, we shall use it for denoting adjacency of words in associative calculi in A. In what follows, $\mathfrak{A} : P \perp Q$ will mean that Q is adjacent to P in the associative calculus $\mathfrak{A}$.

5.1. *If P and Q are words in the alphabet of an associative calculus $\mathfrak{A}$, then $\mathfrak{A} : P \perp Q$ if and only if there exist words R, S, T, U, such that*

$$P \leftrightharpoons RTS \tag{1}$$

$$Q \leftrightharpoons RUS \tag{2}$$

and at least one of the relations

$$T \leftrightarrow U \tag{3}$$

$$U \leftrightarrow T \tag{4}$$

belongs to the defining system of $\mathfrak{A}$.

5.2. *If $\mathfrak{A} : P \perp Q$, then $\mathfrak{A} : Q \perp P$.*

5.3. *If $\mathfrak{A} : P \perp Q$, then $\mathfrak{A} : RP \perp RQ$ and $\mathfrak{A} : PR \perp QR$ for every word R in the alphabet of $\mathfrak{A}$.*

6. Let $\mathfrak{A}$ be an associative calculus in the alphabet A, and let γ be a letter not occurring in A. A γ-system S of words in A will be called an *$\mathfrak{A}$-sequence* if whenever a word X in A is the neighbor in S on the left with a word Y in the same alphabet, i.e., whenever the word $\gamma X \gamma Y \gamma$ occurs in S, we have $\mathfrak{A} : X \perp Y$.

Let P and Q be words in the alphabet A. We shall say that P is equivalent to Q in the calculus $\mathfrak{A}$ if it is possible to construct an $\mathfrak{A}$-sequence S connecting P to Q, i.e., such that P is the first and Q, the last term of S.

Assuming that the symbol $\ll \perp\!\!\!\perp \gg$ is not a letter in the alphabets of the associative calculi under consideration, we shall use this symbol for denoting equivalence in these calculi. In what follows,

$$\mathfrak{A} : P \perp\!\!\!\perp Q \tag{1}$$

will denote that Q is equivalent to P in the associative calculus $\mathfrak{A}$.

The following assertions are obvious:

---

[1] It is frequently called a Thue system. (Translator's note)

6.1. *If* $\mathfrak{A} : P \perp Q$, *then* $\mathfrak{A} : P \perp\!\!\!\perp Q$.

6.2. $\mathfrak{A} : P \perp\!\!\!\perp P$ *for every word* P *in the alphabet of the associative calculus* $\mathfrak{A}$.

6.3. *If* $\mathfrak{A} : P \perp\!\!\!\perp Q$ *then* $\mathfrak{A} : Q \perp\!\!\!\perp P$.

6.4. *If* $\mathfrak{A} : P \perp\!\!\!\perp Q$, *and* $\mathfrak{A} : Q \perp\!\!\!\perp R$, *then* $\mathfrak{A} : P \perp\!\!\!\perp R$.

The following assertion is easily proven with the aid of 5.3:

6.5. *If* $\mathfrak{A} : P \perp\!\!\!\perp Q$, *then* $\mathfrak{A} : RP \perp\!\!\!\perp RQ$ *and* $\mathfrak{A} : PR \perp\!\!\!\perp QR$ *for every word* R *in the alphabet of the calculus* $\mathfrak{A}$.

With the aid of 6.4 and 6.5, it is proven that

6.6. *If* $\mathfrak{A} : P \perp\!\!\!\perp Q$ *and* $\mathfrak{A} : R \perp\!\!\!\perp S$, *then* $\mathfrak{A} : PR \perp\!\!\!\perp QS$.

7. In the literature one finds another definition of associative calculi, essentially equivalent to the one formulated above. It consists in the following.

We shall again consider a system of relations $\mathfrak{S}$ in an alphabet A. *By the associative calculus in the alphabet* A, *defined by the system* $\mathfrak{S}$, we shall mean permission to perform in turn, starting with the relations of $\mathfrak{S}$ and the relation 1(2), the following operations, generating new relations in A:

1) passing from some already available relation 1(1) to the relation

$$\xi T \leftrightarrow \xi U, \tag{1}$$

where ξ is any letter of A;

2) passing from some already available relation 1(1) to the relation

$$T\xi \leftrightarrow U\xi \tag{2}$$

where ξ is any letter of A;

3) passing from some already available relations 1(1) and

$$T \leftrightarrow V \tag{3}$$

to the relation

$$U \leftrightarrow V. \tag{4}$$

In contrast to the definition of an associative calculus given in Section 4, here the totality of initial words is fixed. These are the relations of the system $\mathfrak{S}$ and the relation 1(2). The admissible operations are three operations on relations in A, "generating" (i.e., allowing us to obtain) new relations, starting with available ones. Two of them make it possible to get the relations (1) and (2), starting with an available relation 1(1). The third is applicable to a pair of relations, 1(1) and (3), with equal left sides. Its application to these relations yields the relation (4).

We shall say that a relation is *derivable* in the associative calculus $\mathfrak{A}$ if it can be obtained as a result of a successive application of the admissible operations just singled out, with the stated initial words. We shall then also say that it is a *consequence* of the system $\mathfrak{S}$.

We shall say that the words T and U are *equivalent* in the calculus $\mathfrak{A}$ if the relation 1(1) is derivable in $\mathfrak{A}$. We shall make use of our former notation 6(1) in order to express the equivalence of P and Q in $\mathfrak{A}$.

We thus have two distinct definitions of the associative calculus in the alphabet A, defined by the system of relations $\mathfrak{S}$. To each of them corresponds its own concept of the equivalence of words. It turns out, however, that these two concepts are equivalent, which is conveyed by the following theorem.

7.1. *Let $\mathfrak{S}$ be a system of relations in an alphabet* A*; let $\mathfrak{A}$ be the associative calculus defined by $\mathfrak{S}$ in the sense of Part 4; let $\mathfrak{B}$ be the associative calculus defined by $\mathfrak{S}$ in the sense just indicated. In order for the words P and Q to be equivalent in $\mathfrak{B}$, it is necessary and sufficient that they be equivalent in $\mathfrak{A}$.*

We leave the proof for the reader.

Since the concept of the equivalence of words may be regarded as basic in the theory of associative calculi, the two definitions of an associative calculus formulated above prove to be equivalent in a certain sense. In what follows, when speaking about ≪the associative calculus defined by the system of relations $\mathfrak{S}$ in the alphabet A≫, we shall (in the absence of appropriate stipulations) understand this term in the sense of Part 4.

8. The theory of associative calculi may be regarded as a certain constructive approach to so-called *associative systems*, which are often called *semigroups* in the literature.

In fact, let $\mathfrak{A}$ be an associative calculus in an alphabet A. By virtue of 6.2–6.4, we then have the possibility of applying the abstraction of identification to words in A, identifying words equivalent in $\mathfrak{A}$, i.e., considering words in A up to equivalence in $\mathfrak{A}$. Classes of words in A, regarded in this manner, will play the role of elements of the associative system to be constructed. For the sake of conciseness, we shall call them simply *elements*.

We can further define multiplication of elements as the operation corresponding to the juxtaposition of words representing these elements. The uniqueness of multiplication defined in such a manner is ensured by the fact that the Property 6.6 holds for equivalence; the associativity of multiplication is guaranteed by that of juxtaposition.

Thus, we obtain a certain associative system. We shall say that it is generated by the associative calculus $\mathfrak{A}$.

Associative systems generated by associative calculi will be called F-systems.[2]

It is not difficult to see that every F-system has a unit element. In fact, the element of an F-system, represented by the empty word, plays the role of its unit element.

It is not difficult to see further that every F-system has a finite set of generating element. In fact, let the F-system C be generated by the associative calculus $\mathfrak{A}$ in the alphabet A. As is easily seen, the elements of C represented by

[2] Here the letter F comes from the term "finitely presented system".

letters of A form a system of generating elements of the system C: every non-unit element of C can be represented in the form of a certain product (with possible repetitions of the factors) of these elements.

The elements of the F-system C, represented by the words T and U, are identical if and only if these words are equivalent in $\mathfrak{A}$. If we now understand the term ≪associative calculus≫ in the sense of Part 7, we may say that these words are identical if and only if the relation 1(1) is derivable in our calculus, i.e., is a consequence of $\mathfrak{S}$. This state of affairs may be briefly described by saying that $\mathfrak{S}$ is a complete system of relations in the generating elements of the associative system C.

9. In view of the fact that elements of an F-system are represented by words in the alphabet of an associative calculus generating it, there naturally arises the question of recognizing the identity of elements represented by two different words. These elements are identical if and only if the words representing them are equivalent in the generating associative calculus. Therefore, the recognition of identity for elements of F-systems reduces to the recognition of equivalence for words in associative calculi.

For any words P and Q in the alphabet of an associative calculus $\mathfrak{A}$, the singular problem of recognizing their equivalence in $\mathfrak{A}$ can be raised: is it true that

$$\mathfrak{A} : P \perp\!\!\perp Q?$$

To every given associative calculus $\mathfrak{A}$, there corresponds the class of singular problems of this kind. The mass algorithmic problem corresponding to this class consists in finding a single general constructive method, allowing one to determine for any two given words P and Q, whether they are equivalent in $\mathfrak{A}$. We shall call this problem the *equivalence problem for the calculus* $\mathfrak{A}$.

In view of the fact that the equivalence of two words in the associative calculus $\mathfrak{A}$, generating an F-system C, is equivalent to the identity of the elements of C represented by them, the equivalence problem for $\mathfrak{A}$ may also be regarded as the identity problem for C, i.e., as the problem of constructing a single general constructive method, allowing one to determine for any two elements of C, given by words representing them, whether they are identical. This is what determines the role of the *equivalence problem for an associative calculus* $\mathfrak{A}$ in the study of the associative system generated by this calculus.

Like every mass problem, the equivalence problem for an associative calculus $\mathfrak{A}$ admits a more precise definition in terms of the theory of algorithms—it can be formulated as a normal mass problem [§46]. For this we need only agree to a definite way of writing down the singular equivalence problems for words in the alphabet of $\mathfrak{A}$. Every such singular problem is determined by a pair of words in the alphabet of the calculus. Any symbol which is not a letter of this alphabet can be used for this. We shall assume that the asterisk $*$ is such a symbol, and shall write down the pair of words P and Q in the form

$$P * Q. \tag{1}$$

The equivalence problem for the associative calculus $\mathfrak{A}$ in the alphabet A is then posed as a normal mass problem in the following manner.

An algorithm over the alphabet A$*$ is sought, applicable to every word of the form (1), where P and Q are words in A, and changing it into the empty word if and only if $\mathfrak{A} : P \amalg Q$.

It is this problem which we shall call in what follows the *equivalence problem for the associative calculus* $\mathfrak{A}$.

In the next section, we shall show how to construct an associative calculus whose equivalence problem is unsolvable.

§58 Construction of an Associative Calculus with an Unsolvable Equivalence Problem

1. For the construction of an associative calculus with an unsolvable equivalence problem, we link associative calculi to normal algorithms by means of the following theorem (Markov [2], Chap. VI, §2.1.1):

1.1. *Let π, ρ, σ be letters which are different from each other and do not belong to the alphabet* A. *Then whatever be the normal algorithm $\mathfrak{A}$ in the alphabet* A, *an associative calculus $\mathfrak{B}$ over the alphabet* $A\pi\rho\sigma$ *can be constructed, such that the equality*

$$\mathfrak{A}(P) \simeq Q$$

will hold for the words P and Q in A *if and only if*

$$\mathfrak{B} : \pi\rho P\pi \amalg \pi\sigma Q\pi. \tag{1}$$

In fact, let the algorithm $\mathfrak{A}$ have the scheme

$$\{U_i \to V_i \quad (1 \leqslant i \leqslant N), \tag{2}$$

where the U_i are words in A, while each V_i is either also a word in A or else is obtained from a word in A by adjoining a dot to the left.[3] Let

$$Б \rightleftharpoons A\pi\rho\sigma.$$

We introduce the letters ρ_i $(2 \leqslant i \leqslant N)$, $\sigma_i (1 \leqslant i \leqslant N)$, distinct from each other and from the letters of the alphabet Б; set

$$\rho_1 \rightleftharpoons \rho,$$
$$\sigma_{N+1} \rightleftharpoons \sigma.$$

Denote by P and C the alphabets consisting of the letters ρ_i $(1 \leqslant i \leqslant N)$ and σ_i $(1 \leqslant i \leqslant N+1)$, respectively. Then we set

$$B \rightleftharpoons APC\pi$$

[3] We allow ourselves to speak about the final substitution formulas of the algorithm $\mathfrak{A}$ in this not too precise manner.

and construct the associative calculus $\mathfrak{B}$ in the alphabet B, defined by the following system of relations, written down in an abbreviated form:

$$\begin{cases} \rho_i \xi X \leftrightarrow \xi \rho_i X & (1 \leqslant i \leqslant N; \xi \text{ is a letter of A;} \\ & X \text{ is a word in A}; [\xi X^\delta = l_i; \xi X \neq U_i) \\ \rho_i X \pi \leftrightarrow \sigma_{i+1} X \pi & (1 \leqslant i \leqslant N; X \text{ is a word in A}; X^\delta < l_i) \\ \rho_i U_i \leftrightarrow \sigma_1 V_i & (1 \leqslant i \leqslant N; V_i \text{ is a word in A}) \\ \rho_i U_i \leftrightarrow \sigma_{N+1} V_i' & (1 \leqslant i \leqslant N; V_i \rightleftharpoons \cdot V_i') \\ \pi \sigma_i \leftrightarrow \pi \rho_i & (1 \leqslant i \leqslant N) \\ \xi \sigma_i \leftrightarrow \sigma_i \xi & (1 \leqslant i \leqslant N+1; \xi \text{ is a letter of A}). \end{cases}$$

Here

$$l_i = [U_i^\delta \quad (1 \leqslant i \leqslant N);$$

in the first two groups of relations, X is an arbitrary word in A satisfying the conditions written to the right in parentheses. Since these conditions ensure that the length of X is bounded above, the first two groups of relations contain a finite number of relations; moreover, all these relations can be explicitly written down. The relations of the third group correspond to the simple substitution formulas of the algorithm $\mathfrak{A}$; the relations of the fourth group—to the final substitution formulas. The third and fourth groups together contain a total of N relations. The last two groups of relations need no clarification.

We shall show that the associative calculus $\mathfrak{B}$, constructed in this way, satisfies the condition formulated in Theorem 1.1. For this we introduce certain auxiliary terms and prove some lemmas.

The letter ρ_i $(1 \leqslant i \leqslant N)$, σ_i $(1 \leqslant i \leqslant N+1)$ will be called *operative letters*. Words of the form

$$\pi Q \eta R \pi,$$

where Q and R are words in A and η is an operative letter, will be called *special words*. In other words, we shall call words in B special if they begin and end with the letter π, contain no other occurrences of π and contain exactly one occurrence of an operative letter.

The following properties of any relation in the system (3) is immediately perceived:

1.2. *The left(right) side of a relation contains one and only one occurrence of an operative letter.*

1.3. *The letter π either occurs in neither the left nor the right side of a relation, or else occurs once in each side. In the latter case, either both sides begin with this letter, or else they both end with it.*

1.2 and 1.3 imply

1.4. *Every word adjacent in $\mathfrak{B}$ to any special word is a special word itself.*

By virtue of §57.6, from this follows

1.5. *Every word equivalent in $\mathfrak{B}$ to any special word is a special word itself.*

Let P be a word in the alphabet B. Let us agree to say that the word Q is *adjacent on the right* to the word P if Q can be obtained from P as a result of substituting the right side of one of the relations of the system (3) in place of any occurrence of this relation's left side. We shall make use of the symbol ≪$\vdash$≫ for denoting adjacency on the right. The notation

$$P \vdash Q$$

will mean in the sequel that Q is adjacent on the right to P.

From a comparison of the definitions of adjacency of words in an associative calculus (§57.5) and adjacency on the right, the validity of the following assertion is immediately obvious:

1.6. $\mathfrak{B} : P \perp Q$ *if and only if one of the following holds:* $P \vdash Q$ *or* $Q \vdash P$.

We shall prove the following lemma:

1.7. *Not more than one word is adjacent on the right to a given special word.*

For the proof, it is obviously sufficient to establish that in any special word there occurs not more than one of the left sides of the system (3)'s relations, and not more than once.

Consider any special word T. By definition, $T \rightleftharpoons \pi Q \eta R \pi$, where Q and R are certain words in A and η is an operative letter. Two cases are possible: η is one of the letters ρ_i $(1 \leqslant i \leqslant N)$; η is one of the letters σ_i $(1 \leqslant i \leqslant N+1)$. We shall consider them separately.

a) Let $\eta \rightleftharpoons \rho_j$. Since $\sigma_i \not\rightleftharpoons \rho_j$ $(1 \leqslant i \leqslant N)$, none of the left sides of the relations in the fifth and sixth groups occurs in T. Since $\rho_i \not\rightleftharpoons \rho_j$ when $i \neq j$, among the left sides of the relations in the first four groups, only $\rho_j \xi X$ (ξ is a letter of A; X is a word in A; $[\xi X^\delta = l_j$; $\xi X \not\rightleftharpoons U_j)$ or $\rho_j X \pi$ (X is a word in A; $[X^\delta < l_j)$ or $\rho_j U_j$ can occur in T. The word $\rho_j U_j$ occurs in T only when R begins with the word U_j. A word of the form $\rho_j \xi X$ (ξ is a letter of A; X is a word in A; $[\xi X^\delta = l_j$; $\xi X \not\rightleftharpoons U_j)$ occurs in T only when R does not begin with the word U_j and $[R^\delta \geqslant l_j$; in this case, only one word of the kind indicated occurs in T—the one for which ξX is the initial segment of length l_j of the word R. A word of the form $\rho_j X \pi$ (X is a word in A; $[X^\delta < l_j)$ occurs in T only when $[R^\delta < l_j$; in this case, only one word of the kind indicated occurs in T—the one for which $X \rightleftharpoons R$. Since these three cases are pairwise incompatible, not more than one word of the kinds under consideration occurs in T. The uniqueness of its occurrence in T is obvious.

b) Let $\eta \rightleftharpoons \sigma_j$. Since $\rho_i \not\rightleftharpoons \sigma_j$ $(1 \leqslant i \leqslant N)$, none of the left sides of the relations in the first four groups occurs in T. As for the remaining relations, when $Q \rightleftharpoons \Lambda$ and $1 \leqslant j \leqslant N$, the left side $\pi \sigma_j$ of the jth relation in the fifth group obviously occurs in T and occurs there only once, while the left sides of the relations in the sixth group and the other relations in the fifth group do not occur in T; when $Q \rightleftharpoons \Lambda$ and $j = N+1$, none of the left sides of the system (3)'s relations occurs at all in T; finally, when $Q \not\rightleftharpoons \Lambda$, the left side $\xi \sigma_j$ of one of the relations in the sixth group occurs in T, occurs there only once, but the left sides of the other relations in the sixth group and the left sides of the relations in the fifth group do not occur in T.

With this the lemma is proven.

1.8. *If Q is a word in A, then no word is adjacent on the right to the word $\pi\sigma Q\pi$.*

In fact, none of the left sides of the system (3)'s relations occurs in the word $\pi\sigma_{N+1}Q\pi$, i.e., in $\pi\sigma Q\pi$.

We shall call a $\mathfrak{B}$-sequence S *proper* if the right adjacency $X \vdash Y$ holds for any X and Y such that X is the neighbor of Y on the left in S.

We shall say that P is *properly transformable* into Q if we can construct a proper $\mathfrak{B}$-sequence S, connecting P to Q and such that $[S^\circ \geqslant 2$.[4]

We shall say that P is *transformable into* Q if P is properly transformable into Q or $P \coloneqq Q$. In what follows, the notation

$$P \ltimes Q$$

will mean that P is properly transformable into Q; the notation

$$P \models Q$$

will mean that P is transformable into Q.

1.9. *If P and Q are words in the alphabet* A, *such that the equivalence*

$$\mathfrak{B} : \pi\rho P\pi \perp\!\!\!\perp \pi\sigma Q\pi$$

holds, then

$$\pi\rho P\pi \ltimes \pi\sigma Q\pi.$$

Indeed, let the lemma's hypothesis be satisfied. Then there exists a $\mathfrak{B}$-sequence S connecting $\pi\rho P\pi$ to $\pi\sigma Q\pi$. Since $\sigma \neq \rho$, we have $[S^\circ \geqslant 2$. Let X be the neighbor in S on the left of $\pi\sigma Q\pi$. Then according to 1.6, $X \vdash \pi\sigma Q\pi$ or $\pi\sigma Q\pi \vdash X$. But by virtue of 1.8, $\pi\sigma Q\pi \vdash X$ is impossible. Consequently,

$$X \vdash \pi\sigma Q\pi. \tag{4}$$

If for any words Y and Z, such that Y is the neighbor in S on the left of Z, we have

$$Y \vdash Z, \tag{5}$$

then S is a proper $\mathfrak{B}$-sequence, and by the definition of proper transformability,

$$\pi\rho P\pi \ltimes \pi\sigma Q\pi.$$

[4] See §24.4.

Suppose that not all the right adjacencies (5) hold in S. Using the Least Number Theorem, we easily construct $\mathfrak{B}$-sequences U and V and words Z and Y such that

(6) $$S \rightleftharpoons UY\gamma ZV,$$

(7) $$Z \vdash Y$$

and the $\mathfrak{B}$-sequence γZV is proper.

$\pi\sigma Q\pi$ is the last term of S, and hence also of γZV. Furthermore, it is easy to see that $Z \not\rightleftharpoons \pi\sigma Q\pi$ [(7), 1.8]. Consequently, the $\mathfrak{B}$-sequence V is non-empty. Let T be V's first term. Then

(8) $$V \rightleftharpoons \gamma TV',$$

(9) $$S \rightleftharpoons UY\gamma Z\gamma TV' \quad [(6), (8)]$$

and

(10) $$Z \vdash T.$$

The first term of the $\mathfrak{B}$-sequence S is a special word. Therefore, all the terms of S, including Z, are special [1.5]. From this we conclude on the basis of 1.7 that

(11) $$Y \rightleftharpoons T \quad [(7), (10)].$$

Consequently,

$$S \rightleftharpoons UT\gamma Z\gamma TV' \quad [(9), (11)],$$

and so the system

$$S_1 \rightleftharpoons UTV',$$

being, as is easily seen, a $\mathfrak{B}$-sequence, also connects $\pi\rho P\pi$ to $\pi\sigma Q\pi$. But

$$[S^\circ = [S_1^\circ + 2.$$

Thus, we have replaced the $\mathfrak{B}$-sequence S connecting $\pi\rho P\pi$ to $\pi\sigma Q\pi$ by the shorter $\mathfrak{B}$-sequence S_1. If this sequence still is not proper, then, acting analogously to the way we acted previously, we replace it by an even shorter $\mathfrak{B}$-sequence, S_2, connecting $\pi\rho P\pi$ to $\pi\sigma Q\pi$. Repeating this process of successively shortening a sequence, we finally arrive at a proper $\mathfrak{B}$-sequence connecting $\pi\rho P\pi$ to $\pi\sigma Q\pi$, and by the same token establish that $\pi\rho P\pi \ltimes \pi\sigma Q\pi$.

The lemma is proven.

In the obvious manner, we have

1.10. *If $P \ltimes Q$, then $\mathfrak{B} : P \amalg Q$.*

1.11. *Let S be a proper $\mathfrak{B}$-sequence whose first term is W, and last, $\pi\sigma Q\pi$, where Q is a word in A. Then every proper $\mathfrak{B}$-sequence with the first term W is an initial segment of S.*

Indeed, let the lemma's hypothesis be fulfilled. Then by virtue of 1.5, the word W is special and every $\mathfrak{B}$-sequence beginning with W consists of special words. In order to prove the lemma's conclusion, we shall make use of the method of induction on the construction of a proper $\mathfrak{B}$-sequence.

The basis of the induction is trivial: the sequence $\gamma W \gamma$ is indeed an initial segment of S.

Now let S_1 be a proper $\mathfrak{B}$-sequence with the first term W and the last term X; let

(12) $$X \vdash Y.$$

We shall show that if S_1 is an initial segment of S, then $S_1 Y \gamma$ also possesses this property.

Thus, assume that S_1 is an initial segment of S. Then

(13) $$S \rightleftharpoons S_1 R$$

for some R. Furthermore,

(14) $$S_1 \rightleftharpoons S_2 X \gamma,$$

where S_2 is a proper $\mathfrak{B}$-sequence. Hence

(15) $$S_2 \rightleftharpoons S_3 \gamma$$

for some S_3, and so

(16) $$S \rightleftharpoons S_3 \gamma X \gamma R \quad [(13), (14), (15)].$$

It is easy to see that

$$R \neq \Lambda,$$

since we would otherwise have

$$S \rightleftharpoons S_3 \gamma X \gamma \quad [(16)]$$

and

$$X \rightleftharpoons \pi \sigma Q \pi,$$

which contradicts (12) by virtue of 1.8.

γR is a $\mathfrak{B}$-sequence. Let Z be the first term of this sequence. Then

(17) $$\gamma R \rightleftharpoons \gamma Z \gamma R_1$$

for some R_1. Hence

(18) $$S \rightleftharpoons S_3 \gamma X \gamma Z \gamma R_1 \quad [(16), (17)],$$

and since S is a proper $\mathfrak{B}$-sequence, we have

$$X \vdash Z \quad [(18)],$$

which, together with (12) and by virtue of 1.7, yields

(19) $$Z \rightleftharpoons Y.$$

Consequently,

$$\begin{aligned} S &\rightleftharpoons S_3 \gamma X \gamma Y \gamma R_1 && [(18)] \\ &\rightleftharpoons S_1 Y \gamma R_1 && [(15), (14)]. \end{aligned}$$

Therefore, the $\mathfrak{B}$-sequence $S_1 Y \gamma$ is in fact an initial segment of S. Q.E.D.

1.12. *If the word U_i does not occur in a word P in the alphabet* A, *then*

$$\pi\rho_i P\pi \ltimes \pi\sigma_{i+1}P\pi.$$

Indeed, if $[P^\delta < l_i$, then the relation

$$\rho_i P\pi \leftrightarrow \sigma_{i+1}P\pi$$

belongs to the second group of the system (3)'s relations. Consequently, we then have

$$\pi\rho_i P\pi \vdash \pi\sigma_{i+1}P\pi$$

and so

$$\pi\rho_i P\pi \ltimes \pi\sigma_{i+1}P\pi.$$

Now let $[P^\delta \geqslant l_i$. Since U_i does not occur in P, we have $U_i \not\rightleftharpoons \Lambda$ and, therefore, $l_i > 0$. Hence $P \not\rightleftharpoons \Lambda$ in the given case. Take words Q and R such that

(20) $$P \rightleftharpoons QR$$

and

(21) $$[R^\delta = l_i - 1.$$

Let the letter ξ and the words Y, Z be such that

$$Q \rightleftharpoons Y\xi Z.$$

Then

$$[\xi Z R^\delta \geqslant l_i \quad [(21)]$$

and ξZR does not begin with the word U_i. Consequently, a word of the form $\rho_i\xi X$, which is the left side of one of the relations of the first group in the system (3), occurs in

(22) $$\pi Y\rho_i\xi ZR\pi.$$

Applying the admissible operation of the calculus $\mathfrak{B}$, corresponding to this relation, to the word (22), we obtain

(23) $$\pi Y\rho_i\xi ZR\pi \vdash \pi Y\xi\rho_i ZR\pi.$$

Now, making use of (23) for the inductive step, we can show without difficulty by the usual application of the method of induction that

(24) $$\pi\rho_i QR\pi \ltimes \pi Q\rho_i R\pi.$$

By virtue of (21), the relation

$$\rho_i R\pi \leftrightarrow \sigma_{i+1}R\pi$$

belongs to the second group of relations of the system (3) and, therefore,

$$(25) \qquad \pi Q \rho_i R \pi \vdash \pi Q \sigma_{i+1} R \pi.$$

Finally, bearing in mind the relations in the last group of the system (3), we obtain without difficulty by induction that

$$(26) \qquad \pi Q \sigma_{i+1} R \pi \ltimes \pi \sigma_{i+1} Q R \pi.$$

Thus,

$$\begin{aligned} \pi \rho_i R \pi &\ltimes \pi Q \rho_i R \pi && [(20), (24)] \\ &\vdash \pi Q \sigma_{i+1} R \pi && [(25)] \\ &\ltimes \pi \sigma_{i+1} Q R \pi && [(26)] \\ &\eqcirc \pi \sigma_{i+1} P \pi && [(20)]. \end{aligned}$$

Q.E.D.

1.13. *If none of the words* $U_i (1 \leqslant i \leqslant k-1; k \leqslant N+1)$ *occurs in a word* P *in the alphabet* A, *then*

$$\pi \rho_1 P \pi \ltimes \pi \sigma_k P \pi.$$

In fact, under this condition we have

$$(27) \qquad \pi \rho_i P \pi \ltimes \pi \sigma_{i+1} P \pi \ (1 \leqslant i < k) \quad [1.12].$$

Furthermore, by virtue of the relations in the fifth group of the scheme (3),

$$(28) \qquad \pi \sigma_{i+1} P \pi \vdash \pi \rho_{i+1} P \pi \ (1 \leqslant i < k-1).$$

Consequently,

$$\pi \rho_i P \pi \ltimes \pi \rho_{i+1} P \pi \ (1 \leqslant i < k-1) \quad [(27), (28)],$$

and so,

$$\begin{aligned} \pi \rho_1 P \pi &\ltimes \pi \rho_{k-1} P \pi \\ &\ltimes \pi \sigma_k P \pi \quad [(27)], \end{aligned}$$

from which follows what we are proving.

1.14. *If* P *is a word in* A *and* $\mathfrak{A} : P\rceil$, *then*

$$(29) \qquad \pi \rho P \pi \ltimes \pi \sigma P \pi.$$

Indeed, none of the words U_i $(1 \leqslant i \leqslant N)$ occurs in P in this case, and hence

$$\pi \rho_1 P \pi \ltimes \pi \sigma_{N+1} P \pi,$$

which coincides with (29), since $\rho_1 \eqcirc \rho$, $\sigma_{N+1} \eqcirc \sigma$.

1.15. *If $\mathfrak{A} : P \vdash Q$, then*

$$\pi\rho P\pi \ltimes \pi\rho Q\pi. \tag{30}$$

Indeed, let $\mathfrak{A} : P \vdash Q$. Then one of the simple formulas of the scheme (2) acts during the first step of the algorithm $\mathfrak{A}$'s performance when it is applied to the word P. Let this be the formula

$$U_k \to V_k. \tag{31}$$

Then the words U_i $(1 \leqslant i \leqslant k-1)$ do not occur in P, and V_k is a word in the alphabet A. Moreover, P is a word in A, since $\mathfrak{A}$ is an algorithm in A. Consequently, we may apply Lemma 1.13, according to which

$$\pi\rho_1 P\pi \ltimes \pi\sigma_k P\pi. \tag{32}$$

By virtue of the applicability of the formula (31) to the word P, its left side U_k occurs in P. Let $R * U_k * S$ be the first occurrence of U_k in P. Then

$$P \leftrightharpoons RU_kS, \tag{33}$$

$$Q \leftrightharpoons RV_kS. \tag{34}$$

If $U_k \not\leftrightharpoons \Lambda$, then

$$\pi\rho_k RU_kS\pi \models \pi R\rho_k U_kS\pi \tag{35}$$

[when R is non-empty, relations belonging to the first group of the system (3) are applied] and, therefore,

$$\pi\rho_k P\pi \models \pi R\rho_k U_kS\pi \quad [(35), (33)]. \tag{36}$$

But if $U_k \leftrightharpoons \Lambda$, then $R \leftrightharpoons \Lambda$, $S \leftrightharpoons P$ and so $\pi\rho_k P\pi \leftrightharpoons \pi R\rho_k U_kS\pi$, from which follows (36).

Thus, (36) holds in all cases.

The relation $\rho_k U_k \leftrightarrow \sigma_1 V_k$ belongs to the third group of the system (3), since the formula (31) is simple. Hence

$$\pi R\rho_k U_kS\pi \vdash \pi R\sigma_1 V_kS\pi. \tag{37}$$

Further, using relations in the sixth group of the system (3), we obtain

$$\begin{aligned} \pi R\sigma_1 V_kS\pi &\models \pi\sigma_1 RV_kS\pi \\ &\leftrightharpoons \pi\sigma_1 Q\pi \quad [(34)]. \end{aligned} \tag{38}$$

Finally, relations in the fifth group of the system (3) yield

$$\pi\sigma_k P\pi \vdash \pi\rho_k P\pi, \tag{39}$$

$$\pi\sigma_1 Q\pi \vdash \pi\rho_1 Q\pi. \tag{40}$$

Therefore,

$$\begin{aligned} \pi\rho_1 P\pi &\ltimes \pi\sigma_k P\pi && [(32)] \\ &\vdash \pi\rho_k P\pi && [(39)] \\ &\models \pi R\rho_k U_kS\pi && [(36)] \\ &\vdash \pi R\sigma_1 V_kS\pi && [(37)] \\ &\models \pi\sigma_1 Q\pi && [(38)] \\ &\vdash \pi\rho_1 Q\pi && [(40)], \end{aligned}$$

which is just what proves the proper transformability (30), since $\rho_1 \leftrightharpoons \rho$.

1.16. *If* $\mathfrak{A} : P \vdash\cdot Q$, *then*

$$\pi\rho P\pi \models \pi\sigma Q\pi. \tag{41}$$

Arguing as in the preceding proof, we obtain (32), (39) and (36), where k, R and S retain their previous meaning. Unlike the preceding proof, the formula (31) is now final, and so the fourth group of relations of the system (3) contains the relation

$$\rho_k U_k \leftrightarrow \sigma_{N+1} V_k', \tag{42}$$

where V_k' is a word such that $V_k \coloneqq \cdot V_k'$. The equality (33) holds as before, while instead of (34), we now have

$$Q \coloneqq R V_k' S. \tag{43}$$

Since the relation (42) belongs to the system (3), we have

$$\pi R \rho_k U_k S \pi \vdash \pi R \sigma_{N+1} V_k' S \pi. \tag{44}$$

Finally, relations in the sixth group of the system (3) yield

$$\begin{aligned} \pi R \sigma_{N+1} V_k' S \pi &\models \pi \sigma_{N+1} R V_k' S \pi \\ &\coloneqq \pi \sigma_{N+1} Q \pi \qquad [(43)]. \end{aligned} \tag{45}$$

Therefore,

$$\begin{aligned} \pi\rho_1 P\pi &\ltimes \pi R \rho_k U_k S \pi && [(32), (39), (36)] \\ &\vdash \pi R \sigma_{N+1} V_k' S \pi && [(44)] \\ &\models \pi \sigma_{N+1} Q \pi && [(45)], \end{aligned}$$

which is just what proves the proper transformability (41), since $\rho_1 \coloneqq \rho$, $\sigma_{N+1} \coloneqq \sigma$.

1.17. *If* $\mathfrak{A} : P \models Q$ *then* $\pi\rho P\pi \models \pi\rho Q\pi$.

This follows from 1.15.

1.18. *If* $\mathfrak{A}(P) \coloneqq Q$, *then the equivalence (1) holds.*

Indeed, let $\mathfrak{A}(P) \coloneqq Q$. Then one of the following two cases holds: $\mathfrak{A} : P \models Q]$ or $\mathfrak{A} : P \models\cdot Q$.

In the first case

$$\begin{aligned} \pi\rho P\pi &\models \pi\rho Q\pi && [1.17], \\ \pi\rho Q\pi &\ltimes \pi\sigma Q\pi && [1.14], \end{aligned}$$

whence

$$\pi\rho P\pi \ltimes \pi\sigma Q\pi. \tag{46}$$

Consequently, (1) holds [1.10].

In the second case, there is a word R such that

$$\mathfrak{A} : P \models R \vdash\cdot Q.$$

For it we have

$$\pi\rho P\pi \models \pi\rho R\pi \quad [1.17],$$
$$\pi\rho R\pi \ltimes \pi\sigma R\pi \quad [1.16],$$

from which also follows (46) and (1).

For the completion of the proof of Theorem 1.1, it obviously remains for us to prove the following assertion, which is the converse of Lemma 1.18.

1.19. *If the equivalence* (1) *holds, where P and Q are words in* A, *then* $\mathfrak{A}(P) \coloneqq Q$.

We shall prove this.

Let the equivalence (1) hold for the words P and Q in the alphabet A. Then $\pi\rho P\pi \ltimes \pi\sigma Q\pi$ [1.9], i.e., one can construct a $\mathfrak{B}$-sequence S connecting $\pi\rho P\pi$ to $\pi\sigma Q\pi$.

When the algorithm $\mathfrak{A}$ is applied to the word P, three cases are possible: $\mathfrak{A} : P]$, $\mathfrak{A} : P \vdash P_1$ for some P_1, $\mathfrak{A} : P \vdash\cdot R$ for some R.

If $\mathfrak{A} : P]$, then by virtue of 1.14, the proper transformability

$$\pi\rho P\pi \ltimes \pi\sigma P\pi$$

holds, and so there exists a proper $\mathfrak{B}$-sequence S_1 connecting $\pi\rho P\pi$ to $\pi\sigma P\pi$. By virtue of 1.11, S_1 is an initial segment of S, so that $S \coloneqq S_1 W$, where γW is a proper $\mathfrak{B}$-sequence. It is not hard to see that $W \coloneqq \Lambda$, since otherwise the proper $\mathfrak{B}$-sequence S would have the form

$$S_2\gamma\pi\sigma P\pi\gamma W_1\gamma\pi\sigma Q\pi\gamma,$$

which is obviously impossible [1.8]. Consequently,

$$S \coloneqq S_1,$$

but this implies

$$\pi\sigma Q\pi \coloneqq \pi\sigma P\pi,$$

and hence

$$Q \coloneqq P.$$

Furthermore, $\mathfrak{A}(P) \coloneqq P$, since $\mathfrak{A} : P]$. Consequently, $\mathfrak{A}(P) \coloneqq Q$, which is just what is being asserted.

If $\mathfrak{A} : P \vdash\cdot R$ for some R, then

$$\pi\rho P\pi \ltimes \pi\sigma R\pi$$

and, therefore, there exists a proper $\mathfrak{B}$-sequence S_1 connecting $\pi\rho P\pi$ to $\pi\sigma R\pi$. As in the preceding case, S_1 is an initial segment of S by virtue of 1.11, so that $S \leftrightharpoons S_1 W$, where γW is a proper $\mathfrak{B}$-sequence. Again we have $W \leftrightharpoons \Lambda$ by virtue of the same reasons and, therefore,

$$\pi\sigma R\pi \leftrightharpoons \pi\sigma Q\pi,$$

whence

$$R \leftrightharpoons Q.$$

Furthermore, $\mathfrak{A}(P) \leftrightharpoons R$, since $\mathfrak{A} : P \models\cdot R$. Consequently, $\mathfrak{A}(P) \leftrightharpoons Q$, which is just what is being asserted.

Finally, let $\mathfrak{A} : P \vdash P_1$ for some word P_1. Then by virtue of 1.15, we have

$$\pi\rho P\pi \ltimes \pi\rho P_1\pi$$

and, therefore, there exists a proper $\mathfrak{B}$-sequence S_1 connecting $\pi\rho P\pi$ to $\pi\rho P_1\pi$, such that $[S_1^\circ \geqslant 2$. By virtue of 1.11, S_1 is an initial segment of S, so that $S \leftrightharpoons S_1 W$, where γW is a proper $\mathfrak{B}$-sequence.

S connects $\pi\rho P\pi$ to $\pi\sigma Q\pi$, while S_1 connects $\pi\rho P\pi$ to $\pi\rho P_1\pi$. Hence

$$S \leftrightharpoons \gamma\pi\rho P\pi\gamma Z\gamma\pi\sigma Q\pi\gamma,$$
$$S_1 \leftrightharpoons \gamma\pi\rho P\pi\gamma Z_1\gamma\pi\rho P_1\pi\gamma$$

for some Z and Z_1. Assuming that $W \leftrightharpoons \Lambda$, i.e., that $S \leftrightharpoons S_1$, we get that $\rho \leftrightharpoons \sigma$, whereas $\rho \not\leftrightharpoons \sigma$ by our choice of these letters. Consequently, $W \not\leftrightharpoons \Lambda$ and, therefore, the volume of the $\mathfrak{B}$-sequence γW, i.e., the number of its terms, is greater than zero.

The last term of the sequence γW is the last term of the sequence S, i.e., the word $\pi\sigma Q\pi$. Thus, for some W_1, we have

$$S \leftrightharpoons \gamma\pi\rho P\pi\gamma Z_1\gamma\pi\rho P_1\pi\gamma W_1\gamma\pi\sigma Q\pi\gamma,$$

and so the $\mathfrak{B}$-sequence

$$T_1 \rightleftharpoons \gamma\pi\rho P_1\pi\gamma W_1\gamma\pi\sigma Q\pi\gamma$$

connects the word $\pi\rho P_1\pi$ to the word $\pi\sigma Q\pi$. It is easy to see that the volume of this sequence is strictly less than the volume of the sequence S.

The argument just carried out for the sequence S and the word P is applicable to the sequence T_1 and the word P_1. We conclude from this assertion that one of the two following cases is possible: either $\mathfrak{A}(P_1) \leftrightharpoons Q$ or else there is a word P_2 such that

$$\mathfrak{A} : P_1 \vdash P_2$$

and a proper $\mathfrak{B}$-sequence T_2 connecting the word $\pi\rho P_2\pi$ to the word $\pi\sigma Q\pi$ and such that

$$[T_2^\circ < [T_1^\circ.$$

In the first case, $\mathfrak{A}(P) \coloneqq Q$, since $\mathfrak{A} : P \vdash P_1$ and $\mathfrak{A}(P_1) \coloneqq Q$. In the second case, arguing analogously to the way we did previously, we convince ourselves that either $\mathfrak{A}(P_2) \coloneqq Q$ or else there is a word P_3 such that

$$\mathfrak{A} : P_2 \vdash P_3$$

and a proper $\mathfrak{B}$-sequence T_3 connecting the word $\pi\rho P_3\pi$ to the word $\pi\sigma Q\pi$ and such that

$$[T_3^\circ < [T_2^\circ.$$

This successive construction of words P_i and proper $\mathfrak{B}$-sequences T_i must break off no later than after $[S^\circ - 1$ steps on some word P_k for which

$$\mathfrak{A}(P_k) \coloneqq Q.$$

We then have

$$\mathfrak{A} : P \models P_k$$

and, therefore,

$$\mathfrak{A}(P) \coloneqq Q.$$

Q.E.D.

Thus, Theorem 1.1 is proven.

2. As usual, let A_0 denote the alphabet *ab* and let A_2 denote the alphabet *abcde*. Using Theorem 1.1, we shall prove the following theorem:

2.1. *One can construct an associative calculus $\mathfrak{B}$ over the alphabet A_2, satisfying the following condition: there is no normal algorithm over the alphabet A_0, changing into the empty word those and only those words P in A_0, for which the equivalence*

(1) $$\mathfrak{B} : dcPd \amalg ded$$

does not hold.

We shall in fact construct a normal algorithm $\mathfrak{W}_2$ in the alphabet A_0 according to §51.3.2, i.e., in such a way that the following condition is satisfied: there is no normal algorithm over A_0, annihilating, i.e., changing into the empty word, those and only those words in A_0 which $\mathfrak{W}_2$ does not annihilate. Let us apply Theorem 1.1 to the algorithm $\mathfrak{W}_2$, setting $A \coloneqq A_0$, $\rho \coloneqq c$, $\pi \coloneqq d$, $\sigma \coloneqq e$. In accordance with this theorem, we construct an associative calculus $\mathfrak{B}$ over the alphabet A_0cde, i.e., over A_2, satisfying the following condition: the equality

$$\mathfrak{W}_2(P) \coloneqq Q$$

holds for the words P and Q in A_0 if and only if

$$\mathfrak{B} : dcPd \perp\!\!\!\perp deQd.$$

This calculus is just what we are seeking.

Indeed, $\mathfrak{W}_2$ annihilates by construction those and only those words P in A_0 for which (1) holds. Now if some normal algorithm over A_0 annihilates those and only those words in A_0 for which (1) does not hold, then it annihilates those and only those words in A_0 which $\mathfrak{W}_2$ fails to annihilate. Such a normal algorithm over A_0 would, however, violate the construction of $\mathfrak{W}_2$. Consequently, a normal algorithm over A_0, changing into the empty word those and only those words in A_0 for which the equivalence (1) does not hold, cannot exist. Q.E.D.

Using Theorem §51.3.4 instead of Theorem §51.3.2 in an analogous manner, we obtain the following result:

2.2. *One can construct an associative calculus $\mathfrak{B}$ over the alphabet A_2, satisfying the following condition: there is no normal algorithm over A_0, applicable to every word in A_0 and changing into the empty word those and only those words P in A_0 for which the equivalence (1) holds.*

3. Starting from 2.1, we shall prove the following theorem:

3.1. *One can construct an associative calculus $\mathfrak{B}$ over the alphabet de, satisfying the following condition: there is no normal algorithm over $\mathfrak{B}$'s alphabet, changing those and only those words in this alphabet into the empty word, which are not equivalent in $\mathfrak{B}$ to the word ded.*

We shall in fact construct an associative calculus $\mathfrak{B}$ over the alphabet A_2 in accordance with 2.1, i.e., in such a way that the following condition holds: there is no normal algorithm over A_0, changing into the empty word those and only those words P in A_0 for which the equivalence 2(1) does not hold. We shall show that the calculus $\mathfrak{B}$ is one which we are seeking.

Indeed, let Б denote the alphabet of the calculus $\mathfrak{B}$. Then Б is an extension of A_2, and so Б is an extension of *de*, i.e., $\mathfrak{B}$ is a calculus over *de*. Suppose that $\mathfrak{H}$ is a normal algorithm over Б, changing into the empty word those and only those words in Б which are not equivalent in $\mathfrak{B}$ to the word *ded*.

Let us introduce the letter δ, not belonging to the alphabet Б, and construct a normal algorithm $\mathfrak{C}$ in the alphabet Бδ, presenting it by the scheme

$$\left\{\begin{array}{l} \delta\xi \to \xi\delta \\ \delta \to \cdot\, d \\ \quad \to dc\delta \end{array}\right.$$

(ξ runs through Б).

It is not hard to see that

(1) $$\mathfrak{C}(P) \equiv dcPd$$

for every word P in Б.

We now construct an algorithm $\mathfrak{K}$ as the normal composition of the algorithms $\mathfrak{C}$ and $\mathfrak{H}$:

$$(2) \qquad \mathfrak{K} \rightleftharpoons (\mathfrak{H} \circ \mathfrak{C}).$$

$\mathfrak{K}$ is a normal algorithm over Б and, therefore, over A_0.

$$\mathfrak{K}(P) \simeq \mathfrak{H}(\mathfrak{C}(P))(P \text{ is a word in Б}) \quad [(2)]$$
$$\simeq \mathfrak{H}(dcPd)(P \text{ is a word in Б}) \quad [(1)].$$

It follows from this that the algorithm $\mathfrak{K}$ annihilates the word P in the alphabet A_0 if and only if the algorithm $\mathfrak{H}$ annihilates the word $dcPd$. But $\mathfrak{H}$ annihilates this word if and only if it is not equivalent to the word *ded* in the calculus $\mathfrak{B}$. Therefore, the normal algorithm $\mathfrak{K}$ over the alphabet A_0 changes into the empty word those and only those words in this alphabet for which the equivalence 2(1) does not hold. Such an algorithm violates our construction of the calculus $\mathfrak{B}$. Consequently, there is no normal algorithm $\mathfrak{H}$ over Б, changing into the empty word those and only those words in Б which are not equivalent in $\mathfrak{B}$ to the word *ded*. Q.E.D.

Using Theorem 2.2 in an analogous manner, we obtain the following result:

3.2. *One can construct an associative calculus $\mathfrak{B}$ over the alphabet de, satisfying the following condition: there is no normal algorithm over $\mathfrak{B}$'s alphabet, applicable to every word in this alphabet and changing into the empty word those and only those words in it which are equivalent in $\mathfrak{B}$ to the word ded.*

4. The theorem just established, 3.2, asserts the possibility of constructing an associative calculus $\mathfrak{B}$ and some word B in its alphabet in such a way that the following normal mass problem proves unsolvable: construct a normal algorithm over the alphabet of the calculus $\mathfrak{B}$, applicable to every word in this alphabet and changing into the empty word those and only those words in it which are equivalent in $\mathfrak{B}$ to the word B. We shall call this problem the *problem of equivalence to the word B* in the calculus $\mathfrak{B}$. The following almost obvious theorem establishes a simple connection between problems of this type and previously defined equivalence problems for associative calculi [§57.9].

4.1. *If the equivalence problem for an associative calculus $\mathfrak{A}$ is solvable, then given any word R in its alphabet, the problem of equivalence to this word in the calculus $\mathfrak{A}$ is also solvable.*

Indeed, let A be the alphabet of the calculus $\mathfrak{A}$, let R be a word in A and let $\mathfrak{H}$ be a normal algorithm over A$*$, applicable to every word of the form $P*Q$, where P and Q are words in A, and changing such a word into the empty word if and only if

$$\mathfrak{A} : P \perp\!\!\!\perp Q.$$

We construct the normal algorithm $\mathfrak{B}$ over the alphabet A$*$ as the right joining algorithm of the word $*R$. We have

$$(1) \qquad \mathfrak{B}(P) = P * R. (P \text{ is a word in A}).$$

We construct the algorithm $\mathfrak{K}$ as the normal composition of the algorithms $\mathfrak{B}$ and $\mathfrak{H}$:

$$\mathfrak{K} \rightleftharpoons (\mathfrak{H} \circ \mathfrak{B}). \tag{2}$$

$\mathfrak{K}$ is a normal algorithm over A; moreover, for any word P in A,

$$\begin{aligned} \mathfrak{K}(P) &\simeq \mathfrak{H}(\mathfrak{B}(P))(P \text{ is a word in A}) \quad [(2)] \\ &\simeq \mathfrak{H}(P * R)(P \text{ is a word in A}) \quad [(1)]. \end{aligned} \tag{3}$$

Since the algorithm $\mathfrak{H}$ is applicable to every word of the form $P*Q$, where P and Q are words in A, the algorithm $\mathfrak{K}$ is applicable to every word in A [(3)]. It then follows from (3) that $\mathfrak{K}$ annihilates the word P in A if and only if $\mathfrak{H}$ annihilates the word $P*R$. But this occurs if and only if $\mathfrak{A} : P \perp\!\!\!\perp R$. Therefore, $\mathfrak{K}$ is a normal algorithm over A, applicable to every word in this alphabet and changing into the empty word those and only those words in it which are equivalent in $\mathfrak{A}$ to the word R. By the same token, the problem of equivalence to the word R in the calculus $\mathfrak{A}$ is solved and the theorem is proven.

As an immediate consequence of 4.1, we obtain

4.2. *Let the associative calculus $\mathfrak{A}$ be such that for some word in its alphabet, the problem of equivalence in $\mathfrak{A}$ to this word is unsolvable. Then the equivalence problem for $\mathfrak{A}$ is unsolvable.*

As a consequence of 3.2 and 4.2, we get

4.3. *One can construct an associative calculus whose equivalence problem is unsolvable.*

This theorem admits the following refinement:

4.4. *One can construct an associative calculus $\mathfrak{B}$ in some alphabet Б not containing an asterisk, satisfying the following condition: there is no normal algorithm over Б∗, changing into the empty word those and only those words of the form $P * Q$ (P and Q are words in Б) for which the equivalence*

$$\mathfrak{B} : P \perp\!\!\!\perp Q$$

does not hold.

The proof of this theorem is completely analogous to that of 4.3. One merely has to make use of 3.1 instead of 3.2 and apply the right joining algorithm of the word $*ded$, connecting the algorithms considered in 4.4 to those considered in 3.1.

Having in mind an interpretation of the equivalence problem for an associative calculus as the identity problem for the F-system generated by it [§57.9], we can interpret Theorem 4.3 as a theorem on the possibility of constructing an F-system with an unsolvable identity problem.

5. Only the possibility of constructing an associative calculus is asserted in Theorem 4.3, but the defining system for such a calculus is not explicitly written

down. True, such a system of relations can theoretically be obtained on the basis of all the foregoing. For that we would have to retrace the entire course of the reasoning which led us to Theorem 4.3, concretizing it in the sense of explicitly writing down the schemes of all normal algorithms employed. In particular, we would have to concretize the scheme of a modified universal algorithm $\mathfrak{W}$ for the alphabet A_1 in the role of A and the letter e in the role of δ [§48.2.1]; we would then have to construct the schemes of the normal algorithms $\mathfrak{W}_0$ [§48.2.2], $\mathfrak{W}_1$ [§48.3.1] and $\mathfrak{W}_2$ [§51.3.2]. Finally, starting from the scheme of the algorithm $\mathfrak{W}_2$ and applying the construction indicated in the proof of Theorem 1.1, we would then obtain the scheme of a required associative system. It is clear that as a result we would have obtained something quite unwieldy. But since the problem of an explicit and simplest possible presentation of an associative calculus with an unsolvable equivalence problem is of unquestionable interest, at least as a possible basis for carrying out other constructions, we should seek other ways of solving it.

One such way can be found in Chapter VI of Markov's monograph [2], where the associative calculus $\mathfrak{C}_1$ in the thirteen-letter alphabet

$$abcdefghijklm,$$

whose scheme can be written down in the following abbreviated form:

$$\begin{cases} \varsigma\eta \leftrightarrow \eta\varsigma \\ e\hat{\xi} \leftrightarrow \hat{\xi}e \\ e\xi \leftrightarrow \bar{\xi}e \\ \xi m \leftrightarrow \hat{\xi}m \\ h\xi\bar{\xi} \leftrightarrow \xi h \\ \xi g\hat{\xi} \leftrightarrow g\xi \\ df \leftrightarrow dh \\ fd \leftrightarrow gd \\ hc \leftrightarrow cg, \end{cases}$$

was constructed. Here ξ runs through the two-letter alphabet ab, ς, the four-letter alphabet $abcd$ and η, the five-letter alphabet $fijkl$; moreover,

$$\bar{a} \rightleftharpoons i, \quad \bar{b} \rightleftharpoons j,$$
$$\hat{a} \rightleftharpoons k, \quad \hat{b} \rightleftharpoons l.$$

The problem of equivalence to a certain fixed (true, very long) word, whose construction is indicated, is unsolvable in this calculus. The defining system of the calculus $\mathfrak{C}_1$ consists of 33 relations: the first line includes 20 relations, the next five, two each.

There, Markov also showed that if the four relations

$$\varsigma feam \leftrightarrow feam$$

(ς runs through the alphabet *abcd*) are added to $\mathfrak{C}_1$'s defining system, then this will yield a new associative calculus, $\mathfrak{C}_2$, whose problem of equivalence to the word *feam* will be unsolvable.

6. We present the following "record-breaking" results:

1°. Tseitin [1] (detailed presentation in [2]) proved the unsolvability of the problem of equivalence to a certain fixed word for the associative calculus in the five-letter alphabet *abcde* with the following simple defining system of relations:

$$\begin{cases} ac \leftrightarrow ca \\ ad \leftrightarrow da \\ bc \leftrightarrow cb \\ bd \leftrightarrow db \\ eca \leftrightarrow ce \\ edb \leftrightarrow de \\ cca \leftrightarrow ccae. \end{cases}$$

The associative calculus (in the same alphabet) with the somewhat more complicated defining system

$$\begin{cases} ac \leftrightarrow ca \\ ad \leftrightarrow da \\ bc \leftrightarrow cb \\ bd \leftrightarrow db \\ eca \leftrightarrow ce \\ edb \leftrightarrow de \\ cdca \leftrightarrow cdcae \\ caaa \leftrightarrow aaa \\ daaa \leftrightarrow aaa \end{cases}$$

has an unsolvable problem of equivalence to the word *aaa*.

2°. Matiyasevich [1] constructed an example of an associative calculus in a two-letter alphabet with a defining system of three relations, having an unsolvable problem of equivalence to a certain fixed word. Two of this calculus' three relations are simple. The lengths of the left and right sides of the third relation are equal to 304 and 608, respectively.

The question of the possibility of further decreasing the number of relations remains open for the time being.

3°. A relation $T \leftrightarrow U$ is called *non-cancellable* if the greatest common initial segment and the greatest common terminal segment of the words T and U are empty. Adyan [3] constructed an algorithm solving the equivalence problem for any associative calculus defined by a single non-cancellable relation. The analogous problem for the case of an arbitrary single relation remains unsolved so far.

7. An associative calculus $\mathfrak{A}$ is called a *group calculus* if one can construct a normal algorithm $\mathfrak{C}$ over the alphabet A of the calculus $\mathfrak{A}$, such that for any

word P in the alphabet A, $\mathfrak{C}(P)$ is defined, is a word in the very same alphabet and is such that the following equivalences hold:

$$\mathfrak{A} : P\mathfrak{C}(P) \perp\!\!\!\perp \Lambda$$

and

$$\mathfrak{A} : \mathfrak{C}(P)P \perp\!\!\!\perp \Lambda.$$

This concept yields a constructivization of the concept of a finitely presented group, i.e., a group given by a finite number of generating elements and defining relations. The algorithm $\mathfrak{C}$ figuring in the definition plays the role of the inverse operation.

Novikov [1], [2] constructed the first example of a group associative calculus with an unsolvable equivalence problem. By the same token, the well-known identity problem for groups, formulated back in 1912 by Dehn [1] was settled. Analogous examples were later constructed by Boone [1], Britton [1] and others. The record-breaking result with respect to the number of relations belongs to Borisov [1].

§59 The Problem of Equivalence to the Empty Word

1. A special case of the problem of equivalence to a given word in a given associative calculus [§58.4] is the problem of equivalence to the empty word in this calculus, formulated as follows: for a given associative calculus $\mathfrak{A}$, construct a normal algorithm applicable to every word in the calculus' alphabet and changing into the empty word those and only those words in this alphabet, which are equivalent in $\mathfrak{A}$ to the empty word. We shall see that an associative calculus $\mathfrak{A}$ can be constructed in such a manner that this problem proves unsolvable.

In order to establish this result, we shall show that each problem of equivalence to any given word C in any given associative calculus $\mathfrak{C}$ can always be reduced in a definite way to the problem of equivalence to the empty word in some other associative calculus. We prove the following theorem:

1.1. *Let $\mathfrak{C}$ be an associative calculus in an alphabet Γ, let C be a word in this alphabet, let α and β be letters which are different from each other and do not belong to Γ and let*

$$\text{Б} \rightleftharpoons \Gamma\alpha\beta.$$

We construct an associative calculus $\mathfrak{D}$ in the alphabet Д, defining it by means of the system of relations obtained by adding the relation

$$\alpha C\beta \leftrightarrow \tag{1}$$

to the defining system of the calculus $\mathfrak{C}$. Then for any word P in the alphabet Γ, the equivalence

$$\mathfrak{C} : P \perp\!\!\!\perp C \tag{2}$$

holds if and only if the equivalence

(3) $$\mathfrak{D} : \alpha P \beta \amalg \Lambda$$

holds.

First of all, it is easy to perceive that the equivalence (3) holds for a word P in Γ whenever the equivalence (2) holds for it.

In fact, since all the relations of $\mathfrak{C}$'s defining system belong to $\mathfrak{D}$'s defining system, we have

(4) $$\mathfrak{D} : P \amalg C,$$

(5) $$\mathfrak{D} : \alpha P \beta \amalg \alpha C \beta \quad [(4), §57.6.5],$$

but since the relation (1) belongs to $\mathfrak{D}$'s defining system, we have

(6) $$\mathfrak{D} : \alpha C \beta \amalg \Lambda.$$

By virtue of (5) and (6), we have (3) [§57.6.4].

It remains to prove the converse, i.e., that (2) holds for a word P in Γ whenever (3) holds. Let us introduce some auxiliary concepts and notations for this.

Let us agree to define the *height of a word* Q as the number of occurrences of the letter α in this word, and the *height of a* $\mathfrak{D}$*-sequence* $\mathfrak{Q}$ as the greatest height of this sequence's terms. We shall denote the height of the word Q by the symbol $[Q^H$; the height of the sequence $\mathfrak{Q}$, by the symbol $[\mathfrak{Q}^H$. Let us agree to define the *spread of a sequence* $\mathfrak{S}$ as the number of those terms W of this sequence for which

$$[W^H = [\mathfrak{Q}^H.$$

We shall denote the spread of the sequence $\mathfrak{Q}$ of words by the symbol $[\mathfrak{Q}^S$.

It is obvious that the height of every sequence of words is an natural number, while the spread of every sequence of words is a positive natural number.

In the next lemma, we shall use the terminology introduced in §57.6.

1.2. *If* $\mathfrak{Q}$ *is a* $\mathfrak{D}$*-sequence connecting* V *to* W *and*

(7) $$[V^H < [\mathfrak{Q}^H,$$

(8) $$[W^H < [\mathfrak{Q}^H,$$

then one can construct a $\mathfrak{D}$*-sequence,* $\mathfrak{R}$*, connecting* V *to* W *and such that either*

(9) $$[\mathfrak{R}^H < [\mathfrak{Q}^H$$

or else $[\mathfrak{R}^H = [\mathfrak{Q}^H$ *and* $[R^S < [\mathfrak{Q}^S$.

Indeed, let the $\mathfrak{D}$-sequence $\mathfrak{Q}$ connect V to W and satisfy the conditions (7) and (8). Let N be the volume of $\mathfrak{Q}$, i.e., the number of its terms. Denote the ith term[5] of the sequence $\mathfrak{Q}$ by means of $Q_i (1 \leqslant i \leqslant N)$. Then

(10) $$Q_1 \rightleftharpoons V,$$

(11) $$Q_N \rightleftharpoons W,$$

[5]See §24.8

since $\mathfrak{Q}$ connects V to W. Set

$$h \rightleftharpoons [\mathfrak{Q}^H. \tag{12}$$

According to the definition of a sequence's height, there are numbers i such that $[Q_i^H = h$. They all differ from one and from N, since

$$\begin{aligned} &[Q_1^H < h \quad [(7), (10), (12)], \\ &[Q_N^H < h \quad [(8), (11), (12)]. \end{aligned} \tag{13}$$

Let j denote the smallest of these numbers. Then $1 < j < N$ and, since

$$[Q_i^H \leqslant h \ (1 \leqslant i \leqslant N) \quad [(12)], \tag{14}$$

we have

$$[Q_{j-1}^H < h, \tag{15}$$

while

$$[Q_j^H = h. \tag{16}$$

By virtue of (13), there are numbers i such that $j < i \leqslant N$ and $[Q_i^H < h$. Let k denote the smallest of these numbers. Then $j < k \leqslant N$,

$$[Q_k^H < h, \tag{17}$$

$$[Q_i^H = h \ (j \leqslant i < k) \quad [(14), (16)]. \tag{18}$$

We have

$$\mathfrak{D} : Q_{j-1} \perp Q_j,$$

since $\mathfrak{Q}$ is a $\mathfrak{D}$-sequence. This means that Q_j can be obtained from Q_{j-1} as a result of substituting the right or left side of one of the relations of the calculus $\mathfrak{D}$'s defining system in place of some occurrence of the same relation's other side. Since

$$[Q_{j-1}^H < [Q_j^H \quad [(15), (16)],$$

this relation cannot belong to the defining system of the calculus $\mathfrak{C}$, and the step converting Q_{j-1} into Q_j is the substitution of the left side $\alpha C \beta$ of the relation (1) in place of this relation's right side, i.e., in place of the empty word. Therefore, for some words R and S in the alphabet $Д$,

$$Q_{j-1} \leftrightharpoons RS, \tag{19}$$

$$Q_j \leftrightharpoons R\alpha C\beta S. \tag{20}$$

The words Q_i $(j \leqslant i \leqslant k-1)$, having the same height h [(18)], are terms of the $\mathfrak{D}$-sequence $\mathfrak{Q}$. We have

$$\begin{aligned} &\mathfrak{D} : Q_{i-1} \perp Q_i \ (j < i < k), \\ &[Q_{i-1}^H = [Q_i^H \ (j < i < k). \end{aligned} \tag{21}$$

But since the substitution of the right or left side of the relation (1) in place of an occurrence of the left or right side, respectively, of this relation changes the height, Q_i is obtained from Q_{i-1} as result of a step corresponding to one of the relations of $\mathfrak{C}$'s defining system ($j < i < k$).

We shall show by induction on i that each of the words Q_i ($j \leqslant i < k$) can be represented in the form $R_i \alpha K_i \beta S_i$, where K_i is a word in Γ and R_i, S_i are words in Д; moreover,

$$[R_i^H = [R^H. \tag{22}$$

This follows from (20) for $i = j$: one should put

$$K_j \rightleftharpoons C, \tag{23}$$

$$R_j \rightleftharpoons R, \tag{24}$$

$$S_j \rightleftharpoons S. \tag{25}$$

Suppose that for some i satisfying the conditions $j < i < k - 1$, we have

$$Q_{i-1} \leftrightharpoons R_{i-1} \alpha K_{i-1} \beta S_{i-1}, \tag{26}$$

where K_{i-1} is a word in Γ and R_{i-1}, S_i are words in Д; moreover

$$[R_{i-1}^H = [R^H. \tag{27}$$

We shall show that then

$$Q_i \leftrightharpoons R_i \alpha K_i \beta S_i, \tag{28}$$

where K_i is word in Γ and R_i, S_i are words in Д; moreover, the equality (22) holds.

Q_i is obtained from Q_{i-1} as a result of substituting the right or left side, Y, of one of the relations in the defining system of the calculus $\mathfrak{C}$ in place of some occurrence of the other side, X, of the same relation. X is a word in the alphabet Γ, and so the letters α, β do not occur in X. In view of (26), Q_i has one of the following forms:

$$T \alpha K_{i-1} \beta S_{i-1} \tag{29}$$

(T is the result of substituting Y in place of an occurrence of the word X in R_{i-1}),

$$R_{i-1} \alpha T \beta S_{i-1} \tag{30}$$

(T is the result of substituting Y in place of an occurrence of the word X in K_{i-1}),

$$R_{i-1} \alpha K_{i-1} \beta T \tag{31}$$

(T is the result of substituting Y in place of an occurrence of the word X in S_{i-1}).

Set

$$R_i \rightleftharpoons T,\ K_i \rightleftharpoons K_{i-1},\ S_i \rightleftharpoons S_{i-1} \tag{32}$$

if Q_i has the form (29);

$$R_i \rightleftharpoons R_{i-1},\ K_i \rightleftharpoons T,\ S_i \rightleftharpoons S_{i-1} \tag{33}$$

if Q_i has the form (30);

$$R_i \rightleftharpoons R_{i-1},\ K_i \rightleftharpoons K_{i-1},\ S_i \rightleftharpoons T \tag{34}$$

if Q has the form (31). Then we shall have the equality (28) in all cases. In this equality, K_i either equals K_{i-1} or else is the result of substituting Y for X in K_i [(32)–(34), (30)]. Since Y is a word in Γ, K_i is a word in Γ, just like K_{i-1}. We can perceive in an analogous manner that R_i, S_i are words in Д, just like R_{i-1}, S_{i-1}; moreover,

$$\begin{aligned} [R_i^H &= [R_{i-1}^H \\ &= [R^H \quad [(27)]. \end{aligned}$$

By the same token, our assertion about the form of the words Q_i $(j \leqslant i < k)$ is proven.

Each of the words Q_i $(j \leqslant i < k)$ can obviously be represented in one of these forms uniquely; moreover, the argument given here shows that for each i satisfying the conditions $j < i < k$, we have one of the following three possibilities:

a) $\mathfrak{C}_1 : R_{i-1} \perp R_i, \quad K_{i-1} \leftrightharpoons K_i, \quad S_{i-1} \leftrightharpoons S_i;$

b) $R_{i-1} \leftrightharpoons R_i, \quad \mathfrak{C} : K_{i-1} \perp K_i, \quad S_{i-1} \leftrightharpoons S_i;$

c) $R_{i-1} \leftrightharpoons R_i, \quad K_{i-1} \leftrightharpoons K_i, \quad \mathfrak{C}_1 : S_{i-1} \perp S_i,$

where $\mathfrak{C}_1$ is the associative calculus in the alphabet Д, presented by the same system of relations as $\mathfrak{C}$.

Let us first suppose that for every i satisfying the conditions $j < i < k$, Case b) holds. Then

$$\begin{aligned} R_i &\leftrightharpoons R_j \\ &\leftrightharpoons R\ (j \leqslant i < k) \qquad [(24)], \end{aligned} \tag{35}$$

$$\begin{aligned} S_i &\leftrightharpoons S_j \\ &\leftrightharpoons S\ (j \leqslant i < k) \qquad [(25)], \end{aligned} \tag{36}$$

$$Q_i \leftrightharpoons R\alpha K_i \beta S\ (j \leqslant i < k) \quad [(28), (35), (36)]. \tag{37}$$

We have

$$\mathfrak{D} : Q_{k-1} \perp Q_k,$$

since $\mathfrak{Q}$ is a $\mathfrak{D}$-sequence. Here

$$[Q_k^H < [Q_{k-1}^H \quad [(17), (18)].$$

This means that Q_k is obtained from Q_{k-1} as a result of substituting the right side of the relation (1), i.e., the empty word, for some occurrence of this relation's left side. In view of the fact that $\alpha \neq \beta$ and neither α nor β occurs either in the word K_{k-1} in the alphabet Γ or in the word C in this alphabet, Q_k has one of the forms

$$T\alpha K_{k-1}\beta S$$

(T is the result of substituting Λ in place of an occurrence of the word $\alpha C\beta$ in R),

$$RS,$$

$$R\alpha K_{k-1}\beta T$$

(T is the result of substituting Λ for an occurrence of $\alpha C\beta$ in S). If

(38) $$Q_k \rightleftharpoons RS,$$

then $Q_k \rightleftharpoons Q_{j-1}$ [(38), (19)]. Let

$$\gamma Q_1\gamma X\gamma Q_{j-1}\gamma$$

be the initial segment of the sequence $\mathfrak{Q}$, connecting Q_1 to Q_{j-1}, and let

$$\gamma Q_{k+1}\gamma Y\gamma Q_N\gamma$$

be the terminal segment of this sequence, connecting Q_{k+1} to Q_N. Then the sequence

$$\gamma Q_1\gamma X\gamma Q_{j-1}\gamma Q_{k+1}\gamma Y\gamma Q_N\gamma$$

is, just like $\mathfrak{Q}$, a $\mathfrak{D}$-sequence, and it connects the same words as $\mathfrak{Q}$, i.e., it connects V to W. This is just the sequence which we shall take for $\mathfrak{R}$.

It is clear that

(39) $$[\mathfrak{R}^H \leqslant [\mathfrak{Q}^H$$

and that for

(40) $$[\mathfrak{R}^H = [\mathfrak{Q}^H$$

we have

(41) $$[\mathfrak{R}^S = [\mathfrak{Q}^S - (k-j) \quad [(18), (12), (17)],$$

(42) $$[\mathfrak{R}^S < [\mathfrak{Q}^S \quad [(41)].$$

Thus, in this case the sequence $\mathfrak{R}$ possesses all the required properties.

Let us now suppose that

$$Q_k \rightleftharpoons T\alpha K_{k-1}\beta S, \tag{43}$$

where T is the result of substituting the empty word for some occurrence of $\alpha C\beta$ in R. Then put

$$P_{j-1} \rightleftharpoons TS, \tag{44}$$

$$T_i \rightleftharpoons T\alpha K_i \beta S \ (j \leqslant i < k). \tag{45}$$

Let $\mathfrak{P}$ be the system of words whose volume is $k-j+1$ and whose ith term is the word P_{j+i-2}. We take the sequence

$$\gamma Q_1 \gamma X \gamma Q_{j-1} \mathfrak{P} Q_{k+1} \gamma Y \gamma Q_N \gamma$$

for $\mathfrak{R}$. Since $j > 1$ and

$$P_{k-1} \rightleftharpoons Q_k \quad [(45), (43)], \tag{46}$$

the sequence $\mathfrak{R}$ connects (even when $k = N$) the same words as $\mathfrak{Q}$, i.e., it connects V to W.

We further have

$$\mathfrak{D} : Q_r \perp Q_{r-1} \ (1 \leqslant r \leqslant j-2); \tag{47}$$

$$\mathfrak{D} : P_{k-1} \perp Q_{k+1}, \ \mathfrak{D} : Q_s \perp Q_{s+1} \ (k+1 \leqslant s \leqslant N-1), \tag{48}$$

since $\mathfrak{Q}$ is a $\mathfrak{D}$-sequence and the equality (46) holds;

$$\mathfrak{D} : R \perp T, \tag{49}$$

since T is the result of substituting the right side of the relation (1) in place of an occurrence of its left side in R;

$$\mathfrak{D} : Q_{j-1} \perp P_{j-1} \quad [(49), (19), (44), §57.5.3]. \tag{50}$$

By hypothesis, we also have

$$\mathfrak{C} : K_{i-1} \perp K_i \ (j < i < k),$$

whence

$$\mathfrak{D} : K_{i-1} \perp K_i \ (j < i < k), \tag{51}$$

$$\mathfrak{D} : P_{i-1} \perp P_i \ (j < i < k) \quad [(51), (45), §57.5.3]. \tag{52}$$

Finally, we have

$$P_j \rightleftharpoons T\alpha C\beta S \quad [(45), (23)],$$

from which, according to (44),

(53) $$\mathfrak{D} : P_{j-1} \perp P_j.$$

The adjacencies (47), (50), (53), (52), (48) show that $\mathfrak{R}$ is a $\mathfrak{D}$-sequence.
Comparing Q_i and P_i for $j \leqslant i < k$ and noting that

(54) $$[T^H < [R^H,$$

we perceive that

$$[P_i^H < [Q_i^H \quad [(45), (37), (54)],$$

whence

(55) $$[P_i^H < h \quad (j \leqslant i < k) \quad [(18)].$$

We also have

(56) $$\begin{aligned}[P_{j-1}^H &< [Q_{j-1}^H \quad [(44), (19), (54)]\\ &< h \qquad [(15)].\end{aligned}$$

Thus, the height of each term of the sequence $\mathfrak{P}$ is less than h, i.e., less than the height of the sequence $\mathfrak{P}$. Consequently,

$$[\mathfrak{R}^H \leqslant [\mathfrak{Q}^H.$$

Since passing from $\mathfrak{Q}$ to $\mathfrak{R}$ consists, according to (46), in replacing the $k-j$ words Q_i $(j \leqslant i \leqslant k-1)$ of height h [(18)] by the words P_i $(j-1 \leqslant i \leqslant k-2)$ of smaller height [(55), (56)], we have the inequality (42) when $[\mathfrak{R}^H = [\mathfrak{Q}^H$. This shows that the sequence $\mathfrak{R}$ possesses all the required properties.

The desired sequence $\mathfrak{R}$ is constructed in an entirely analogous manner in case

$$Q_k \rightleftharpoons R\alpha K_{k-1}\beta T,$$

where T is the result of substituting the empty word for some occurrence of $\alpha C\beta$ in S.

We have therefore completed our analysis of the case when the condition b) is satisfied for every i such that $j < i < k$.

Let us now suppose that for some such i's, these conditions are not all satisfied, and let us denote the smallest of these i's by m. Then by what was proven above, either the condition a) or the condition c) holds for $i = m$, while the condition b) holds for every i such that $j < i < m$. The case when the condition a) holds for $i = m$ and the case when the condition c) holds for $i = m$ can be handled in an entirely analogous manner. It will therefore be sufficient to carry out the construction of a desired sequence R for the first of these cases.

Thus, let

(57) $$\mathfrak{C}_1 : R_{m-1} \perp R_m,$$

(58) $$K_{m-1} \rightleftharpoons K_m,$$

(59) $$S_{m-1} \rightleftharpoons S_m,$$

while the condition b) is satisfied for every i such that $j < i < m$. Moreover, m is a number such that $j < m < k$.

We then have

$$R_i \equiv R_j$$

(60) $$\equiv R \ (j \leqslant i < m) \qquad [(24)],$$

$$S_i \equiv S_j$$

(61) $$\equiv S \ (j \leqslant i \leqslant m) \qquad [(59), (25)],$$

(62) $$Q \equiv R\alpha K_i \beta S \ (j \leqslant i < m) \quad [(28), (60), (61)].$$

Set

(63) $$P_{j-1} \rightleftharpoons P_m S,$$

(64) $$P_i \rightleftharpoons R_m \alpha K_i \beta S \ (j \leqslant i < m)$$

and take as $\mathfrak{R}$ the sequence whose ith term is Q_i for $1 \leqslant i < j$, P_{i-1} for $j \leqslant i < m+1$ and Q_i for $m+1 \leqslant i \leqslant N$.

Since $j > 1$ and $m < k \leqslant N$, the sequence $\mathfrak{R}$ connects the same words as $\mathfrak{Q}$, i.e., connects V to W.

We have

(65) $$\mathfrak{D} : Q_i \perp Q_{i+1} \qquad (1 \leqslant i < j-1),$$

(66) $$\mathfrak{D} : P_{m-1} \perp Q_{m+1}, \ \mathfrak{D} : Q_i \perp Q_{i+1} \quad (m+1 \leqslant i \leqslant N-1),$$

since $\mathfrak{Q}$ is a $\mathfrak{D}$-sequence and

$$P_{m-1} \equiv Q_m \quad [(64), (58), (61), (28)].$$

We further have

$$\mathfrak{C} : K_{i-1} \perp K_i \quad (j < i < m),$$

whence

(67) $$\mathfrak{D} : K_{i-1} \perp K_i \quad (j < i < m),$$

(68) $$\mathfrak{D} : P_{i-1} \perp P_i \quad (j < i < m) \quad [(64), (67), §57.5.3].$$

We also have

(69) $$P_j \equiv R_m \alpha C \beta S \quad [(64), (23)],$$

(70) $$\mathfrak{D} : P_{j-1} \perp P_j \qquad [(63), (69)].$$

Finally, we have

(71) $$\mathfrak{C}_1 : R \perp R_m \quad [(57), (60)],$$

(72) $$\mathfrak{D} : R \perp R_m \quad [(71)],$$

(73) $$\mathfrak{D} : Q_{j-1} \perp P_{j-1} \quad [(72), (19), (63), §57.5.3].$$

The adjacencies (65), (73), (70), (68) and (66) show that $\mathfrak{R}$ is a $\mathfrak{D}$-sequence. We further have

$$
\begin{aligned}
(74) \qquad & [R_m^H = [R^H && [(22)], \\
& [P_{j-1}^H = [Q_{j-1}^H && [(74), (63), (19)] \\
(75) \qquad & \qquad < h && [(15)], \\
& [P_i^H = [Q_i^H \ (j \leqslant i < m) && [(74), (64), (62)] \\
(76) \qquad & \qquad = h && [(18)].
\end{aligned}
$$

It can be seen from a comparison of the sequences $\mathfrak{R}$ and $\mathfrak{Q}$ that $\mathfrak{R}$ is obtained from $\mathfrak{Q}$ as a result of replacing the words Q_i $(j \leqslant i \leqslant m)$ by the words P_i $(j-1 \leqslant i \leqslant m-1)$. Since the words Q_i $(j \leqslant i \leqslant m)$ have height h [(18)], but among the words P_i $(j-1 \leqslant i \leqslant m-1)$, the first has a height less than h [(75)], while the rest have height h [(76)], we have the inequality (39); moreover, in case of equality (40), we have

$$
\begin{aligned}
[\mathfrak{R}^S &= [\mathfrak{Q}^S - 1 \\
&< [\mathfrak{Q}^S.
\end{aligned}
$$

Therefore, the sequence we have constructed, $\mathfrak{R}$, possesses all the required properties. With this, the proof of Lemma 1.2 is completed.

1.3. *If the hypotheses of Lemma 1.2 hold, one can construct a $\mathfrak{D}$-sequence $\mathfrak{R}$ connecting V to W and satisfying the condition (9).*

Let, in fact, the hypotheses of Lemma 1.2 hold. We then construct, according to this lemma, a $\mathfrak{D}$-sequence $\mathfrak{R}_1$ connecting V to W and satisfying either the condition

$$(77) \qquad [\mathfrak{R}_1^H < [\mathfrak{Q}^H$$

or else the pair of conditions

$$(78) \qquad [\mathfrak{R}_1^H = [\mathfrak{Q}^H,$$

$$(79) \qquad [\mathfrak{R}_1^S < [\mathfrak{Q}^S.$$

If the condition (77) holds, set $\mathfrak{R} \rightleftharpoons \mathfrak{R}_1$; the sequence $\mathfrak{R}$ obviously possesses the required properties.

But if the conditions (78) and (79) hold, then apply Lemma 1.2 to the words V, W and the sequence $\mathfrak{R}_1$ (in the role of $\mathfrak{Q}$). This is possible, since

$$
\begin{aligned}
&[V^H < [\mathfrak{R}_1^H \quad [(7), (78)], \\
&[W^H < [\mathfrak{R}_1^H \quad [(8), (78)].
\end{aligned}
$$

Construct, according to 1.2, a $\mathfrak{D}$-sequence $\mathfrak{R}_2$ connecting V to W and satisfying either the condition

$$(80) \qquad [\mathfrak{R}_2^H < [\mathfrak{R}_1^H$$

or else the conditions

$$[\mathfrak{R}_2^H = [\mathfrak{R}_1^H, \tag{81}$$
$$[\mathfrak{R}_2^S < [\mathfrak{R}_1^S. \tag{82}$$

If the condition (80) holds, set $\mathfrak{R} \rightleftharpoons \mathfrak{R}_2$; but if the conditions (81) and (82) hold, apply Lemma 1.2 again. If you keep on acting in this way, you will continue obtaining $\mathfrak{D}$-sequences connecting V to W and such that their heights are equal, while their spreads are strictly decreasing.

This process must obviously break off (after at most $[\mathfrak{Q}^S$ steps), but it can only break off when a $\mathfrak{D}$-sequence is obtained, connecting V to W and having a height less than that of $\mathfrak{Q}$. Such a $\mathfrak{D}$-sequence can therefore be constructed.

Q.E.D.

1.4. *If $\mathfrak{D} : V \amalg W$, then one can construct a $\mathfrak{D}$-sequence $\mathfrak{R}$ connecting V to W and satisfying at least one of the inequalities*

$$[V^H \geqslant [\mathfrak{R}^H, \tag{83}$$
$$[W^H \geqslant [\mathfrak{R}^H. \tag{84}$$

Indeed, let $\mathfrak{D} : V \amalg W$ Then there exists a $\mathfrak{D}$-sequence $\mathfrak{Q}$ connecting V to W.

If it satisfies the conditions (7) and (8), we construct, according to 1.3, a $\mathfrak{D}$-sequence $\mathfrak{Q}_1$ connecting V to W and such that $[\mathfrak{Q}_1^H < [\mathfrak{Q}^H$. If

$$[V^H < [\mathfrak{Q}_1^H$$

and

$$[W^H < [\mathfrak{Q}_1^H,$$

we construct, according to 1.3, a $\mathfrak{D}$-sequence $\mathfrak{Q}_2$ connecting V to W and such that $[\mathfrak{Q}_2^H < [\mathfrak{Q}_1^H$. This process of successively constructing $\mathfrak{D}$-sequences connecting V to W and having strictly decreasing heights must obviously break off (after at most $[\mathfrak{Q}^H$ steps) and this, according to 1.3, will happen only when we obtain a $\mathfrak{D}$-sequence $\mathfrak{Q}_n$ connecting V to W and such that at least one of the inequalities

$$[V^H < [\mathfrak{Q}_n^H,$$
$$[W^H < [\mathfrak{Q}_n^H$$

is satisfied. This sequence $\mathfrak{Q}_n$ may obviously be taken as $\mathfrak{R}$.

We can now complete the proof of Theorem 1.1.

Let us assume that the equivalence (3) i.e., equivalence

$$\mathfrak{D} : \alpha P \beta \amalg \Lambda$$

holds for a word P in the alphabet Γ. We shall show that then the equivalence (2) i.e., equivalence

$$\mathfrak{C} : P \perp\!\!\!\perp C$$

holds.

Since

(85) $$[\alpha P \beta^H = 1$$

and $[\Lambda^H = 0$, one can construct, according to 1.4, a $\mathfrak{D}$-sequence $\mathfrak{Q}$ connecting $\alpha P \beta$ to Λ and satisfying the condition

$$[\mathfrak{Q}^H \leqslant 1$$

i.e., the condition

(86) $$[Q_i^H \leqslant 1 \; (1 \leqslant i \leqslant N),$$

where Q_i is the ith term of the sequence $\mathfrak{Q}$ and N is its volume.

We have

(87) $$Q_1 \rightleftharpoons \alpha P \beta,$$

$$Q_N \rightleftharpoons \Lambda,$$

(89) $$[Q_N^H = 0 \qquad [(88)].$$

Let j denote the least number i for which $[Q_i^H = 0$. By virtue of (89), such numbers exist and $j \leqslant N$. By virtue of (85) and (87), $j > 1$. Consider the terms Q_i $(1 \leqslant i \leqslant j)$ of the $\mathfrak{D}$-sequence $\mathfrak{Q}$.

By definition of the number j, we have

(90) $$[Q_i^H = 1 \; (1 \leqslant i < j) \quad [(86)],$$

(91) $$[Q_j^H = 0.$$

If we keep on arguing as in the proof of Lemma 1.2, we shall convince ourselves that each word Q_i $(1 \leqslant i < j)$ can be uniquely represented in the form $R_i \alpha K_i \beta S_i$, where

(92) $$K_1 \rightleftharpoons P,$$

R_i, K_i and S_i are words in Γ and for every i $(1 < i < j)$, we have either $\mathfrak{C} : K_{i-1} \perp K_i$ or else $K_{i-1} \rightleftharpoons K_i$. Hence we have

(93) $$\mathfrak{C} : K_1 \perp\!\!\!\perp K_{j-1}$$

On the other hand, we have

$$Q_{j-1} \rightleftharpoons R_{j-1} \alpha K_{j-1} \beta S_{j-1},$$

$$\mathfrak{D} : Q_{j-1} \perp Q_j,$$

from which it follows by virtue of (90) and (91) that

(94) $$K_{j-1} \rightleftharpoons C.$$

By virtue of (92)–(94), the equivalence (2) holds. Q.E.D.

2. Theorems §58.3.1 and 1.1 make it possible to establish the following result (Markov [6]):

2.1. *One can construct an associative calculus $\mathfrak{D}$ satisfying the following condition: there is no normal algorithm over the alphabet of the calculus $\mathfrak{D}$, changing into the empty word those and only those words in this alphabet, which are not equivalent in $\mathfrak{D}$ to the empty word.*

Indeed, let us construct an associative calculus $\mathfrak{B}$ over the alphabet *de* according to §58.3.1. We apply Theorem 1.1 to $\mathfrak{B}$. In this connection, let the role of $\mathfrak{C}$ be played by $\mathfrak{B}$, the role of Γ—the alphabet of the calculus $\mathfrak{B}$, the role of C—the word *ded* in this alphabet, the roles of α and β—any two letters not belonging to Γ. According to 1.1, we construct an associative calculus $\mathfrak{D}$ in the alphabet Д, equal to $\Gamma\alpha\beta$, in such a way that the equivalence

$$\mathfrak{C} : P \perp\!\!\!\perp C$$

will hold for a word P in Γ if and only if the equivalence

$$\mathfrak{D} : \alpha P \beta \perp\!\!\!\perp \Lambda$$

will hold. The calculus $\mathfrak{D}$ will then satisfy the condition formulated in Theorem 2.1.

Suppose that $\mathfrak{H}$ really is a normal algorithm over Д annihilating those and only those words in Д which are not equivalent to the empty word in $\mathfrak{D}$.

Construct the normal algorithm $\mathfrak{F}$ in the alphabet Д with the scheme

$$\begin{cases} \beta\xi \to \xi\beta \\ \beta \to \cdot\beta \\ \quad \to \alpha\beta, \end{cases}$$

where ξ runs through the alphabet Γ.

It is clear that the graphic equality

(1) $$\mathfrak{F}(P) \eqcirc \alpha P \beta$$

holds for any word P in Γ.

Construct an algorithm $\mathfrak{K}$ as the normal composition of the algorithms $\mathfrak{F}$ and $\mathfrak{H}$:

(2) $$\mathfrak{K} \rightleftharpoons (\mathfrak{H} \circ \mathfrak{F}).$$

$\mathfrak{K}$ is a normal algorithm over Γ [(2)]; moreover, for any word P in Γ, we have

$$\begin{aligned} \mathfrak{K}(P) &\simeq \mathfrak{H}(\mathfrak{F}(P)) \quad [(2)] \\ &\simeq \mathfrak{H}(\alpha P \beta) \quad [(1)]. \end{aligned}$$

Consequently, $\mathfrak{K}$ annihilates the word P in Γ if and only if $\mathfrak{H}$ annihilates the word $\alpha P \beta$, i.e., if and only if $\alpha P \beta$ is not equivalent to the empty word in $\mathfrak{D}$. But according to the construction of the calculus $\mathfrak{D}$, the equivalence

$$\mathfrak{D} : \alpha P \beta \perp\!\!\!\perp \Lambda$$

holds if and only if the word P is not equivalent to *ded* in the calculus $\mathfrak{B}$. Therefore, $\mathfrak{K}$ annihilates those and only those words in Γ which are not equivalent to *ded* in $\mathfrak{B}$. Such a normal algorithm $\mathfrak{K}$ over Γ, however violates the construction of the calculus $\mathfrak{B}$. Consequently, there is no normal algorithm $\mathfrak{H}$ over Д, annihilating those and only those words in Д which are not equivalent to the empty word in $\mathfrak{D}$.

Analogously, with the aid of theorems §58.3.2 and 1.1, one establishes

2.2. *One can construct an associative calculus $\mathfrak{D}$ satisfying the following condition: there is no normal algorithm over the alphabet of the calculus $\mathfrak{D}$, applicable to every word in this alphabet and changing into the empty word those and only those words in it which are equivalent in $\mathfrak{D}$ to the empty word.*

3. Analogously to the way in which the equivalence problem in a given associative calculus $\mathfrak{A}$ can be interpreted as the identity problem in the F-system generated by it [§57.9], the problem of equivalence to a given word in the calculus $\mathfrak{A}$ can be interpreted as the problem of identity to a given element of this F-system. In particular, the problem of equivalence to the empty word in a calculus $\mathfrak{A}$ can be interpreted as the problem of identity to the unit element of the F-system generated by the calculus $\mathfrak{A}$, i.e., as the problem of constructing a normal algorithm recognizing the elements which are identical to the F-system's unit element by the words representing these elements. In accordance with this, the theorem just proven may be regarded as the establishment of the possibility of constructing an F-system with an unsolvable problem of identity to the unit element.

§60 The Method of Computable Invariants

We conclude this chapter with a presentation of a general method, proposed by Markov (see [14], [15]) and having a very natural application to the problems considered by us.

1. We shall hold some alphabet A fixed and consider *binary relations* between words in it. By this, as is customary, we shall mean two-place verbal predicates whose free variables assume words in the alphabet A as their values. For the sake of conciseness, we shall sometimes omit any reference to the alphabet and shall speak simply of ≪binary relations≫.

Our main interest in the sequel will be *binary relations, such as equivalence*, which are reflexive, symmetric and transitive. We shall confine ourselves to a consideration of precisely such relations.

2. Thus, let $\mathfrak{R}$ be any binary relation of the type under consideration. We shall call a normal algorithm $\mathfrak{I}$ over the alphabet A a *computable invariant* of this relation if $\mathfrak{I}$ is applicable to every word in A and changes any two words connected by this relation into one and the same word.

We shall say that the computable invariant $\mathfrak{I}$ *distinguishes* the words X and Y if

$$\mathfrak{I}(X) \neq \mathfrak{I}(Y).$$

We shall call two words X and Y *indistinguishable by invariants of the relation* $\mathfrak{R}$ if for any computable invariant $\mathfrak{I}$ of this relation, we have the graphic equality

$$\mathfrak{I}(X) = \mathfrak{I}(Y).$$

Of course, any two words connected by a relation $\mathfrak{R}$ are indistinguishable by this relation's invariants. The converse, as we shall later be convinced, is not always true.

3. To each binary relation $\mathfrak{R}$, there corresponds in a natural way the mass algorithmic problem $P_{\mathfrak{R}}$ of recognizing those pairs of words $X * Y$ for which X and Y are connected by the relation $\mathfrak{R}$.

It is comparatively easy to show that

3.1. *If the problem $P_{\mathfrak{R}}$ is solvable in the sense that one can find a normal algorithm solving it, then one can construct a computable invariant $\mathfrak{I}_{\mathfrak{R}}$ of the relation $\mathfrak{R}$, distinguishing any two words, X and Y, not connected by this relation.*

Leaving the construction's details to the reader, we restrict ourselves to presenting only the general plan of the construction required here.

We first arrange the words in the alphabet A in a computable sequence with the aid of some ≪good≫ method. (For example, we can write down words in the order of increasing length, and for equal lengths—in lexicographic order.)

We construct the invariant $\mathfrak{I}_{\mathfrak{R}}$ in such a way that it assumes natural numbers as its values and is a non-decreasing function of its variable.

We assign the number 0 to the first term of the sequence constructed by us of words in the alphabet A.

Now let N be a natural number greater than one, and let the values of $\mathfrak{I}_{\mathfrak{R}}$ be defined for all terms of the sequence with numbers less than N. We give the value of $\mathfrak{I}_{\mathfrak{R}}$ at the Nth term of the sequence in the following way. Using a (normal) algorithm solving the problems $P_{\mathfrak{R}}$, we determine whether there are any terms connected to the Nth by the relation $\mathfrak{R}$ among those terms preceding it in the sequence. If there are no such terms, we increase the value of $\mathfrak{I}_{\mathfrak{R}}$ by one in comparison with the previous values. But if there are such terms then we take the value $\mathfrak{I}_{\mathfrak{R}}$ at the first such term as its newly defined value.

Having presented the algorithm so constructed in the form of a normal algorithm (the possibility of doing so is suggested by the normalization principle), we then establish without any particular trouble that it is a computable invariant with respect to $\mathfrak{R}$ and that it possesses the property indicated in the statement of Theorem 3.1.

Therefore, in order to prove the unsolvability of the problem $P_{\mathfrak{R}}$, it is sufficient to find words X and Y which are not connected by the relation $\mathfrak{R}$, but are not distinguishable by its invariants.

4. To any associative calculus, there corresponds in a natural way a certain reflexive, symmetric and transitive relation—the relation of equivalence of words in this calculus—and so one can speak about this relation's computable invariants.

We shall now show that

4.1. *One can construct an associative calculus $\mathfrak{B}$ and words X, Y in its alphabet in such a way that the following conditions will hold:*

1°. *X and Y are non-equivalent in $\mathfrak{B}$.*

2°. *X and Y are indistinguishable by invariants of the relation of equivalence of words in $\mathfrak{B}$.*

In fact, take the normal algorithm $\mathfrak{G}$ constructed for the proof of Theorem §53.1.1 over the alphabet A_0 in such a way that 1) whenever $\mathfrak{G}$ is applicable to some word P in A_0, $\mathfrak{G}(P) \eqcirc a$ or $\mathfrak{G}(P) \eqcirc b$ and 2) $\mathfrak{G}$ is not completable with respect to A_0.

Let Γ be the alphabet of the algorithm $\mathfrak{G}$ and let the letters π, ρ, σ not belong to Γ and be distinct from each other. We apply Theorem §58.1.1 to $\mathfrak{G}$ and construct, according to this theorem, an associative calculus $\mathfrak{B}$ over the alphabet $\Gamma\pi\rho\sigma$ in such a way that the equality $\mathfrak{G}(P) \eqcirc Q$ for words P and Q in the alphabet Γ is fulfilled if and only if the equivalence

$$\mathfrak{B} : \pi\rho P\pi \perp\!\!\!\perp \pi\sigma Q\pi$$

holds.

Let us put

$$X \rightleftharpoons \pi\sigma a\pi, \tag{1}$$

$$Y \rightleftharpoons \pi\sigma\beta\pi \tag{2}$$

and show that the calculus $\mathfrak{B}$ and the words X and Y in its alphabet satisfy the conditions 1° and 2°.

Indeed, suppose that

$$\mathfrak{B} : X \perp\!\!\!\perp Y,$$

i.e., that

$$\mathfrak{B} : \pi\sigma a\pi \perp\!\!\!\perp \pi\sigma b\pi \quad [(1), (2)]. \tag{3}$$

Let P be an arbitrary word in A_0. Assume $\mathfrak{G}$ is applicable to P. Then by virtue of our choice of $\mathfrak{G}$, either $\mathfrak{G}(P) \eqcirc a$ or else $\mathfrak{G}(P) \eqcirc b$.

Suppose that $\mathfrak{G}(P) \eqcirc a$. Then

$$\mathfrak{B} : \pi\rho P\pi \perp\!\!\!\perp \pi\sigma a\pi. \tag{4}$$

But the equivalence (3) holds, and so

$$\mathfrak{B} : \pi\rho P\pi \perp\!\!\!\perp \pi\sigma b\pi \quad [(4), (3)].$$

But then $\mathfrak{G}(P) \eqcirc b$, and so $a \eqcirc b$, which is not true. Consequently, the equality $\mathfrak{G}(P) \eqcirc a$ is impossible.

Analogously, neither is the equality $\mathfrak{G}(P) \eqcirc b$ possible.

Hence, the algorithm $\mathfrak{G}$ is not applicable to any word P in A_0, but then it is completable with respect to A_0 [see §53.1.1°], which contradicts the choice of $\mathfrak{G}$. Therefore, the condition 1° of Theorem 4.1 is satisfied.

Let us now verify the condition 2°.

Let $\mathfrak{I}$ be any computable invariant of the relation of equivalence to the words in $\mathfrak{B}$, and let the words $\mathfrak{I}(\pi\sigma a\pi)$ and $\mathfrak{I}(\pi\sigma b\pi)$ be distinct. Using our combining theorems and, in particular, the Branching Theorem, we construct[6] a normal

[6] We leave the construction's details to the reader, counting on her or his accumulated experience.

algorithm $\mathfrak{L}$ over the alphabet A_0 in such a way that for any word P in this alphabet, the graphic[7] equality

$$\mathfrak{L}(P) \rightleftharpoons \begin{cases} a \text{ if } & \mathfrak{I}(\pi\rho P\pi) \rightleftharpoons \mathfrak{I}(\rho\sigma a\pi), \\ b \text{ if } & \mathfrak{I}(\pi\rho P\pi) \not\rightleftharpoons \mathfrak{I}(\pi\sigma a\pi). \end{cases} \tag{5}$$

is fulfilled. The algorithm $\mathfrak{L}$ is applicable to any word in A_0, and is therefore complete with respect to A_0.

Now let $\mathfrak{G}$ be applicable to P in A_0. Then either $\mathfrak{G}(P) \rightleftharpoons a$ or else $\mathfrak{G}(P) \rightleftharpoons b$. Suppose $\mathfrak{G}(P) \rightleftharpoons a$. Then

$$\mathfrak{B} : \pi\rho P\pi \perp\!\!\!\perp \pi\sigma a\pi,$$

and so

$$\mathfrak{I}(\pi\rho P\pi) \rightleftharpoons \mathfrak{I}(\pi\sigma a\pi), \tag{6}$$

but then $\mathfrak{L}(P) \rightleftharpoons a$ [(5), (6)] and $\mathfrak{L}(P) \rightleftharpoons \mathfrak{G}(P)$.

Suppose that $\mathfrak{G}(P) \rightleftharpoons b$. Then

$$\mathfrak{B} : \pi\rho P\pi \perp\!\!\!\perp \pi\sigma b\pi$$

and hence

$$\mathfrak{I}(\pi\rho P\pi) \rightleftharpoons \mathfrak{I}(\pi\sigma b\pi).$$

But $\mathfrak{I}(\pi\sigma b\pi) \not\rightleftharpoons \mathfrak{I}(\pi\sigma a\pi)$, and so

$$\mathfrak{I}(\pi\rho P\pi) \not\rightleftharpoons \mathfrak{I}(\pi\sigma a\pi), \tag{7}$$

so that $\mathfrak{L}(P) \rightleftharpoons b$ [(5), (7)] and $\mathfrak{L}(P) \rightleftharpoons \mathfrak{G}(P)$. Thus, the algorithm $\mathfrak{L}$ is a completion of $\mathfrak{G}$ with respect to A_0, which contradicts our choice of $\mathfrak{G}$.

Consequently , the words $\mathfrak{I}(\pi\sigma a\pi)$ and $\mathfrak{I}(\pi\sigma b\pi)$ cannot be distinct, and since the invariant $\mathfrak{I}$ is arbitrary, the words $\pi\sigma a\pi$ and $\pi\sigma b\pi$, i.e., X and Y, are indistinguishable by invariants of the relation of equivalence of words in $\mathfrak{B}$, which is just what remained to be proven.

REMARK: What was said at the end of Part 3 allows us to conclude that the equivalence problem for the calculus $\mathfrak{B}$ is unsolvable, and in this sense Theorem 4.1 is a strengthening of Theorem §58.4.3. True, it is not known whether this strengthening is essential. Nagorny [5] constructed an example of an enumerable, but undecidable [see §49.1] binary equivalence relation for words, such that any two words not connected by this relation are distinguishable by its invariants. The question of the possibility of obtaining an analogous result for equivalence relations in associative calculi remains open.

---

[7] We call attention to the fact that the algorithm $\mathfrak{I}$ is applicable to every word in the alphabet $\Gamma\pi\rho\sigma$.

§61 The Problems of Recognizing Properties of Associative Calculi

As we have already noted, from the algebraic point of view, associative calculi are *means* of presenting associative systems. Properties of associative systems, arising as a result of various sorts of algebraic classifications, generate the corresponding properties of associative calculi in a natural manner. It is natural to be interested in the mass algorithmic problems of recognizing these properties. In this section, we give a general result pertaining to the problems of recognizing these properties. Unfortunately, because of a lack of space, we shall be forced to make our presentation concise, restricting ourselves to references to sources in the literature containing complete proofs of the assertions we make.

1. Let $\mathfrak{A}$ and $\mathfrak{B}$ be associative calculi in the alphabets A and Б, respectively. We shall say that a normal algorithm $\mathfrak{C}$ over the alphabet $\text{A} \cup \text{Б}$ is a *homomorphism of* $\mathfrak{A}$ *into* $\mathfrak{B}$ if it satisfies the following conditions:

H.1. The algorithm $\mathfrak{C}$ changes each word in the alphabet A into some word in the alphabet Б.

H.2. If $\mathfrak{A} : P \perp\!\!\!\perp Q$, then $\mathfrak{B} : \mathfrak{C}(P) \perp\!\!\!\perp \mathfrak{C}(Q)$.

H.3. For any P and Q in A, the following equivalence holds:

$$\mathfrak{B} : \mathfrak{C}(PQ) \perp\!\!\!\perp \mathfrak{C}(P)\mathfrak{C}(Q).$$

The following theorem is easily proven:

1.1. *If $\mathfrak{C}$ is a homomorphism of $\mathfrak{A}$ into $\mathfrak{B}$ and $\mathfrak{D}$ is a homomorphism of $\mathfrak{B}$ into $\mathfrak{E}$, then the normal composition $(\mathfrak{D} \circ \mathfrak{C})$ is a homomorphism of $\mathfrak{A}$ into $\mathfrak{E}$.*

2. Let $\mathfrak{A}$ and $\mathfrak{B}$ be associative calculi in alphabets A and Б, respectively, and let $\mathfrak{C}$ be a homomorphism of $\mathfrak{A}$ into $\mathfrak{B}$. We shall say that $\mathfrak{C}$ is an *isomorphism of* $\mathfrak{A}$ *into* $\mathfrak{B}$ if the following condition is fulfilled:

H.4. if P and Q are words in A, such that $\mathfrak{B} : \mathfrak{C}(P) \perp\!\!\!\perp \mathfrak{C}(Q)$, then $\mathfrak{A} : P \perp\!\!\!\perp Q$.

It is easy to verify that we have the assertion

2.1. *If $\mathfrak{C}$ is an isomorphism of $\mathfrak{A}$ into $\mathfrak{B}$ and $\mathfrak{D}$ is an isomorphism of $\mathfrak{B}$ into $\mathfrak{E}$, then $(\mathfrak{D} \circ \mathfrak{C})$ is an isomorphism of $\mathfrak{A}$ into $\mathfrak{E}$.*

3. Let $\mathfrak{A}$ and $\mathfrak{B}$ be associative calculi in alphabets A and Б, respectively, and let $\mathfrak{C}$ be a *homomorphism* of $\mathfrak{A}$ into $\mathfrak{B}$. We shall say that $\mathfrak{C}$ is a *homomorphism of* $\mathfrak{A}$ *onto* $\mathfrak{B}$ if the following condition is fulfilled:

H.5. There is a normal algorithm $\mathfrak{G}$ over $\text{A} \cup \text{Б}$, changing each word in the alphabet Б into some word in the alphabet A and such that for any Q in Б, the following equivalence holds:

(1) $$\mathfrak{B} : \mathfrak{C}(\mathfrak{G}(Q)) \perp\!\!\!\perp Q.$$

3.1. *If $\mathfrak{C}$ is a homomorphism of $\mathfrak{A}$ onto $\mathfrak{B}$ and $\mathfrak{D}$ is a homomorphism of $\mathfrak{C}$ onto $\mathfrak{E}$, then $(\mathfrak{D} \circ \mathfrak{C})$ is a homomorphism of $\mathfrak{A}$ onto $\mathfrak{E}$.*

It is not hard to show that

3.2. *If $\mathfrak{C}$ is a homomorphism of $\mathfrak{A}$ onto $\mathfrak{B}$, then*

$$\mathfrak{B} : \mathfrak{C}(\Lambda) \perp\!\!\!\perp \Lambda.$$

4. By an *isomorphism of $\mathfrak{A}$ onto $\mathfrak{B}$*, we shall understand a homomorphism of $\mathfrak{A}$ onto $\mathfrak{B}$, which is simultaneously an isomorphism of $\mathfrak{A}$ into $\mathfrak{B}$.

From 2.1 and 3.1 follows our next assertion:

4.1. *If $\mathfrak{C}$ is an isomorphism of $\mathfrak{A}$ onto $\mathfrak{B}$ and $\mathfrak{D}$ is an isomorphism of $\mathfrak{B}$ onto $\mathfrak{E}$, then $(\mathfrak{D} \circ \mathfrak{C})$ is an isomorphism of $\mathfrak{A}$ onto $\mathfrak{E}$.*

It is obvious that

4.2. *The identity normal algorithm in the alphabet of an associative calculus $\mathfrak{A}$ is an isomorphism of $\mathfrak{A}$ onto itself.*

It is not difficult to verify the following assertion.

4.3. *Let $\mathfrak{A}$ and $\mathfrak{B}$ be associative calculi in alphabets* А *and* Б, *respectively, let $\mathfrak{C}$ be an isomorphism of $\mathfrak{A}$ onto $\mathfrak{B}$ and let $\mathfrak{G}$ be a normal algorithm over* А ∪ Б, *changing each word in the alphabet* Б *into some word in the alphabet* А *and satisfying the condition* 3(1). *Then $\mathfrak{G}$ is an isomorphism of $\mathfrak{B}$ onto $\mathfrak{A}$.*

5. Let $\mathfrak{A}$ and $\mathfrak{B}$ be associative calculi. We shall say that $\mathfrak{A}$ is *embeddable* in $\mathfrak{B}$ if there is an isomorphism of $\mathfrak{A}$ into $\mathfrak{B}$. We shall say that $\mathfrak{A}$ is *isomorphic* to $\mathfrak{B}$ if there is an isomorphism of $\mathfrak{A}$ onto $\mathfrak{B}$.

From 2.1 follows

5.1. *If a calculus $\mathfrak{A}$ is embeddable in $\mathfrak{B}$ and $\mathfrak{B}$ is embeddable in $\mathfrak{E}$, then $\mathfrak{A}$ is embeddable in $\mathfrak{E}$.*

Bearing in mind that every isomorphism of $\mathfrak{A}$ onto $\mathfrak{B}$ is an isomorphism of $\mathfrak{A}$ into $\mathfrak{B}$, we obtain

5.2. *If a calculus $\mathfrak{A}$ is isomorphic to $\mathfrak{B}$, then $\mathfrak{A}$ is embeddable in $\mathfrak{B}$.*

It follows from 4.2 that

5.3. *Every associative calculus is isomorphic to itself.*

It follows from 4.1 that

5.4. *If a calculus $\mathfrak{A}$ is isomorphic to $\mathfrak{B}$ and $\mathfrak{B}$ is isomorphic to $\mathfrak{E}$, then $\mathfrak{A}$ is isomorphic to $\mathfrak{E}$.*

With the aid of 4.3, it is not hard to verify that

5.5. *If a calculus $\mathfrak{A}$ is isomorphic to $\mathfrak{B}$, then $\mathfrak{B}$ is isomorphic to $\mathfrak{A}$.*

The following simple assertion will be helpful for what is to come:

5.6. *Let* A *be an arbitrary alphabet and let n be a positive natural number. Then each associative calculus in an n-letter alphabet is isomorphic to some calculus in an alphabet having no letters in common with* A *and also consisting of n letters.*

As is easily understood, after choosing a suitable alphabet, a desired calculus' defining system of relations can be obtained in the given case from the initial calculus' system of relations by a simple renaming of the letters occurring in it.

6. Let F be any property of associative calculi. We shall say that F is *invariant* if every associative calculus isomorphic to a calculus with the property F possesses it, itself. We shall say that F is *hereditary* if every associative calculus embeddable in a calculus with the property F possesses it, itself.

We obviously have

6.1. *Every hereditary property of associative calculi is invariant.*

Let us give some examples of hereditary properties.

We shall say that an associative calculus $\mathfrak{A}$ is *unitary* if every two words in its alphabet are equivalent in $\mathfrak{A}$ (or if—which is the same thing— every word in its alphabet is equivalent in $\mathfrak{A}$ to the empty word).

It can be proven without any trouble that

6.2. *Unitariness is a hereditary property of associative calculi.*

6.3. *A unitary associative calculus is embeddable in every other associative calculus.*

6.4. *Every unitary associative calculus is isomorphic to any other unitary associative calculus.*

6.5. *If $\mathfrak{A}$ is a unitary associative calculus, then for the unitariness of $\mathfrak{B}$, it is necessary and sufficient that $\mathfrak{B}$ be isomorphic to $\mathfrak{A}$.*

We shall say that an associative calculus $\mathfrak{A}$ is *finite* if there is a non-empty system of words in this calculus' alphabet, such that each word in its alphabet is equivalent in $\mathfrak{A}$ to one of the terms of this system.

It can be shown without any particular trouble that

6.6. *Finiteness is a hereditary property of associative calculi.*

We shall say that an associative calculus $\mathfrak{A}$ in an alphabet A is a *semigroup calculus*[8] if for any P, Q and R in the alphabet A, the following conditions are fulfilled:

1. if $\mathfrak{A} : PR \perp\!\!\!\perp QR$, then $\mathfrak{A} : P \perp\!\!\!\perp Q$;
2. if $\mathfrak{A} : RP \perp\!\!\!\perp RQ$, then $\mathfrak{A} : P \perp\!\!\!\perp Q$.

It is established with no particular trouble that

[8]Semigroup associative calculi are often called ≪finitely presented cancellation semigroups≫ in the literature.

6.7. *The property of being a semigroup associative calculus is a hereditary property of associative calculi.*

We shall say that an associative calculus $\mathfrak{A}$ in an alphabet A is a *group calculus* (see §58.7) if there is a normal algorithm $\mathfrak{F}$ over A, changing each word in the alphabet A into a word in this same alphabet and such that for every P in A the following equivalences hold:

$$\mathfrak{A} : P\mathfrak{F}(P) \perp\!\!\!\perp \Lambda$$

and

$$\mathfrak{A} : \mathfrak{F}(P)P \perp\!\!\!\perp \Lambda.$$

The property of being a group associative calculus is not a hereditary property of associative calculi. In fact, the calculus in the alphabet ab, presented by the system of relations

$$\begin{cases} ab \leftrightarrow \\ ba \leftrightarrow, \end{cases}$$

is, as can be seen without difficulty, a group calculus, but the associative calculus in the alphabet a presented by the empty system of relations, is embeddable in it and is not a group calculus.

It is not hard to show that

6.8. *The property of being a group associative calculus is invariant.*

6.9. *Every group calculus is a semigroup calculus.*

It follows from 6.9 and 6.7 that

6.10. *Every associative calculus embeddable in a group calculus is a semigroup calculus.*

With the aid of 5.1, one easily gets that

6.11. *The embeddability of an associative calculus in a group calculus is a hereditary property.*

Assume that $*$ is not a letter of the alphabet A. We shall call an associative calculus $\mathfrak{A}$ in the alphabet A *decidable* if there is a normal algorithm over A$*$, recognizing whether or not a word in this alphabet is of the form $P * Q$ where P and Q are words in A for which the equivalence

$$\mathfrak{A} : P \perp\!\!\!\perp Q$$

holds.

It is not difficult to show that

6.12. *The decidability of associative calculi is a hereditary property of associative calculi.*

We shall call an associative calculus $\mathfrak{A}$ Abelian if for any two words P and Q in its alphabet, we have the equivalence

$$\mathfrak{A} : PQ \perp\!\!\!\perp QP.$$

It is easily perceived that

6.13. *The property of being Abelian is a hereditary property of associative calculi.*

7. From the point of view of algebra, invariant properties of associative calculi are of major interest. The algebraicist will be inclined to identify isomorphic associative calculi, regarding them as merely different means of presenting one and the same "abstract associative system". Given such an approach, non-invariant properties of associative systems are not particularly interesting, being properties not of abstract associative systems, themselves, but only of the means of presenting them. Granted, however, that the method of presenting abstract associative systems by means of associative calculi is valuable, the algebraicist poses the problem of finding general methods for recognizing invariant properties of presentations.

To this problem there corresponds the following formulation of a normal mass problem.

Let F be a certain invariant property of associative calculi and let A be some alphabet. We introduce *representations* of associative calculi in A, similarly to the way in which this was done for normal algorithms in §42. It is required to construct a normal algorithm over a two-letter extension of the alphabet A (one letter is needed for separating the left sides of the relations from the right sides, and another one, in order to have the possibility of separating relations from each other), applicable to the representation $\mathfrak{A}^{\text{Rp}}$ of each associative calculus $\mathfrak{A}$ in A and changing $\mathfrak{A}^{\text{Rp}}$ into the empty word if and only if $\mathfrak{A}$ possesses the property F.

We shall call this normal mass problem the *problem of recognizing the property F for the alphabet* A.

The unsolvability of this problem was proven (Markov [7], [8]; see also [9]; a complete presentation is in the monograph [2], Chapter VI, §11) under quite general conditions imposed on F and A. Namely, the following theorem was proven:

7.1. *Let F be an invariant property of associative calculi. If there is a calculus with this property, as well as a calculus which is not embeddable in any calculus with this property, then one can find an alphabet for which the problem of recognizing the property F is unsolvable (in the sense that a normal algorithm sought in this problem cannot exist). If, moreover, the alphabet of a calculus with the property F consists of n letters, then for any alphabet with more than $n+3$ letters, the problem of recognizing the property F is unsolvable. In particular, if there is a unitary calculus with the property F, then for any alphabet with more than 3 letters, the problem of recognizing the property F is unsolvable.*

The reader will find a detailed proof of this theorem in the monograph [2] (Chap. VI, §11, Parts 8–11).

In applications to hereditary properties, Theorem 7.1 can be considerably simplified. Indeed, suppose that F is a hereditary property of associative calculi and that there is a calculus $\mathfrak{A}$ with the property F, as well as a calculus $\mathfrak{B}$ without it. Then the unitary associative calculus in the alphabet a, presented by the system of relations

$$\{a \leftrightarrow,$$

is embeddable in $\mathfrak{A}$ [6.3], and so by virtue of the fact that F is hereditary, it possesses this property. On the other hand, the calculus $\mathfrak{B}$, not possessing the property F, is not embeddable in any associative calculus with this property (once again by virtue of F's being hereditary). Consequently, the hypotheses of the third assertion in Theorem 7.1 are fulfilled, and so this assertion is applicable. We therefore obtain the following theorem:

7.2. *Let F be a hereditary property of associative calculi. If there is a calculus with the property F, as well as a calculus without it, then for every alphabet with more than 3 letters, the problem of recognizing the property F is unsolvable.*

Theorem 7.2 may be applied to the hereditary properties considered above. This yields the following result:

7.3. *For every alphabet containing more than three letters, the problems of recognizing unitariness, finiteness, being a semigroup calculus, embeddability in a group calculus, decidability, being Abelian are unsolvable.*

In fact, every unitary calculus possesses the six hereditary properties just listed. On the other hand, there is a non-finite calculus (the calculus in the alphabet a, presented by the empty system of relations) which, by the same token, is not unitary; there is a non-semigroup calculus (the calculus in the alphabet a, presented by the system consisting of the single relation $aa \leftrightarrow a$) which, by the same token, is not embeddable in a group calculus [6.10]; there is an undecidable calculus [§58.4.4] and, finally, there is a non-Abelian calculus (the calculus in the alphabet ab, presented by the empty system of relations).

Theorem 7.2 is not applicable to an associative calculus' property of being a group calculus, since this property is not hereditary. We nevertheless have

7.4. *For every alphabet containing more than three letters, the problem of recognizing the property of being a group calculus is unsolvable.*

Indeed, every unitary associative calculus, as can be seen without difficulty, is a group calculus. On the other hand, there is a calculus which is not embeddable in any group calculus. Consequently, the third assertion of Theorem 7.1 is applicable to the property of being a group calculus.

8. In conclusion, let us consider the *isomorphism problem for associative calculi*, consisting in seeking an algorithm allowing one to recognize, for any two associative calculi in given alphabets, whether or not they are isomorphic. This problem can be posed as a normal mass problem in the following manner.

Let A and Б be arbitrary alphabets. It is required to construct a normal algorithm applicable to every word of the form $\mathfrak{A}^{\mathrm{Rp}} * \mathfrak{B}^{\mathrm{Rp}}$, where $\mathfrak{A}$ and $\mathfrak{B}$ are associative calculi in the alphabets A and Б, respectively, and changing such a word into the empty word if and only if $\mathfrak{A}$ is isomorphic to $\mathfrak{B}$.

We shall call this problem the *isomorphism problem for the alphabets* A *and* Б.

The following assertion holds (Markov [8]; see detailed presentation in [2] on pp. 369–370).

8.1. *If at least one of the alphabets* A, Б *contains more than three letters, then the isomorphism problem for these alphabets is unsolvable.*

Indeed, suppose for definiteness that Б contains more than three letters. Construct a unitary calculus 𝔄 in A. According to 6.5, an associative calculus 𝔅 in Б is unitary if and only if 𝔄 is isomorphic to 𝔅. But the problem of recognizing unitariness for the alphabet Б is unsolvable, from which the assertion 8.1 of interest to us is easily derived.[9]

9. Being interested only in invariant properties of associative calculi, one can, along with *representations* of associative calculi, introduce their *records* in such a way that complete information about a calculus, itself, be contained in its record. Theorem 5.6 shows that in the case under consideration, it is important for us to know not the alphabet, itself, but only the number of letters of which it consists. But the representation of an associative calculus can be encoded by a word in a two-letter alphabet, using a method analogous to the one which we applied to normal algorithms in §45. The record of a calculus will consist precisely of these two components.

Speaking set-theoretically, the assertion 8.1 shows that the set of pairs of records of isomorphic associative calculi is *undecidable*. There arises the question of what can be said affirmatively about this set. It turns out that this set is *enumerable*. Namely, the following theorem is valid (Nagorny [4]).

9.1. *One can construct an algorithm applicable to the pair of records of associative calculi* 𝔄 *and* 𝔅 *if and only if* 𝔄 *is isomorphic to* 𝔅, *and in this case, changing this pair of records into the record of an isomorphism of* 𝔄 *onto* 𝔅.

The method applied for the proof of this assertion consists in the following. Finite lists of words of a certain simple type are considered. For any two associative calculi, the concept of their *T-isomorphism* is defined in such a manner that a T-isomorphism of calculi 𝔄 and 𝔅 is a list of the type indicated, and for any list of this type, one can verify algorithmically whether it is a T-isomorphism of the calculi 𝔄 and 𝔅. Moreover, it turns out that a T-isomorphism of 𝔄 and 𝔅 exists if and only if 𝔄 is isomorphic to 𝔅. Thus, writing down the lists of this type one after another and carrying out an appropriate verification, we will find a T-isomorphism of 𝔄 and 𝔅, if it exists, and then, starting with it, we will construct an isomorphism of 𝔄 onto 𝔅. But if 𝔄 is not isomorphic to 𝔅, then the process of checking out the lists under consideration will never come to an end.

This same device can be made use of in other similar situations. Thus, with its aid, one can prove the *enumerability* of the (undecidable by virtue of results given above) sets of unitary, Abelian and group associative calculi.

---

[9]E. I. Greendlinger [1] solved the "isomorphism problem" defined by imposing the following restriction on the associative calculi to be considered: every occurrence of a relation's (left or right) side in a relation's (left or right) side is both an initial occurrence and a terminal occurrence. In [2] she constructed an undecidable associative calculus satisfying this restriction. (Translator's note).

The set-theoretic terminology which we made use of in this part was not, of course, used here in any essential way, but rather out of consideration for convenience. The analysis we gave can be kept entirely within the bounds of constructive ideas.

10. It is natural to try to carry over the problems considered in Part 7 to group associative systems, which are, from the point of view of the algebraicist, a means of presenting certain special («finitely presented») groups. Results analogous to Theorem 7.1 were obtained by Adyan [1], [2] and Rabin [1] for such calculi.

CHAPTER IX

ALGORITHMS AND MATHEMATICAL ANALYSIS

The question of the effectiveness of the fundamental concepts and constructions of mathematical analysis, as is well-known, attracted the attention of a number of the most prominent mathematicians practically from the very beginning of research in the set-theoretical foundations of analysis. However, a discussion of these problems *in exact terms* became truly possible only after a precise concept of an algorithm had been formulated. It is interesting that the concept of a computable real number already appears in the title on the first, basic paper of Turing [1], in which was given the precise formulation of the concept of an algorithm that later received the name ≪Turing machine≫.

Ever since then, many questions of principle in the foundations of mathematical analysis have been clarified using tools from the theory of algorithms. The constructive content of a number of analysis' fundamental results, justifying many methods which are of practical value, were also revealed. Important investigations were carried out on the determination of initial data sufficient for actually finding the value of one or another unknown quantity. The complexity dependence between values of initial and resulting quantities were investigated. Finally, there was developed and realized a detailed program of constructing mathematical analysis on the basis of the theory of algorithms and constructive logic (so-called ≪constructive analysis≫), not appealing to conceptions from set-theory. These investigations acquire particular significance in connection with the needs of the general theory of computing methods. Unfortunately, the results of these investigations are practically unknown outside a narrow circle of specialists. Including in this monograph a number of fragments clarifying the significance of a precise concept of an algorithm for mathematical analysis and computational mathematics, we would like to attract the reader's attention to the range of questions just described.

A detailed presentation of the concepts and methods of constructive analysis, and also a comparison of them with the classical set-theoretic approach, can be found by the reader in, for example, Shanin's fundamental paper [2][1] and

[1]The volume of the ≪Proceedings of the Steklov Institute of Mathematics≫ in which this paper appears is devoted in its entirety to problems of constructive analysis.

Kushner's book [1].

§62 Constructive Real Numbers and Constructive Real Functions

We shall show in this section how the fundamental concepts of constructive mathematical analysis—the concept of a constructive (computable) real number and the concept of a constructive (computable) real function—can be defined on the basis of the concept of an algorithm (and with no recourse whatsoever to set-theoretic conceptions).

By P we denote the three-letter alphabet $\mathsf{I} - /$ in which we agreed to construct the rational numbers in §1. Let P^+ be any two-letter extension of this alphabet. As "arbitrary" normal algorithms over P, we shall take normal algorithms in this fixed alphabet P^+.

We shall call a normal algorithm $\mathfrak{A}$ in the alphabet P^+ a *constructive sequence of rational* (respectively *natural*) numbers if it is applicable to every natural number and changes it into some rational (respectively, natural) number. We shall call the number $\mathfrak{A}(N)$ the Nth *term of the sequence* $\mathfrak{A}$.

Let $\mathfrak{A}$ be a constructive sequence of rational numbers. We shall call a constructive sequence $\mathfrak{B}$ of natural numbers a *convergence regulator for the sequence* $\mathfrak{A}$ if for any natural numbers L, M and N, it follows from $M, N \geqslant \mathfrak{B}(L)$ that

$$|\mathfrak{A}(M) - \mathfrak{A}(N)| < 2^{-L}.$$

We shall call a constructive sequence of rational numbers *fundamental* if one can construct a convergence regulator for it.

We shall call every word of the form $\mathfrak{A}^{\mathrm{Rc}}\delta\mathfrak{A}^{\mathrm{Rc}}$, where $\mathfrak{A}$ is a fundamental constructive sequence of rational numbers, $\mathfrak{B}$ is a convergence regulator for it and δ is a separating symbol, a *constructive real number*.

2. We call the reader's attention to the fact that a constructive real number is a certain *word* in the alphabet $ab\delta$. This word contains information about two algorithms, the first of which generates a "good" (computable) fundamental sequence of rational numbers, while the second allows one to find, for any given degree of accuracy, a place in this sequence where the given degree of accuracy begins being guaranteed.

In its general outlines, this definition "mimics" Cantor's definition of a real number in terms of fundamental sequences of rational numbers, but differs from it advantageously as regards the possibility of effectively approximating real numbers by means of rationals.

Let us note (see §53.1 for comparison) that we define the concept of a *constructive real number independently,* without any reference to the general concept of a *real number*. The constructive continuum may therefore be considered quite independently, in isolation from the classical continuum (although, of course, it generates in a natural manner a certain ≪good≫ subset of the latter).

3. We shall call the two constructive real numbers $\mathfrak{A}_1^{\mathrm{Rc}}\delta\mathfrak{B}_1^{\mathrm{Rc}}$ and $\mathfrak{A}_2^{\mathrm{Rc}}\delta\mathfrak{B}_2^{\mathrm{Rc}}$ *equal* if for any natural number N,

$$|\mathfrak{A}_1(\mathfrak{B}_1(N)) - \mathfrak{A}_2(\mathfrak{B}_2(N))| \leqslant 2^{-(N-1)}.$$

For the relation of equality between constructive real numbers, we shall make use of the customary symbol«=».

In such a form, the concept of a constructive real number was first introduced in the literature by Shanin in [2]. There one can find a detailed presentation, with commentary on the history of the development of this and related concepts.

4. We now turn to a formulation of the definition of a constructive real function. For this we fix any two-letter extension Д of the alphabet $ab\gamma\delta$, in which we shall write out γ-systems of constructive real numbers. We shall call two such systems equal if they have an equal number of terms and if their terms with the same numbers are equal constructive real numbers.

We shall call a normal algorithm $\mathfrak{A}$ in the alphabet Д a constructive real function of N variables if for any two equal N-term systems, S and T, of constructive real numbers, whenever $\mathfrak{A}(S)$ is defined, $\mathfrak{A}(T)$ is also defined and $\mathfrak{A}(S)$, $\mathfrak{A}(T)$ are equal constructive real numbers.

Thus, a constructive real function assigns constructive real numbers to systems of constructive real numbers with the observance of a natural compatibility requirement: a function is simultaneously defined or undefined at two equal systems, and assumes equal values there in case it is defined.

This concept of a constructive real function, a constructive analogue of the concept (going back to Dirichlet) of a pointwise mapping, was introduced by Markov ([10], [11]). The history of this question and analysis of related concepts can be found in the paper of Shanin [2] which was already mentioned.

5. Let us introduce several more concepts which will be needed for what is to follow.

Let r be any fixed rational number. By $\mathfrak{A}_r$ we denote the normal algorithm in the alphabet P^+, such that for any natural number N,

$$\mathfrak{A}_r(N) \simeq r.$$

The algorithm $\mathfrak{A}_r$ is a constructive sequence of rational numbers and is easily seen to be fundamental, because the algorithm $\mathfrak{A}_0$ is a convergence regulator for $\mathfrak{A}_r$. Therefore, the word $\mathfrak{A}_r^{\mathrm{Rc}}\delta\mathfrak{A}_0^{\mathrm{Rc}}$ is a constructive real number. We shall call it the *real image of the rational number r*. As our notation for this number, we pick the symbol $\hat{r}$.

Let x be a constructive real number. We shall call it *rational*[2] if one can find a rational number r such that $x = \hat{r}$. We shall call x *irrational* if it is not rational, and *pseudorational* if it is not irrational. We shall write $x = r$ instead of $x = \hat{r}$ in the sequel.

It is easy to see that every rational constructive real number is pseudorational. We shall consider the question of the extent to which the converse is valid in Part 5 of the next section.

---

[2] We call the reader's attention to the fact that a rational constructive real number is not a rational number in the sense in which we defined it in §1.6. The distinction between rational numbers and rational real numbers is, of course, also drawn in classical analysis.

§63 Specker's Example

1. In this section, we present a result which has played a fundamental role in the foundations of analysis: Specker's example [1], which sheds light on the effectiveness of the operation of taking the least upper bound of a bounded set of numbers (within the framework of set-theoretical ideas, the existence of such a least upper bound is, as the reader well remembers from his analysis course, ensured by Weierstrass' well-known theorem).

Namely, we show that

1.1. *One can construct a bounded non-decreasing constructive sequence of rational numbers for which there is no convergence regulator.*

Several different proofs of this theorem are known at the present time. The one we give is close to Specker's original proof.

In order to present the conception behind Specker's construction as clearly as possible, we take the liberty of not constructing certain normal algorithms required in the course of the proof, but merely refer to the possibility of constructing them. Having armed her or himself with patience, the reader can, without any special effort, fill in the gaps which will therefore arise in our proof. Thus, let $\mathfrak{A}$ be any normal algorithm. We denote this algorithm's alphabet by means of A. Let us enumerate the words in A by any "good" method whatsoever (for example, let us order them according to length, and for equal lengths—lexicographically). We shall denote the word with the number N by the symbol P_N. It is clear that there are uniform constructive methods (algorithms) for finding the number of a given word and for re-establishing a word, starting with its number. The normalization principle §27.6 suggests that these algorithms can be presented in the form of normal algorithms (and this is actually so, as the reader can convince her or himself without assistance).

We shall now hold a natural number L fixed and apply the algorithm $\mathfrak{A}$ to the word P_L, computing in the course of this application the arithmetical function $f_{\mathfrak{A},L}$ whose values for an arbitrary natural number N are defined by the following conditions:

1) $f_{\mathfrak{A},L}(N) = 0$ if the process of applying $\mathfrak{A}$ to P_L has not ended in less than N steps;

2) $f_{\mathfrak{A},L}(N) = 1$ if the process of applying $\mathfrak{A}$ to P_L has ended in less than N steps.

It is easy to see that there is a uniform constructive method, an algorithm, for computing the values of this everywhere defined, monotone function $f_{\mathfrak{A},L}$ assuming the values 0 and 1. The normalization principle suggests that there then also exists a *normal algorithm* computing this function's values. We shall assume in what follows that some such normal algorithm $\mathfrak{F}_{\mathfrak{A},L}$ is held fixed.

It is easy to comprehend that

1.2. *The algorithm $\mathfrak{A}$ is applicable to the word P in* A *if and only if there exists a natural number N such that $\mathfrak{F}_{\mathfrak{A},L}(N) = 1$, where L is the number of the word P.*

For any natural number N, we now set

$$s_{\mathfrak{A}}(N) \rightleftharpoons \sum_{i=0}^{N} \frac{\mathfrak{F}_{\mathfrak{A},i}(N)}{2^i}.$$

The rational-valued function $s_{\mathfrak{A}}$ is defined for any natural number and there is an algorithm for computing its values. The normalization principle again suggests that the algorithm we are considering can be presented in the form of a normal algorithm. Assuming the construction required here to have been carried out, we denote the normal algorithm so obtained by means of $\mathfrak{S}_{\mathfrak{A}}$. Thus, for any natural N, the equality

$$\mathfrak{S}_{\mathfrak{A}}(N) = \sum_{i=0}^{N} \frac{\mathfrak{F}_{\mathfrak{A},i}(N)}{2^i}$$

holds. Hence, $\mathfrak{S}_{\mathfrak{A}}$ changes each natural number into a rational number. According to the Reduction Theorem (§41.7.1), we may assume that $\mathfrak{S}_{\mathfrak{A}}$ is a normal algorithm in P^+ (see §62.1). Therefore, $\mathfrak{S}_{\mathfrak{A}}$ is a constructive sequence of rational numbers.

A simple verification shows that for any N, the inequalities

$$\mathfrak{S}_{\mathfrak{A}}(N) \leqslant \mathfrak{S}_{\mathfrak{A}}(N+1)$$

and

$$0 \leqslant \mathfrak{S}_{\mathfrak{A}}(N) \leqslant \sum_{i=0}^{N} \frac{1}{2^i} < 2$$

are fulfilled.

We shall now establish that

1.3. *If there exists a convergence regulator for the sequence $\mathfrak{S}_{\mathfrak{A}}$, then the problem of recognizing the applicability of the algorithm $\mathfrak{A}$ to words in the alphabet* A *is solvable.*

Indeed, let $\mathfrak{B}$ be a convergence regulator for $\mathfrak{S}_{\mathfrak{A}}$. We construct a normal algorithm $\mathfrak{C}$ in the alphabet P^+ in such a way that for any N, the equality

$$\mathfrak{C}(N) \simeq \max(N, \mathfrak{B}(N))$$

is fulfilled (we again leave the details of the construction to the reader). Since for any N we have

$$\mathfrak{C}(N) \geqslant \mathfrak{B}(N),$$

$\mathfrak{C}$ is also a convergence regulator for $\mathfrak{S}_{\mathfrak{A}}$.

We now take any word P in A and let N be its number. Let $K \geqslant \mathfrak{C}(N)$. Then

$$|\mathfrak{S}_{\mathfrak{A}}(K) - \mathfrak{S}_{\mathfrak{A}}(\mathfrak{C}N))| < \frac{1}{2^N}$$

and, since the sequence $\mathfrak{S}_{\mathfrak{A}}$ is monotonic, the absolute value sign may be removed:

$$\mathfrak{S}_{\mathfrak{A}}(K) - \mathfrak{S}_{\mathfrak{A}}(\mathfrak{C}(N)) < \frac{1}{2^N}.$$

Consequently,

$$\sum_{i=0}^{K} \frac{\mathfrak{F}_{\mathfrak{A},i}(K)}{2^i} - \sum_{i=0}^{\mathfrak{C}(N)} \frac{\mathfrak{F}_{\mathfrak{A},i}(\mathfrak{C}(N))}{2^i} < \frac{1}{2^N},$$

and so

$$\sum_{i=0}^{\mathfrak{C}(N)} \frac{\mathfrak{F}_{\mathfrak{A},i}(K) - \mathfrak{F}_{\mathfrak{A},i}(\mathfrak{C}(N))}{2^i} < \frac{1}{2^N}.$$

All the summands of this sum are non-negative. Furthermore, $N \leqslant \mathfrak{C}(N)$ by virtue of our choice of $\mathfrak{C}$. Therefore

$$0 \leqslant \frac{\mathfrak{F}_{\mathfrak{A},N}(K) - \mathfrak{F}_{\mathfrak{A},N}(\mathfrak{C}(N))}{2^N} < \frac{1}{2^N}$$

and hence

$$0 \leqslant \mathfrak{F}_{\mathfrak{A},N}(K) - \mathfrak{F}_{\mathfrak{A},N}(\mathfrak{C}(N)) < 1.$$

Recalling that the function represented by the algorithm $\mathfrak{F}_{\mathfrak{A},N}$ is non-decreasing and assumes 0 and 1 as its values, we conclude from this that for any $K \geqslant \mathfrak{C}(N)$, the equality

$$\mathfrak{F}_{\mathfrak{A},N}(K) = \mathfrak{F}_{\mathfrak{A},N}(\mathfrak{C}(N))$$

holds.

We now compute $\mathfrak{F}_{\mathfrak{A},N}(\mathfrak{C}(N))$. If it turns out that $\mathfrak{F}_{\mathfrak{A},N}(\mathfrak{C}(N)) = 1$, then we can assert that the algorithm $\mathfrak{A}$ is applicable to the word P [1.2]. But if it turns out that $\mathfrak{F}_{\mathfrak{A},N}(\mathfrak{C}(N)) = 0$, then $\mathfrak{F}_{\mathfrak{A},N}(K) = 0$ for all K: for $K \geqslant \mathfrak{C}(N)$—by virtue of what was just proven; for$K < \mathfrak{C}(N)$—by virtue of the monotonicity of $\mathfrak{F}_{\mathfrak{A},N}$. Therefore, there exists no M in this case, such that $\mathfrak{F}_{\mathfrak{A},N}(M) = 1$, and so $\mathfrak{A}$ is not applicable to the word P [1.2].

Thus, we have developed an algorithm allowing one to determine, for an arbitrary given word P in A, whether or not the algorithm $\mathfrak{A}$ is applicable to P. Presenting this algorithm in the form of a normal algorithm, we complete the proof of the assertion 1.3.

Theorem 1.1 is now obtained without any trouble. Indeed, it is sufficient to take any normal algorithm $\mathfrak{A}$ with an unsolvable problem of recognizing applicability [§48.2.4] and form the constructive sequence $\mathfrak{S}_{\mathfrak{A}}$ of rational numbers corresponding to it.

2. Theorem 1.1 shows that so-called ≪pure existence theorems≫ of classical mathematical analysis can sometimes turn out to be very unreliable guides in conditions of real computational practice.

§64 The Problem of Recognizing Equality for Real Numbers

1. A mathematician adhering to the traditional set-theoretic point of view regards the disjunction

$$\ll x = 0 \text{ or it is not the case that } x = 0 \gg, \tag{1}$$

where x is an *arbitrary* real number, as automatically true. For her or him, it is a simple consequence of a general logical law—the ≪law of the excluded middle≫.

Nevertheless, the problem of actually determining, for a given x, precisely which term of the disjunction (1) is true, can turn out to be difficult even under very favorable conditions.

For example, let x be given in the form of an infinite decimal fraction, such that there is an *algorithm* allowing one to compute any of its digits (such an algorithm, of course, often fails to exist). Since we have no right to demand any further information whatsoever, nothing else comes to mind than to try computing x with continually increasing accuracy. But regardless of how many digits of x we may have computed, we are in no position to say anything about x if they are all equal to zero: it is quite possible that the next digit will turn out to be different from zero. Thus, if $x = 0$, this "direct" method does not lead to our goal. We shall attempt to show that neither is any other method capable of solving the problem in which we are interested.

But nevertheless, in many arguments and computational procedures used in analysis, one finds branching points where one is invited to "see" whether the value of a certain quantity is equal to zero and to argue or act accordingly, depending on this. A typical example is the plan for finding a root of a continuous function with alternating signs ≪method of bisecting a segment≫. Here one is invited from the very beginning to "see" whether the value of a function is equal to zero at the midpoint of a segment, and depending on this, to continue or break off a process. Since there is no actual way to clarify the question of interest to us in this situation, the first step of the process already turns out in general to be unfulfillable.

We shall return to our investigation of this example in §68, but now we shall prove the following assertion:

1.1. *No normal algorithm can be applicable to every constructive real number and changes such a number into the empty word if and only if it equals zero.*

2. Before turning directly to the proof of this theorem, we carry out the following analysis.

Let us hold some normal algorithm $\mathfrak{A}$ fixed and denote its alphabet by the letter A. Let P be an arbitrary word in A. We shall develop the process of applying $\mathfrak{A}$ to P, and in the course of this process, we shall compute an arithmetical function $f_{\mathfrak{A},P}$, putting[3]

$$f_{\mathfrak{A},P}(N) \rightleftharpoons \begin{cases} N & \text{if the application of } \mathfrak{A} \text{ to } P \text{ has not ended by the time that } N \text{ steps were performed;} \\ (\mathfrak{A} : P) & \text{if the application of } \mathfrak{A} \text{ to } P \text{ ended with a total number of steps not exceeding } N \end{cases}$$

As we see, there is an algorithm for computing the values of the function $f_{\mathfrak{A},P}$. A corresponding *normal* algorithm $\mathfrak{F}_{\mathfrak{A},P}$ can also be constructed with no particular trouble. Moreover, one can construct a normal algorithm which changes the pair consisting of $\mathfrak{A}$'s record and the word P into the record of $\mathfrak{F}_{\mathfrak{A},P}$. Taking the normalization principle into account, we consider this construction fulfilled.

It is not hard now to convince oneself that the following assertion holds:

2.1. *The algorithm $\mathfrak{A}$ is not applicable to the word P if and only if $\mathfrak{F}_{\mathfrak{A},P}(N) \simeq N$ for every N.*

3. There exists an algorithm which changes each natural number N into the rational number $2^{-\mathfrak{F}_{\mathfrak{A},P}(N)}$. This algorithm can of course be normalized; moreover, P^+ can be taken as the alphabet of the resulting normal algorithm. In what follows, we shall assume that such a normal algorithm has been constructed and shall denote it by the symbol $\mathfrak{G}_{\mathfrak{A},P}$.

Recalling the definition, we see that this algorithm is a constructive sequence of rational numbers. Furthermore, given any $M, N \geqslant L$, the inequality

$$|\mathfrak{G}_{\mathfrak{A},P}(M) - \mathfrak{G}_{\mathfrak{A},P}(N)| < \frac{1}{2^L}$$

holds for any P[4], so that the identity normal algorithm $\mathfrak{I}$ in the alphabet P^+ is a convergence regulator for this sequence. Consequently, the word

$$\mathfrak{G}_{\mathfrak{A},P}^{\mathrm{Rc}} \delta \mathfrak{I}^{\mathrm{Rc}}$$

is a constructive real number for any P. We denote it by the symbol $x_{\mathfrak{A},P}$. Here the construction of $x_{\mathfrak{A},P}$, starting with $\mathfrak{A}$ and P, can again be normalized, and we shall assume that this normalization has been carried out.

It is not difficult to see that

3.1. $\mathrm{x}_{\mathfrak{A},P} = 0$ *if and only if the algorithm $\mathfrak{A}$ is not applicable to P.*

4. It is now becoming clear that if there exists a normal algorithm recognizing the equality of constructive real numbers to zero, then there also exists a normal algorithm recognizing to which words in the alphabet A the algorithm $\mathfrak{A}$ is inapplicable.

---

[3]Let us recall that $(\mathfrak{A} : P)$ denotes the number of steps performed by the algorithm $\mathfrak{A}$ when applied to the word P [see §27.3 and §25.8].

[4]Here it is not necessary to examine separately the case when $\mathfrak{A}$ is and is not applicable to the word P.

Indeed, let some normal algorithm $\mathfrak{C}$ recognize the equality of constructive real numbers to zero. We take a word P in the alphabet A, write down the constructive real number $x_{\mathfrak{A},P}$ and apply the algorithm $\mathfrak{C}$ to it. If it turns out that $x_{\mathfrak{A},P} = 0$, then the algorithm $\mathfrak{A}$ is not applicable to the word P. But if $x_{\mathfrak{A},P} \neq 0$, then this is false.

We have thus obtained an algorithm solving the problem of recognizing the inapplicability of the algorithm $\mathfrak{A}$ to words in its alphabet. The normalization principle suggests that this algorithm can be normalized, and this is indeed the case. But this yields a contradiction to Theorem §48.2.3, according to which there exists a normal algorithm with an unsolvable problem of recognizing inapplicability to words in its alphabet.

Thus, there is no normal algorithm recognizing the equality of constructive real numbers to zero.

Theorem 1.1 is proven.

5. Analyzing our method of defining the numbers $x_{\mathfrak{A},P}$, we see that they are quite simple in structure. In fact, if the algorithm $\mathfrak{A}$ is not applicable to the word P, then $x_{\mathfrak{A},P} = 0$ [3.1] and, therefore, $x_{\mathfrak{A},P}$ is rational in this case. It is also not difficult to see that if $\mathfrak{A}$ is applicable to P and $(\mathfrak{A} : P) = k$, then $x_{\mathfrak{A},P} = 2^{-k}$, and so $x_{\mathfrak{A},P}$ is rational in this case, too. Therefore, we have the following implication:

(1) ≪if $\mathfrak{A}$ is applicable to P or if $\mathfrak{A}$ is not applicable to P, then the constructive real number $x_{\mathfrak{A},P}$ is rational≫.

A mathematician acknowledging the ≪law of the excluded middle≫ as a universal logical law draws the conclusion from this that every number $x_{\mathfrak{A},P}$ is rational. It seems to us, however, that the following argument is more careful:

Take any of the numbers $x_{\mathfrak{A},P}$ and assume that it is irrational, i.e., not rational. Then the premise of the implication (1) is false, and both its disjunctive terms are therefore false, i.e., 1) $\mathfrak{A}$ is not applicable to P and 2) it is not the case that $\mathfrak{A}$ is not applicable to P. We have thus obtained a contradiction. Consequently, none of the numbers $x_{\mathfrak{A},P}$ is irrational and hence, according to the definition, they are all pseudorational.

6. Let us now assume that there is a normal algorithm $\mathfrak{R}$, applicable to every pseudorational constructive real number and changing it into the rational number[5] which it equals. Let us take an arbitrary word P in A, construct the number $x_{\mathfrak{A},P}$ and apply the algorithm $\mathfrak{R}$ to it. If $\mathfrak{R}(x_{\mathfrak{A},P}) = 0$, then $x_{\mathfrak{A},P} = 0$ and, therefore, $\mathfrak{A}$ is not applicable to P [3.1]. But if $\mathfrak{R}(x_{\mathfrak{A},P}) \neq 0$, then it is not the case that $\mathfrak{A}$ is not applicable to P [3.1]. Thus, we have developed an algorithm solving for $\mathfrak{A}$ the problem of recognizing inapplicability to words in its alphabet. This algorithm can of course be normalized, and since $\mathfrak{A}$ is an arbitrary normal algorithm, this yields a contradiction to Theorem §48.2.3.

Therefore a *normal* algorithm $\mathfrak{R}$, satisfying the conditions formulated in the beginning of this part, cannot exist. But then there is no basis for counting

[5]In the sense of §1.6, i.e., an integer or a "fraction" of the form M/N, where M is an integer, while N is a natural number different from zero.

on the existence of any *algorithm,* i.e., uniform general method, whatsoever with these properties.

After this remark, we suggest that the reader answer the following question by her or himself: is it true that every pseudorational constructive real number is rational?

7. Our examination of the numbers $x_{\mathfrak{A},P}$ gives us a convenient excuse to note another interesting circumstance.

Let us agree to say that the number $\mathfrak{A}_1^{\mathrm{Rc}}\delta\mathfrak{B}_1^{\mathrm{Rc}}$ *majorizes* the number $\mathfrak{A}_2^{\mathrm{Rc}}\delta\mathfrak{B}_2^{\mathrm{Rc}}$ if for every natural number N, the inequality

$$\mathfrak{A}_1(\mathfrak{B}_1(N)) - \mathfrak{A}_2(\mathfrak{B}_2(N)) \geqslant -2^{-(n-1)}$$

holds.

In order to denote this relation, which is an analogue of the usual non-strict order in the set of real numbers, we shall, according to tradition, make use of the customary symbol≪$\geqslant$≫, using the symbol ≪$\leqslant$≫ in the natural sense. After this, the relation ≪$>$≫ can be introduced in the ordinary way[6].

Of course, the implication

$$\ll \text{ if } x > y \text{ or } x = y \text{ then } x \geqslant y \gg$$

holds for any two constructive real numbers x and y. We shall study the question of how valid the converse implication is. (Let us emphasize that the majorizing symbol $\geqslant$ is understood here in the sense of the definition just formulated and there is no a priori basis for regarding it as the disjunction ≪greater or equal≫.)

It obviously follows from the definition of majorizing that

7.1. *Given any P in* A, $x_{\mathfrak{A},P} \geqslant 0$[7].

Let us now suppose that there is a normal algorithm $\mathfrak{C}$, applicable to every constructive real number x majorizing 0 and such that $x > 0$ if $\mathfrak{C}(x) \equiv \Lambda$ and $x = 0$ if $\mathfrak{C}(x) \not\equiv \Lambda$. We take an arbitrary word P in A, construct the number $x_{\mathfrak{A},P}$ and apply the algorithm $\mathfrak{C}$ to it. If $\mathfrak{C}(x_{\mathfrak{A},P}) \equiv \Lambda$, then $x_{\mathfrak{A},P} > 0$ and, therefore, $\mathfrak{A}$ is applicable to P [3.1]. But if $\mathfrak{C}(x_{\mathfrak{A},P}) \not\equiv \Lambda$, then $x_{\mathfrak{A},P} = 0$ and, therefore, $\mathfrak{A}$ is not applicable to P [3.1].Thus, we have obtained an algorithm recognizing the applicability of the normal algorithm $\mathfrak{A}$ to words in its alphabet. Normalizing this algorithm—which, of course, is possible—we obtain, in view of the fact that $\mathfrak{A}$ is arbitrary, a contradiction to Theorem §48.2.4.

Thus, we have proven that

7.2. *There is no normal algorithm $\mathfrak{C}$ applicable to every constructive real number x majorizing 0 and such that $x > 0$ if $\mathfrak{C}(x) \equiv \Lambda$ and $x = 0$ if $\mathfrak{C}(x) \not\equiv \Lambda$.*

In view of the normalization principle, there remains little hope to obtain any other (not a normal) algorithm whatsoever with the properties listed in the

[6] It can also be introduced independently—for example, in the following manner: $\mathfrak{A}_1^{\mathrm{Rc}}\delta\mathfrak{B}_1^{\mathrm{Rc}} > \mathfrak{A}_2^{\mathrm{Rc}}\delta\mathfrak{B}_2^{\mathrm{Rc}}$ means that there exists a natural number N such that $\mathfrak{A}_1(\mathfrak{B}_1(N)) - \mathfrak{A}_2(\mathfrak{B}_2(N)) > 2^{-(N-1)}$.

[7] Naturally, we shall understand the notation $x \geqslant r$ $(x > r)$ as $x \geqslant \hat{r}$ $(x > \hat{r})$.

formulation of Theorem 7.2. We now let the reader decide for her or himself how valid the disjunctive understanding of the majorizing relation is.

8. Basing ourselves on the definitions of equality and majorizing, we can show without difficulty that the following assertion holds:

8.1. *Given any constructive real numbers x and y, $x = y$ if and only if $x \geqslant y$ and $y \geqslant x$.*

§65 Recognizing Majorization. Arithmetical Operations. Piecewise Representation of Functions

1. Let x be an arbitrary constructive real number. Consider the disjunction

$$\ll x \leqslant 0 \text{ or } x \geqslant 0 \gg . \tag{1}$$

It is not hard to show that for no x can both terms of this disjunction be false. The problem therefore arises of seeking an algorithm which would allow one, starting with x, to find the true term of the disjunction (1). (Note that both terms of the disjunction under consideration, unlike the disjunction §64.1(1), can turn out to be true.)

2. Before beginning to solve the problem we formulated, we shall carry out the following construction.

Fix any normal algorithm $\mathfrak{A}$ over the alphabet A_0, such that whenever it is applicable to any word P in A_0, it changes it either into a or into b. Let P be an arbitrary word in the alphabet of the algorithm $\mathfrak{A}$. By means of the method given in §64.3, first construct the normal algorithm $\mathfrak{F}_{\mathfrak{A},P}$, and then the constructive sequence of rational numbers $\mathfrak{G}_{\mathfrak{A},P}$.

It is easily perceived from the conditions satisfied by the algorithm $\mathfrak{F}_{\mathfrak{A},P}$ [see §64.2] that the equality $\mathfrak{G}_{\mathfrak{A},P}(N+1) = \mathfrak{G}_{\mathfrak{A},P}(N)$ holds for any N if and only if $\mathfrak{A}$ is applicable to P.

We now generate a constructive sequence of rational numbers, $\mathfrak{H}_{\mathfrak{A},P}$, in such a way that the equality

$$\mathfrak{H}_{\mathfrak{A},P}(0) = \mathfrak{G}_{\mathfrak{A},P}(0)$$

is fulfilled and

$$\mathfrak{H}_{\mathfrak{A},P}(N+1) = \begin{cases} \mathfrak{G}_{\mathfrak{A},P}(N+1) & \text{if } \mathfrak{G}_{\mathfrak{A},P}(N+1) \neq \mathfrak{G}_{\mathfrak{A},P}(N), \\ -\mathfrak{G}_{\mathfrak{A},P}(N+1) & \text{if } \mathfrak{G}_{\mathfrak{A},P}(N+1) = \mathfrak{G}_{\mathfrak{A},P}(N) \text{ and } \mathfrak{A}(P) \simeq a, \\ \mathfrak{G}_{\mathfrak{A},P}(N+1) & \text{if } \mathfrak{G}_{\mathfrak{A},P}(N+1) = \mathfrak{G}_{\mathfrak{A},P}(N) \text{ and } \mathfrak{A}(P) \simeq b. \end{cases}$$

Regardless of what P one takes, a normal algorithm $\mathfrak{J}$ such that $\mathfrak{J}(N) \simeq N+1$ is a convergence regulator for $\mathfrak{H}_{\mathfrak{A},P}$, and so the word $\mathfrak{H}_{\mathfrak{A},P}^{\mathrm{Rc}} \delta \mathfrak{J}^{\mathrm{Rc}}$ is a constructive real number. We shall agree to denote this number by the symbol $y_{\mathfrak{A},P}$.

The normalization principle suggests that we can construct a normal algorithm $\mathfrak{K}$ such that $\mathfrak{K}(\mathfrak{A}^{\mathrm{Rc}} * P) \simeq y_{\mathfrak{A},P}$. This really is so, and in what follows, we shall assume that this construction has been carried out.

3. Let us now suppose that there is a normal algorithm $\mathfrak{C}$, applicable to every constructive real number x and such that if $\mathfrak{C}(x) \coloneqq \Lambda$, then $x \leqslant 0$, and if $\mathfrak{C}(x) \not\coloneqq \Lambda$, then $x \geqslant 0$. We define the algorithm $\mathfrak{L}$ in the alphabet A_0 by the following procedure. Given an arbitrary word P in A_0, let $y_{\mathfrak{A},P}$ be the number constructed with the aid of the algorithm $\mathfrak{K}$ from Part 2 and let the algorithm $\mathfrak{C}$ be applied to it. If it turns out that $\mathfrak{C}(y_{\mathfrak{A},P}) \coloneqq \Lambda$, then set $\mathfrak{L}(P) \rightleftharpoons a$. But if $\mathfrak{C}(y_{\mathfrak{A},P})$ will be different from Λ, set $\mathfrak{L}(P) \rightleftharpoons b$. Taking the normalization principle into account, we shall regard the algorithm $\mathfrak{L}$ as normalized, i.e., we shall assume that it is a normal algorithm over A_0.

One can perceive from the description of $\mathfrak{L}$ that it is applicable to every word P in A_0 and that $\mathfrak{L}(P)$ is either a or b. We shall show that $\mathfrak{L}$ is a completion of $\mathfrak{A}$ with respect to A_0. In fact, let P be an arbitrary word in A_0 and let $\mathfrak{A}$ be applicable to P. Then $\mathfrak{A}(P) \coloneqq a$ or $\mathfrak{A}(P) \coloneqq b$.

In the first case, $y_{\mathfrak{A},P} = -2^{-(\mathfrak{A}:P)}$, i.e., $y_{\mathfrak{A},P} < 0$. But then $\mathfrak{C}(y_{\mathfrak{A},P}) \coloneqq \Lambda$, since it would follow from $\mathfrak{C}(y_{\mathfrak{A},P}) \not\coloneqq \Lambda$ that $y_{\mathfrak{A},P} \geqslant 0$. Therefore, $\mathfrak{L}(P) \coloneqq a$. Thus $\mathfrak{L}(P) \coloneqq \mathfrak{A}(P)$ in this case.

In the second case, $\mathfrak{A}(P) \coloneqq b$ and so $y_{\mathfrak{A},P} = 2^{-(\mathfrak{A}:P)}$, i.e., $y_{\mathfrak{A},P} > 0$. But then $\mathfrak{C}(y_{\mathfrak{A},P}) \not\coloneqq \Lambda$ and, therefore, $\mathfrak{L}(P) \coloneqq b$. Thus, $\mathfrak{L}(P) \coloneqq \mathfrak{A}(P)$ in the second case.

Consequently, $\mathfrak{L}(P) \coloneqq \mathfrak{A}(P)$ in both cases, and hence $\mathfrak{L}$ really is a completion of $\mathfrak{A}$ with respect to A_0.

The upshot of our analysis is that if there exists a normal algorithm $\mathfrak{C}$ with the properties given at the beginning of this part, then every normal algorithm $\mathfrak{A}$ with the properties listed in the beginning of Part 2 is completable with respect to A_0, which contradicts Theorem §53.1.1.

Thus, we have proven that the following assertion holds (Tseitin [3]):

3.1. *There is no normal algorithm $\mathfrak{C}$, applicable to every constructive real number x and such that $x \leqslant 0$ if $\mathfrak{C}(x) \coloneqq \Lambda$ and $x \geqslant 0$ if $\mathfrak{C}(x) \not\coloneqq \Lambda$.*

4. In the sequel, we shall need a number of ordinary arithmetical operations on constructive real numbers. We now turn to their definition.

Let $\mathfrak{A}_1^{\text{Rc}}\delta\mathfrak{B}_1^{\text{Rc}}$ and $\mathfrak{A}_2^{\text{Rc}}\delta\mathfrak{B}_2^{\text{Rc}}$ be two constructive real numbers. We shall call the constructive sequences of rational numbers $\mathfrak{A}_3$, $\mathfrak{A}_4$ and $\mathfrak{A}_5$ the *sum, difference* and *product,* respectively of the sequences $\mathfrak{A}_1$ and $\mathfrak{A}_2$ if for every natural number N, we have the following qualities:

$$\mathfrak{A}_3(N) = \mathfrak{A}_1(N) + \mathfrak{A}_2(N),$$
$$\mathfrak{A}_4(N) = \mathfrak{A}_1(N) - \mathfrak{A}_2(N),$$
$$\mathfrak{A}_5(N) = \mathfrak{A}_1(N) \cdot \mathfrak{A}_2(N).$$

It is easy to show that the sequences $\mathfrak{A}_3$, $\mathfrak{A}_4$ and $\mathfrak{A}_5$ are fundamental. Namely, the normal algorithm $\mathfrak{B}$ in P^+, such that

$$\mathfrak{B}(N) = \max(\mathfrak{B}_1(N+1), \mathfrak{B}_2(N+1))$$

for every natural number N is a convergence regulator for the sequences $\mathfrak{A}_3$ and $\mathfrak{A}_4$. If, in addition, M is a natural number such that $|\mathfrak{A}_1(N)| < 2^M$ and

$|\mathfrak{A}_2(N)| < 2^M$, them the normal algorithm $\mathfrak{C}$ in P^+, such that

$$\mathfrak{C}(N) = \max(\mathfrak{B}_1(N+M+1), \mathfrak{B}_2(N+M+1))$$

for every N is a convergence regulator for the sequence $\mathfrak{A}_5$.

Therefore, the words $\mathfrak{A}_3^{\mathrm{Rc}}\delta\mathfrak{B}^{\mathrm{Rc}}$, $\mathfrak{A}_4^{\mathrm{Rc}}\delta\mathfrak{B}^{\mathrm{Rc}}$ and $\mathfrak{A}_5^{\mathrm{Rc}}\delta\mathfrak{B}^{\mathrm{Rc}}$ are constructive real numbers. We shall call them the sum, difference and product, respectively, of the numbers $\mathfrak{A}_1^{\mathrm{Rc}}\delta\mathfrak{B}_1^{\mathrm{Rc}}$ and $\mathfrak{A}_2^{\mathrm{Rc}}\delta\mathfrak{B}_2^{\mathrm{Rc}}$.

It is not hard to show that one can construct normal algorithms changing any pair of constructive real numbers x and y into their *sum* $x+y$, *difference* $x-y$ and *product* $x \cdot y$.

Let $\mathfrak{A}^{\mathrm{Rc}}\delta\mathfrak{B}^{\mathrm{Rc}}$ be a constructive real number. The constructive sequence of rational numbers, $\mathfrak{A}_1$, such that

$$\mathfrak{A}_1(N) = |\mathfrak{A}(N)|$$

for any N, is easily seen to be fundamental: the algorithm $\mathfrak{B}$ is its convergence regulator. We shall call the constructive real number $\mathfrak{A}_1^{\mathrm{Rc}}\delta\mathfrak{B}_1^{\mathrm{Rc}}$ the *absolute value* of the number $\mathfrak{A}^{\mathrm{Rc}}\delta\mathfrak{B}^{\mathrm{Rc}}$.

One can construct a suitable normal algorithm for finding the absolute value $|x|$ of a constructive real number x.

After this, we can define the *maximum* and *minimum* of two constructive real numbers x and y:

$$\max(x,y) \rightleftharpoons \frac{1}{2} \cdot (x + y + |x-y|),$$
$$\min(x,y) \rightleftharpoons \frac{1}{2} \cdot (x + y - |x-y|).$$

in order to find $\max(x,y)$ and $\min(x,y)$, suitable normal algorithms can be constructed with no trouble.

The reader has probably already noticed that $|x|$, and also $\max(x,y)$ and $\min(x,y)$ are defined by us, because of Theorem 3.1, without recourse to the usual device in classical analysis of ≪analyzing cases≫.

Analogously to the way in which we defined the sum, difference and product of two constructive real numbers, we can also define the *quotient of two such numbers*. A slight difficulty arises here, related to the possibility of the divisor's being equal to zero. However, it is easily overcome (see Shanin([2]; §6) or Kushner ([1]; Chap. 2, §4)).

5. Let us consider one more auxiliary question—that of a piecewise presentation of a constructive real function.

In view of Theorem 3.1, it would seem at first sight that a definite obstacle might be concealed here. However, when the functions to be pasted together satisfy suitable compatibility conditions, it could be overcome in the following manner.

Let f_1 and f_2 be constructive real functions, and a, a constructive real number such that $f_1(a) = f_2(a)$. Then a constructive real function, f, given[8] in

[8] We leave the construction's details to the reader.

such a way that the conditional equality

$$f(x) \simeq f_1(\min(x,a)) + f_2(\max(x,a)) - f_1(a)$$

holds for arbitrary x, satisfies the condition[9]

$$f(x) \cong \begin{cases} f_1(x) & \text{if } x \leqslant a, \\ f_2(x) & \text{if } x \geqslant a. \end{cases}$$

§66 Cauchy's Theorem on a Zero of a Continuous Function with Alternating Signs

In the present section, we shall consider the question of the possible influence of the experience accumulated in the theory of algorithm on the posing of problems in numerical analysis.

1. Cauchy's Theorem on a zero of a continuous function with alternating signs plays an important role in classical mathematical analysis. It asserts that there exists a functional, F, defined on the set of all functions f which are continuous on the interval $[0,1]$ and satisfy the condition $f(0) \cdot f(1) < 0$, such that $F(f) \in (0,1)$ and $f(F(f)) = 0$ for any function f from this set.

There arises the natural question of this functional's effectiveness, i.e., the question as to how effectively we can solve the equation

$$f(x) = 0$$

for arbitrary functions from the given set. It is known that there are no good numerical methods for the solution of this problem. The analysis carried out below uncovers, in a certain sense, the reason for this situation.

2. First of all, we must introduce the concept of a continuous constructive real function. We shall do this in the manner which is typical for constructive analysis.

Let the constructive real function f be defined at the point x. We shall call a constructive sequence $\mathfrak{A}$ of natural numbers a *continuity regulator* for the function f at the point x if for any constructive real number y such that $f(y)$ is defined and for any natural number N, one has the implication

$$\ll \text{if } |x-y| < 2^{\mathfrak{A}(N)}, \text{ then } |f(x) - f(y)| < 2^{-N} \gg .$$

We shall say that f is *continuous at the point* x if one can construct a continuity regulator for it at the point x.

---

[9]By analogy with the symbol $\simeq$, we understand the symbol $\cong$, used in the sequel, in such a manner that both sides connected by it make sense for one and the same x, and in case they do make sense, their values are equal constructive real numbers.

The function f is called *continuous in the interval* $[a, b]$, where a and b are constructive real numbers such that $a \leqslant b$, if one can find a normal algorithm changing each constructive real number x such that $a \leqslant x \leqslant b$ into a continuity regulator for f at the point x.

These definitions, as we see, differ from the ordinary definitions customary in classical analysis in that it is required that δ be found, starting with x and ε, in an *effective (computable)* manner. Therefore, a continuous constructive function is continuous in a certain stronger sense than a continuous function of classical analysis.

We note that in fact every constructive real function is continuous at any point where it is defined (Tseitin [4]). However, the proof of this assertion is very complicated and we shall not make use of it, preferring to include requirements which are automatically fulfilled in a number on theorems given below.

A somewhat weaker property of constructive real functions, that is, their non-discontinuity, was detected earlier by Markov (see [10], [11]). Developing Markov's approach, Tseitin brought it to a logical conclusion with the result just mentioned. An analogous result was later obtained by Moschovakis [1]. Not having the possibility of presenting Markov's result in its complete form due to lack of space, we shall, nevertheless, illustrate it by means of a simple example. We shall prove that

2.1. *No constructive real function f can satisfy the following conditions:* 1) $f(0) = 1$, 2) $f(x) = 0$ *for every constructive real number x different from 0.*

Indeed, under the assumption that such a function exists, taking an arbitrary constructive real number x and computing the value of $f(x)$ with a sufficient degree of accuracy (for example, with an accuracy of $\frac{1}{4}$), we find out which of the numbers, 0 or 1, $f(x)$ equals. But then we shall be able to determine which of the disjunctive terms,

$$\ll x = 0 \text{ or } x \neq 0 \gg,$$

holds. Normalizing this method, we get a contradiction to the assertion §64.1.1.

Figure 1.

3. Now let c be an arbitrary constructive real number. Consider a constructive real function f_c, defined in such a manner that its graph in the interval $[0, 1]$ has the form depicted in Figure 1 (we take the construction from Martin-Löf; see [1], p. 74).

This function can be constructed for any c with the aid of the general method of pasting, described in Part 5 of the preceding section. It is easy to convince oneself that it is continuous at any point of the interval $[0, 1]$ (of course, we are speaking of constructive real numbers x satisfying the condition $0 \leqslant x \leqslant 1$).

Let us assume, finally, that there is a normal algorithm applicable to the record of each constructive real function f_c which is continuous in the interval $[0, 1]$ and takes on values of opposite signs at the points 0 and 1, changing this record into a constructive real number, x_f, lying between 0 and 1 and such that

$$f(x_f) = 0.$$

We take an arbitrary constructive real number c, construct the function f_c and apply this algorithm to its record. This gives us a constructive real number, x_{f_c}, for which

$$f_c(x_{f_c}) = 0.$$

Computing the number x_{f_c} with sufficient accuracy, we find the true term of the disjunction

$$\ll x_{f_c} > \frac{1}{3} \text{ or } x_{f_c} < \frac{2}{3} \gg .$$

If its first term is true, then $c \leqslant 0$. But if the second term of this disjunction is true, then $c \geqslant 0$. Thus, we have developed an algorithm allowing one, for any constructive real number c, to find a true term of the disjunction

$$\ll c \leqslant 0 \text{ or } c \geqslant 0 \gg .$$

Normalizing this algorithm, we obtain a contradiction to Theorem §65.1.1.

Therefore, the following assertion holds (Tseitin [3]):

3.1. *No normal algorithm can be applicable to the record of every continuous constructive real function f changing sign in the interval $[0, 1]$ and transform any such record into a constructive real number which is a root of the equation $f(x) = 0$ and lies between 0 and 1.*

Nevertheless, it is not hard to show (Tseitin [3]) that

3.2. *No continuous constructive real function which changes sign in the interval $[0, 1]$ can be different from zero at every constructive real point of the interval $[0, 1]$.*

The proof of this assertion will be carried out in §68.1.

§67 Markov's Principle

1. Here we have for the first time come across the necessity of applying reasoning whose use we have been avoiding so far and whose admissibility is a subject on which different mathematicians apparently disagree. The question is the demonstration of the applicability of a normal algorithm to initial data by means

of an apagogical proof[10]. For a mathematician adhering to a classical, set-theoretic point of view, this method is undoubtedly applicable, since it is a simple consequence of a general logical law—the ≪law of the excluded middle≫. For a mathematician sharing a constructive point of view, this method presents a certain problem, requiring careful consideration in each particular case.

We shall now present our point of view on this question, which we believe is shared by a significant group of mathematicians adhering to constructivism.[11]

2. The theory of normal algorithms is constructed in this monograph within the scope of the abstraction of *potential* feasibility. Words in an alphabet under consideration, normal algorithms—all these are *potentially* feasible constructive objects. We regard the very process of applying a normal algorithm to a given word as a potentially feasible constructive process. We do not consider it necessary for the process of applying an algorithm $\mathfrak{A}$ to a word P to be it actually carried out in front of us from beginning to end in order for us to be certain that $\mathfrak{A}$ is applicable to P. But how can one be sure of the applicability of an algorithm to a word?

3. We consider it possible to apply here the method of apagogical proof,[12] i.e., to assert that the algorithm $\mathfrak{A}$ is applicable to the word P if the assumption that the process of applying $\mathfrak{A}$ to P will continue endlessly has been refuted by means of a reductio ad absurdum.

We see no reasonable grounds for refuting this method of reasoning, since it does not take us beyond the bounds of the constructive point of view: The abstraction of actual infinity is not applied, existence continues to be understood as the potential feasibility of a construction. If we assert, on the basis of the proven impossibility of a determined process' unlimited continuability, that this process will end, then this gives us a perfectly definite means of performing the desired construction: we continue the process until it is completed. The fact that we might not have an "a priori" bound on the number of steps to be performed would require no essential change in anything said here. Besides, the requirement that this number be bounded beforehand can hardly be formulated precisely or objectively.

It is not difficult to see that the above method of proving the applicability of an algorithm yields the possibility of justifying the following mode of reasoning.

Let there be a normal algorithm allowing one to determine, for the one-place predicate $\mathfrak{P}$ in the alphabet A and for any given word P in A, whether or not P satisfies $\mathfrak{P}$. If the assumption that no word satisfies the predicate $\mathfrak{P}$ has been refuted, then there exists a word satisfying $\mathfrak{P}$. This word can then be found by means of a successive testing of words, beginning with the empty word, in which case, for each word under consideration, we determine with the aid of the available algorithm, whether it satisfies the predicate $\mathfrak{P}$. By virtue of this, we call the given mode of reasoning the method of constructive selection, and the

---

[10]See footnote on p. 176. (Translator's note)

[11]Our treatment basically follows Markov's paper [3]. Also see his paper [10].

[12]See footnote on p. 176. (Translator's note)

logical principle declaring this method permissible, Markov's Principle.[13]

4. There are a variety of important applications of the Markov's Principle in constructive mathematics, and especially in constructive analysis. As an example, let us prove the following assertion with its aid:

4.1. *Given any constructive real numbers x and y, if it is not the case that $x = y$, then $x > y$ or $y > x$.*

Indeed, let $x = \mathfrak{A}_1^{\mathrm{Rc}} \delta \mathfrak{B}_1^{\mathrm{Rc}}$, $y = \mathfrak{A}_2^{\mathrm{Rc}} \delta \mathfrak{B}_2^{\mathrm{Rc}}$. Consider the following arithmetical predicate with the variable N for natural numbers:

$$|\mathfrak{A}_1(\mathfrak{B}_1(N)) - \mathfrak{A}_2(\mathfrak{B}_2(N))| > 2^{-(N-1)}. \tag{1}$$

It is not difficult to convince oneself that there exists a normal algorithm recognizing those N which satisfy this predicate. Now assume that there does not exist an N satisfying (1). Then for any N,

$$|\mathfrak{A}_1(\mathfrak{B}_1(N)) - \mathfrak{A}_2(\mathfrak{B}_2(N))| \leqslant 2^{-(N-1)}.$$

But this means that $x = y$ [§62.3], which contradicts our assertion's hypothesis. Therefore, according to Markov's Principle, there exists an N satisfying the predicate (1). One of the following inequalities holds for this N:

$$\mathfrak{A}_1(\mathfrak{B}_1(N)) - \mathfrak{A}_2(\mathfrak{B}_2(N)) > 2^{-(N-1)},$$
$$\mathfrak{A}_2(\mathfrak{B}_2(N)) - \mathfrak{A}_1(\mathfrak{B}_1(N)) > 2^{-(N-1)}.$$

In the first case, $x > y$ and in the second case $y > x$.

Theorem 4.1 is proven.

§68 Cauchy's Theorem on a Zero of a Continuous Function with Alternating Signs (Continuation of §66)

1. Let us now turn to the proof of Theorem §66.3.2. (Note that it is a non-trivial assertion: among the functions of the type indicated, we might have expected to find one whose continuous extension to the classical continuum has zeros which all turn out to be non-constructive.)

Thus, let the constructive real function f be continuous in $[0, 1]$ and change sign in this segment (we shall assume for the sake of definiteness that $f(0) < 0$ and $f(1) > 0$). By assumption, f is different from zero at every constructive point x of the segment under consideration. Therefore, according to Theorem §67.4.1, $f(x) > 0$ or $f(x) < 0$, and we have available a normal algorithm determining just which of these cases holds. Using this algorithm, we construct normal algorithms $\mathfrak{X}$ and $\mathfrak{Y}$ in the alphabet P^+, such that the following equalities are fulfilled:

$$\mathfrak{X}(0) \rightleftharpoons 0,$$
$$\mathfrak{Y}(0) \rightleftharpoons 1,$$

[13] Here and elsewhere the authors use the Russian equivalent of the term "constructive selection principle". (Translator's note)

$$\mathfrak{X}(N+1) \rightleftharpoons \begin{cases} \dfrac{\mathfrak{X}(N)+\mathfrak{Y}(N)}{2} & \text{if } f\left(\dfrac{\mathfrak{X}(N)+\mathfrak{Y}(N)}{2}\right) < 0, \\ \mathfrak{X}(N) & \text{if } f\left(\dfrac{\mathfrak{X}(N)+\mathfrak{Y}(N)}{2}\right) > 0, \end{cases}$$

$$\mathfrak{Y}(N+1) \rightleftharpoons \begin{cases} \mathfrak{Y}(N) & \text{if } f\left(\dfrac{\mathfrak{X}(N)+\mathfrak{Y}(N)}{2}\right) < 0, \\ \dfrac{\mathfrak{X}(N)+\mathfrak{Y}(N)}{2} & \text{if } f\left(\dfrac{\mathfrak{X}(N)+\mathfrak{Y}(N)}{2}\right) > 0 \end{cases}$$

(here N is an arbitrary natural number).

The algorithms $\mathfrak{X}$ and $\mathfrak{Y}$ are constructive sequences of rational numbers, and also, as is easily seen, fundamental sequences. A simple calculation shows that

$$|\mathfrak{X}(N) - \mathfrak{Y}(N)| = 2^{-N},$$

so that if we denote by x and y, respectively, the constructive real numbers determined by the sequences $\mathfrak{X}$ and $\mathfrak{Y}$, then $x = y$. Let us also note that $f(\mathfrak{X}(N)) < 0$ and $f(\mathfrak{Y}(N)) > 0$. Now, using the continuity of f, it is not hard to show that $f(x) \leqslant 0$ and $f(y) \geqslant 0$, but from this we can conclude on the basis of §64.8.1 that $f(x) = 0$, which contradicts the theorem's hypothesis.

2. The reader will now have no trouble convincing her or himself by the method of ≪bisecting a segment≫ that the following assertion is valid:

2.1. *Given any constructive real function which is continuous and changes sign in the segment* $[0, 1]$ *and given any rational number* $\varepsilon > 0$, *one can construct a constructive real number* x *lying between 0 and 1 and such that* $|f(x)| < \varepsilon$.

Indeed, let the theorem's hypothesis be fulfilled and suppose, for the sake of definiteness, that $f(0) < 0$, while $f(1) > 0$. We compute $|f(1/2)|$ with sufficient accuracy to enable us to single out a true term of the disjunction

$$\ll |f(1/2)| < \varepsilon \text{ or } |f(1/2)| > \varepsilon/2 \gg .$$

If it turns out that $|f(1/2)| < \varepsilon$, then we take $1/2$ as x and put an end to the process. But if it turns out that $|f(1/2)| > \varepsilon/2$, then $f(1/2) > \varepsilon/2$ or $f(1/2) < -\varepsilon/2$. In the first case, we take the segment $[0, 1/2]$, and in the second, the segment $[1/2, 1]$, and perform the analogous next step of the process. According to Markov's Principle, this process must break off, since its unlimited continuation would give us a constructive real number which would be a point of discontinuity of the function f, which is continuous by hypothesis. Normalizing this argument, we obtain the proof of Theorem 2.1.

3. Summarizing the contents of §66 and §68, we may assert that for any continuous constructive real function f in $[0, 1]$ with alternating signs, the following situation holds: 1) it cannot turn out to have no constructive zero[14] [§66.3.2]; 2) information about this function, itself, is in general not enough for actually

[14] i.e., a constructive root of the equation $f(x) = 0$.

finding one of its zeros [§66.3.1]; 3) the function f and a given degree of accuracy, $\varepsilon > 0$, constitute sufficient information for finding a constructive point x at which $|f(x)| \leqslant \varepsilon$ [2.1].

An experienced specialist in numerical methods should, of course, have a strong feeling for the situation described here. However, we believe that the possibility of expressing this feeling in the form of precise mathematical assertions is not without interest.

LITERATURE

Adyan, S. I. (Адян, С. И.)

1 Конечно-определенные группы и алгоритмы. ДАН СССР **117** (1957) 9–12.

"Finitely presented groups and algorithms." (Translantion published in *Soviet Math. Dokl.* Providence, R.I.)

2 Об алгоритмических проблемах в эффективно-полных классах групп. ДАН СССР **123** (1958) 13–16.

"On algorithmic problems in effectively complete classes of groups." (Translation published in *Soviet Math. Dokl.,* Providence, R.I.)

3 Определяющие соотношения и алгоритмические проблемы для групп и полугрупп. Тр. матем. ин-та АН СССР им. В. А. Стеклова, **85** М. Наука (1966).

"Defining relations and algorithmic problems for groups and semigroups." (Translation published in *Proc. Steklov Inst. Math.,* Providence, R.I.)

Barzdin, Ya. M. (Барздинь, Я. М.)

1 Сложность программ, распознающих принадлежность натуральных чисел, не превышающих n, рекурсивно перечислимому множеству. ДАН СССР, **182** (1968) 1249–1252.

"Complexity of programs recognizing whether natural numbers not exceeding n belong to a recursively enumerable set." (Translation published in *Soviet Math. Dokl.,* Providence, R.I.)

Boone, W. W.

1 "The word problem". *Ann. of Math.* **70** (1959) 207–265.

Borisov, V. V. (Борисов, В. В.)

1 Простые примеры групп с неразрешимой проблемой тождества. Мат. заметки **6** (1969) 521–523.

"Simple examples of groups with an unsolvable identity problem." (Translation published in *Math. Notes*, New York.)

Britton, J. L.

1 "The word problem." *Ann. of Math.,* **77** (1963) 16–32.

Brouwer, L. E. J.

1 "De onbetrouwbaarheid der logische principes." *Tijdschrift voor wijsbegeerte*, **2** (1908) 152–158.

Church, A.

1 "An unsolvable problem of elementary number theory." *Amer. J. Math.,* **58** (1936) 345–363.

2 "A note on the Entscheidungsproblem." *J. Symbolic Logic,* **1** (1936) 40–41; Correction, Ibid., 101–102.

Dehn, M.

1 "Über unendliche diskontinuierliche Gruppen." *Math. Ann.,* **71** (1912) 116–144.

Detlovs, V. K. (Детловс, В. К.)

1 Нормальные алгорифмы и рекурсивные функции. ДАН СССР **90** (1953) 723–725
"Normal algorithms and recursive functions." (Translation published in *Soviet Math. Dokl.,* Providence, R.I.)

2 Эквивалентность нормальных алгорифмов и рекурсивных функций. Тр. матем. ин-та АН СССР им. В. А. Стеклова, **52** М.-Л.: Изд. АН СССР (1958) 75–139.

Gödel, K.

1 "Über formal unentscheidbare Sätze der Principia Mathematica und verwandter Systeme I." *Monatsh. für Math. und Physik,* **38** (1931) 173–198.

Greendlinger, E. I. (Гриндлингер, Е. И.)

1 Решение проблемы изоморфизма для одного класса полугрупп. Сиб. матем. журн. **5** (1964) 788–792.
"Solution of the isomorphism problem for a class of semigroups." (Translation published in *Sib. Matem. J.,* Providence, R.I.)

2 О неразрешимости проблемы тождества слов для класса полугрупп с разрешимой проблемой изоморфизма. ДАН СССР **171** (1966).
"On the unsolvability of the word problem for a class of semigroups with a solvable isomorphism problem." (Translation published in *Soviet Math. Dokl.,* Providence, R.I.)

Heyting, A. (Гейтинг, А.)

1 *Intuitionism. An introduction.* North Holland Publ. Co. Amsterdam, 1956. (Russian translation, edited by A. A. Markov and with his commentaries, published in Moscow: Интуиционизм. Введение.)

Hilbert, D., and Bernays, P.

1 *Grundlagen der Mathematik. I, 2* Aufl. Springer-Verlag, Berlin, Heidelberg, New York, 1968.
(Russian translation by N. M. Nagorny published in Moscow: Основания математики. Логические исчисления и формализация арифметики.)

2 *Grundlagen der Mathematik. II, 2* Aufl. Springer-Verlag, Berlin, Heidelberg, New York, 1970.
Russian translation by N. M. Nagorny published in Moscow: Основания математики. Теория доказательств.)

Kleene, S. C. (Клини, С. К.)

1 "General recursive functions of natural numbers." *Math. Ann.,* **112** (1936) 727–742.

2 "Recursive predicates and quantifiers." *Trans. Amer. Math. Soc.* **53** (1943) 41–73.

3 "On the interpretation of intuitionistic number theory." *J. Symbolic Logic* **10** (1945) 109–124.
4 "A symmetric form of Gödel's theorem." *Koninklijke Nederlandsche Akademie van Wetenschappen*, Proc. of the section of sciences, **53** (1950) 800–802.
5 *Introduction to metamathematics.* Van Nostrand Co. Inc., New York, Toronto, 1952.
(Russian translation published in Moscow: Введение в метаматематику.)
6 "On the sources of the theory of recursive functions." *Annals of the History of Computing,* **3**.
(Russian translation published in Novosibirsk: Об истоках теории рекурсивных функций. Алгоритмы в современной математике и ее приложениях. Часть II.)

Kolmogorov, A. N. (Колмогоров, А. Н.)
1 О понятии алгоритма. Успехи матем. наук **8** (1953) 175–176.
"On the concept of an algorithm." (Translation published in *Russian Math. Surveys*, London).
2 Три подхода к определению понятия количества информации. Проблемы передачи информации, **I** (1965) 3–11.
"Three approaches to the definition of the concept of the 'amount of information'." (Translation published in *Problems of Information Transmission*, New York.)

Kolmogorov, A. N., and Uspenskii, V. A. (Колмогоров, А. Н., и Успенский, В. А.)
1 К определению алгоритма. Успехи матем. наук **13** (1958) 3–28.
"Towards the definition of an algorithm." (Translation published in *Russian Math. Surveys*, London.)

Kushner, B. A. (Кушнер, Б. А.)
1 Лекции по конструктивному математическому анализу. М. Наука 1973.
"Lectures on constructive mathematical analysis." (Translation published in Providence, R. I.)

Linnik, Yu. V., and Shanin, N. A. (Линник, Ю. В., и Шанин, Н. А.)
2 Андрей Андреевич Марков. К пятидесятилетию со дня рождения. Успехи Матем. наук **9** (1954) 145–149
"Andrei Andreevich Markov. On his fiftieth birthday". (Translation published in *Russian Math. Surveys*, London).

Maltsev, A. I. (Мальцев, А. И.)
1 Алгоритмы и рекурсивные функции. М. Наука, 1965.
Algorithms and recursive functions. (Translation published in Amsterdam).

Manin, Yu. I. (Манин, Ю. И.)
1 Вычислимое и невычислимое. М. Советское радио, 1980.
The computable and the non-computable. (Translation published in Berlin, Heidelberg, New York).

Markov, A. A. (Марков, А. А.)
1 Теория алгорифмов. Тр. матем. ин-та АН СССР им. В. А. Стеклова, **38** М. Изд. АН СССР (1951) 176–189.

"The theory of algorithms." (Translation published in *Proc. Steklov Inst. Math.,* Providence, R. I.)

2 Теория алгорифмов. Тр. Матем. ин-та АН СССР им. В. А. Стеклова, **42,** М.-Л. Изд. АН СССР (1954).
"The theory of algorithms." (Translation published in Jerusalem and New York).

3 О конструктивной математике. Тр. матем. ин-та АН СССР им. В. А. Стеклова, **67** М.-Л.: Изд. АН СССР (1962).
"On constructive mathematics." (Translation published in *Proc. Steklov Inst. Math.*, Providence, R. I.)

4 О логике конструктивной математики. М. Знание, 1972.
"On the logic of constructive mathematics."

5 Невозможность некоторых алгорифмов в теории ассоциативных систем. ДАН СССР **55** (1947) 587–590.
"The impossibility of certain algorithms in the theory of associative systems." (Translation published in *Soviet Math. Dokl.*, Providence, R. I.)

6 Невозможность некоторых алгорифмов в теории ассоциативных систем. II. ДАН СССР **58** (1947) 353–356.
"The impossibility of certain algorithms in the theory of associative systems. II." (Translation published in *Soviet Math. Dokl.*, Providence, R. I.)

7 Невозможность некоторых алгоритмов в теории ассоциативных систем. ДАН СССР **77** (1951) 19–20.
"The impossibility of certain algorithms in the theory of associative systems." (Translation published in *Soviet Math. Dokl.*, Providence, R. I.)

8 Невозможность алгорифмов распознавания некоторых свойств ассоциативных систем. ДАН СССР **77** (1951) 953–956.
"The impossibility of algorithms recognizing certain properties of associative systems." (Translation published in *Soviet Math. Dokl.*, Providence, R. I.)

9 О неразрешимых алгорифмических проблемах. Матем. сборник, **31** **(73)** (1952) 34–42.
"On the unsolvability of algorithmic problems." (Translation published in *Math. USSR Sb.*, Providence, R. I.)

10 О непрерывности конструктивных функций. Успехи матем. наук, **9** (1954) 226–229.
"On the continuity of constructive functions." (Translation published in *Russian Math. Surveys*, London).

11 О конструктивных функциях. Тр. матем. ин-та АН СССР им. В. А. Стеклова, **52** М.–Л.: Изд. АН СССР (1958) 315–348.
"On constructive functions." (Translation published in *Proc. Steklov Inst. Math.*, Providence, R. I.)

12 Неразрешимость проблемы гомеоморфии. ДАН СССР **121** (1958) 218–220.
"The unsolvability of the homeomorphy problem." (Translation published in *Soviet Math. Dokl.*, Providence, R. I.)

13 О неразрешимости некоторых проблем топологии. ДАН СССР **123** (1958) 978–980.

"On the unsolvability of certain problems in topology." (Translation published in *Soviet Math. Dokl.*, Providence, R. I.)

14 О вычислимых инвариантах. ДАН СССР **146** (1962) 1017–1020.
"On computable invariants." (Translation published in *Soviet Math. Dokl.*, Providence, R. I.)

15 О неотличимости по инвариантам. Изд. АН СССР сер. матем. **27** (1963) 407–436.
"On indistinguishability by invariants." (Translation published in *Math. USSR–Izv.*, Providence, R. I)

16 О нормальных алгорифмах, вычисляющих булевы функции. ДАН СССР **157** (1964) 262–264.
"On normal algorithms computing Boolean functions." (Translation published in *Soviet Math. Dokl.*, Providence, R. I.)

17 О нормальных алгорифмах, связанных с вычислением булевых функций. Изв. АН СССР сер. матем. **31** (1967) 161–208.
"On normal algorithms related to the computation of Boolean functions." (Translation published in *Math. USSR–Izv.*, Providence, R. I.)

18 О языке $Я_0$. ДАН СССР **214** (1974) 40–43.
"On the language L_o. (Translation published in *Soviet Math. Dokl.*, Providence, R. I.)

19 О языке $Я_1$. ДАН СССР **214** (1974) 279–282.
"On the language L_1." (Translation published in *Soviet Math. Dokl.*, Providence, R. I.)

20 О языке $Я_2$. ДАН СССР **214** (1974) 513–516.
"On the language L_2." (Translation published in *Soviet Math. Dokl.*, Providence, R. I.)

21 О языке $Я_3$. ДАН СССР **214** (1974) 765–768.
"On the language L_3. (Translation published in *Soviet Math. Dokl.*, Providence, R. I.)

22 О языках $Я_4, Я_{5,\ldots}$. ДАН СССР **214** (1974) 1031–1034.
"On the languages $L_4, L_{5,\ldots}$. (Translation published in *Soviet Math. Dokl.*, Providence, R. I.)

23 О языке $Я_\omega$. ДАН СССР **214** (1974) 1262–1264.
"On the language L_ω." (Translation published in *Soviet Math. Dokl.*, Providence, R. I.)

24 О языке $Я_{\omega|}$. ДАН СССР **215** (1974) 57–60.
"On the language $L_{\omega|}$." (Translation published in *Soviet Math. Dokl.*, Providence, R. I.)

25 О полноте классического исчисления предикатов в конструктивной математической логике. ДАН СССР **215** (1974) 266–269.
"On the completeness of classical predicate calculus in constructive mathematical logic." (Translation published in *Soviet Math. Dokl.*, Providence, R. I.)

26 "Essai de construction d'une logique de la mathématique constructive." *Revue intern. de Philosophie*, **98**, (1971)

27 Попытка построения логики конструктивной математики. Исследования по теории алгорифмов и математической логике, М. Изд. ВЦ АН СССР (1976) 3–31.
"An attempt to construct a logic of constructive mathematics." (Russian original of [26].)

Markov, A., and Nagorny, N. (Марков, А. и Нагорный, Н.)

1 Об одном языке для описания работы вычислительных машин. Проблемы кибернетики, вып. **19** М. Наука 5–38.
"On a language for describing the performance of computers." (Translation published in *Probleme der Kybernetik*, Berlin.)

Martin-Löf, P. (Мартин-Лёф, П.)

1 *Notes on constructive mathematics.* Almquist and Wiksell, Stockholm, 1970.
(Russian translation published in Moscow: Очерки по конструктивной математике).

Matiyasevich, Yu. V. (Матиясевич, Ю. В.)

1 Простые примеры неразрешимых ассоциативных исчислений. ДАН СССР **173** (1967) 1264–1266.
"Simple examples of unsolvable associative calculi." (Translation published in *Soviet Math. Dokl.*, Providence, R. I.)

2 Диофантовость перечислимых множеств. ДАН СССР **191** (1970) 279–282.
"The Diophanticity of the enumerable sets." (Translation published in *Soviet Math. Dokl.*, Providence, R. I.)

Mendelson, E. (Мендельсон, Э).

1 *Introduction to mathematical logic.* Van Nostrand Co., Inc., Princeton, Toronto, New York, London, 1964.
(Russian translation published in Moscow: Введение в математическую логику).

Moschovakis, Y. N.

1 "Recursive metric spaces." *Fund. Math.*, **55** (1964).

Nagorny, N. M. (Нагорный, Н. М.)

1 К усилению теоремы приведения теории алгорифмов. ДАН СССР **90** (1953) 341–342.
"Towards the strengthening of the Reduction Theorem in the theory of algorithms." (Translation published in *Soviet Math. Dokl.*, Providence, R. I.)

2 О минимальном алфавите алгорифмов над данным алфавитом. Тр. матем. ин-та АН СССР им. В. А. Стеклова, **52** М.-Л. Изд. АН СССР (1958) 66–74.
"On a minimal alphabet of algorithms over a given alphabet." (Translation published in *Proc. Steklov Inst. Math.*, Providence, R. I.)

3 Некоторые обобшения понятия нормального алгорифма. Тр. матем. ин-та АН СССР им. В. А. Стеклова, **52** М.–Л.: Изд. АН СССР (1958) 7–65.
"Some generalizations of the concept of a normal algorithm." (Translation published in *Proc. Steklov Inst. Math.*, Providence, R. I.)

4 О разыскании изоморфизмов ассоциативных исчислений. *Z. Math.* "On the search of the isomorphisms of associative calculi." *Logic und Grundl. Math.*, **6** (1960) 319–324.

5 Об отделимости по инвариантам. Исследования по теории алгорифмов и математической логике. М. Изд. ВЦ АН СССР (1973) 205–210. "On separability by invariants."

Nagorny, N. M., and Shanin, N. A. (Нагорный Н. М., и Шанин, Н. А.)

1 Андрей Андреевич Марков. К шестидесятилетию со дня рождения. Успехи. матем. наук, **19** (1964) 207–223. "Andrei Andreevich Markov. On his sixtieth birthday." (Translation published in *Russian Math. Surveys*, London.)

Novikov, P. S. (Новиков, П. С.)

1 Об алгоритмической неразрешимости проблемы тождества. ДАН СССР **85** (1952) 709–712. "On the algorithmic unsolvability of the identity problem." (Translation published in *Soviet Math. Dokl.*, Providence, R. I.)

2 Об алгоритмической неразрешимости проблемы тождества слов в теории групп. Тр. матем. ин-та АН СССР им. В. А. Стеклова, **44** М.–Л. Изд. АН СССР (1955). "On the algorithmic unsolvability of the problem of the identity of words in the theory of groups." (Translation published in *Proc. Steklov Inst. Math.*, Providence, R. I.)

Orlovskii, E. S. (Орловский, Э. С.)

1 Некоторые вопросы теории алгорифмов. Тр. матем. ин-та АН СССР им. В. А. Стеклова, **52** М.–Л. Изд. АН СССР (1958). "Some questions in the theory of algorithms." (Translation published in *Proc. Steklov Inst. Math.*, Providence, R. I.)

Ostroukhov, D. A.(Остроухов, Д. А.)

1 Об оценке сложности нормальных алгорифмов. ДАН СССР **184** (1969) 1292–1294. "On the estimation of the complexity of normal algorithms." (Translation published in *Soviet Math. Dokl.*, Providence, R. I.)

2 Линеаризация конструктивных последовательностей нормальных алгорифмов. Исследования по теории алгорифмов и математической логике. т. 1, М. Изд. ВЦ АН СССР (1973) 97–133. "The linearization of constructive sequences of normal algorithms."

Péter, R. (Петер, Р.)

1 *Rekursive Funktionen.* Budapest, 1951. (Russian translation published in Moscow: Рекурсивные функции.)

Petri, N. V. (Петри, Н. В.)

1 Об алгорифмах, связанных с предикатами и булевыми функциями. ДАН СССР **185** (1969) 37–39. "On algorithms related to predicates and Boolean functions." (Translation published in *Soviet Math. Dokl.*, Providence, R. I.)

Post, E. L. (Пост, Э. Л.)

1 "Finite combinatory processes–formulation I." *J. Symbolic Logic*, **1** (1936) 103–105.

(Russian translation published in Uspenskii's book [3].)
2 "Recursive unsolvability of a problem of Thue." *J. Sumbolic Logic*, **12** (1947) 1–11.

Rabin, M. O.
1 "Recursive unsolvability of group theoretic problems." *Ann. of Math.*, **67** (1958) 172–194.

Rice, H. G.
1 "Classes of recursively enumerable sets of positive integers and their decision problems." *Trans. Amer. Math. Soc.*, **74** (1953) 358–366.

Rogers, H.
1 *Theory of recursive functions and effective computability.* McGraw–Hill Book Co., New York, St. Louis, San Francisco, Toronto, London, Sydney, 1967.
(Russian translation published in Moscow: Теория рекурсивиых функций и эффективная вычислимость.)

Schönhage, A.
1 "Universelle Turing Speicherung." In: *Automatentheorie und formale Sprachen.* Dorr, J., Hotz, G., eds. Mannheim, 1970, 369–383.
2 "Storage modification machines." *SIAM Journal on Computing*, **9** (1980) 490–508.

Shanin, N. A. (Шанин, Н. А.)
1 О конструктивном понимании математических суждений. Тр. матем. ин–та им. В. А. Стеклова, **52** М.–Л. Изд. АН СССР (1958) 226–311.
"On the constructive understanding of mathematical judgements." (Translation published in *Proc. Steklov Inst. Math.*, Providence, R. I.)
2 Конструктивные вещественные числа и конструктивные функциональные пространства. Тр. матем. ин–та АН СССР им. В. А. Стеклова, **67** М.–Л. Изд. СССР (1962) 15–294.
"Constructive real numbers and constructive functional spaces." (Translation published in *Proc. Steklov Inst. Math.*, Providence, R. I.)

Specker, E.
1 "Nicht konstruktiv beweisbare Sätze der Analysis." *J. Symbolic Logic,* **14** (1949) 145–158.

Syrkin, G. I. (Сыркин, Г. И.)
1 Критерий справедливости теоремы перевода в теории нормальных алгорифмов. ДАН СССР **173** (1967) 270–272.
"Criteria for the validity of the translation theorem in the theory of algorithms." (Translation published in *Soviet Math. Dokl.*, Providence, R. I.)

Thue, A.
1 "Probleme über Veränderungen von Zeichenreihen nach gegebenen Regeln." *Videnskapsselskapets Skrifter. I. Mat. Naturv. Kl.* **10** (1914).

Tseitin, G. S. (Цейтин, Г. С.)
1 Ассоциативное исчисление с неразрешимой проблемой эквивалентности. ДАН СССР **107** (1956) 370–371.
"An associative calculus with an unsolvable equivalence problem." (Translation published in *Soviet Math. Dokl.*, Providence, R. I.)

2 Ассоциативное исчисление с неразрешимой проблемой эквивалентности. Тр. матем. ин-та им. В. А. Стеклова АН СССР **52** М.-Л. Изд. АН СССР (1958) 172–189.

"An associative calculus with an unsolvable equivalence problem." (Translation published in *Proc. Steklov Inst. Math.*, Providence, R. I.)

3 Теоремы о среднем значении в конструктивном анализе. Тр. матем. ин-та АН СССР им. В. А. Стеклова **67** М.-Л. Изд. АН СССР (1962) 362–384.

"Mean value theorems in constructive analysis." (Translation published in *Proc. Steklov Inst. Math.*, Providence, R. I.)

4 Алгоритмические операторы в конструктивных метрических пространствах. Тр. матем. ин-та АН СССР. им В. А. Стеклова **67** М.-Л. Изд. АН СССР (1962) 295–361.

"Algorithmic operators in constructive metric spaces." (Translation published in *Proc. Steklov Inst. Math.*, Providence, R. I.)

5 Нижняя оценка числа шагов для обращающего нормального алгорифма и других аналогичных алгорифмов. Зап. научн. семинаров Ленингр. отд. матем. ин-та АН СССР **20** (1971) 243–262.

"A lower bound for the number of steps taken by the reversing normal algorithm and other analgous algorithms." (Translation published in *J. Soviet Math.*, New York).

Turing, A. M.

1 "On computable numbers, with an application to the Entscheidungsproblem." *Proc. London Math. Soc.*, ser. 2, **42** (1936) 230–265.

2 "On computable numbers, with an application to the Entscheidungsproblem. A correction." *Proc. London Math. Soc.*, ser. 2,**43** (1937) 544–546.

Uspenskii, V. A. (Успенский, В. А.)

1 Теорема Гёделя и теория алгоритмов. ДАН СССР **91** (1953) 737–740.

"Gödel's Theorem and the theory of algorithms." (Translation published in *Soviet Math. Dokl.*, Providence, R. I.)

2 Лекции о вычислимых функциях. М. Физматгиз, 1960.

"Lectures on computable functions." (French translation published in Paris).

3 Машина Поста. М. Наука, 1979.

"Post machines."

Uspenskii, V. A., and Semenov, A. L. (Успенский, В. А., и Семенов, А. Л.)

1 "What are the gains of the theory of algorithms: basic developments connected with the concept of algorithm and with its application in mathematics." In: Ershov, A. P., Knuth, D. E., eds. *Algorithms in modern mathematics and computer science*, Springer, 1981 (*LN in computer science*, **122** 100–234.

2 Теория алгоритмов: ее основные открытия и приложения. В кн: Алгоритмы в современной математике и ее приложениях, часть 1, Новосибирск, (1982) 99–342.

"The theory of algorithms: its basic discoveries and applications, Part 1."

Zharov, V. G. (Жаров, В. Г.)

1 Об оценке сложности членов конструктивной последовательности нормальных алгорифмов. ДАН СССР **203** (1972) 746–748.
"On bounds for the complexity of the therms of constructive sequences of normal algorithms." (Translation published in *Soviet Math. Dokl.*, Providence, R. I.)

2 О сложности универсального нормального алгорифма. Теория алгорифмов и математическая логика. М. Изд. ВЦ АН СССР (1974) 34–54.
"On the complexity of a universal normal algorithm."

PRINCIPAL NOTATIONS

Λ

$P \asymp Q$

$P \neq Q$

$[P, Q]$

$T \rightleftharpoons S$

$(P \rightharpoondown Q)$

$(P \leftharpoondown Q)$

$[P^{\mathbf{A}}$

$[P^{\delta}$

$[P^{\tau}$

$[P^{\smile}$

$[P^{-}$

$[X^{\circ}$

$[Y^{\triangle}$

$P * Q * R$

$\Sigma(P, Q, R)$

$A \subset Б$

$(A \cup Б)$

$(A \setminus Б)$

A_0

A_1

A_2

$\mathfrak{A} : P \vdash Q$

$\mathfrak{A} : P \vdash\!\cdot\, Q$

$\mathfrak{A} : P]$

$\mathfrak{A} : P \vDash Q$

$\mathfrak{A} : P \vDash\!\cdot\, Q$

$\mathfrak{A} : P \vDash Q]$

$\mathfrak{A} : P \Rightarrow Q$

$\mathfrak{A}(P)$

$!\,\mathfrak{A}(P)$

$(\mathfrak{A} : P)$

$\mathfrak{A}^{\cdot}$

$(\mathfrak{A} \circ \mathfrak{B})$

$(\mathfrak{A} \frown \mathfrak{B})$

$(\mathfrak{A} \mathbf{Y} \mathfrak{B} \mathbf{|} \mathfrak{C})$

$(\mathfrak{A} \circlearrowleft \mathfrak{C})$

$\mathfrak{A}^{\mathrm{Rp}}$

$\mathfrak{A}^{\mathrm{Rc}}$

$\widetilde{\mathfrak{A}}_P$

$\mathfrak{A} : P \perp Q$

$\mathfrak{A} : P \perp\!\!\perp Q$

$P \vdash Q$

$P \vDash Q$

$P \sqsubseteq Q$

$[\mathfrak{Q}^{H}$

$[\mathfrak{Q}^{S}$

NAME INDEX

SUBJECT INDEX

The manufacturer's authorised representative in the EU is Springer Nature Customer Service Centre GmbH, Europaplatz 3, 69115 Heidelberg, Germany. If you have any concerns regarding our products, please contact ProductSafety@springernature.com

Printed and bound by CPI Group (UK) Ltd, Croydon, CR0 4YY
30/09/2026
02225505-0015